CAMBRIDGE LIBRARY COLLECTION

Books of enduring scholarly value

Technology

The focus of this series is engineering, broadly construed. It covers technological innovation from a range of periods and cultures, but centres on the technological achievements of the industrial era in the West, particularly in the nineteenth century, as understood by their contemporaries. Infrastructure is one major focus, covering the building of railways and canals, bridges and tunnels, land drainage, the laying of submarine cables, and the construction of docks and lighthouses. Other key topics include developments in industrial and manufacturing fields such as mining technology, the production of iron and steel, the use of steam power, and chemical processes such as photography and textile dyes.

Reports of the Late John Smeaton

Celebrated for his construction of the Eddystone Lighthouse near Plymouth, John Smeaton (1724–92) established himself as Britain's foremost civil engineer in the eighteenth century. A founder member of the Society of Civil Engineers, he was instrumental in promoting the growth of the profession. After his death his papers were acquired by the president of the Royal Society, Sir Joseph Banks, Smeaton's friend and patron. Using these materials, a special committee decided to publish 'every paper of any consequence' written by Smeaton, as a 'fund of practical instruction' for current and future engineers. These were published in four illustrated volumes between 1812 and 1814. Volume 2 contains Smeaton's reports on engineering works for bridges, including a proposal for the widening and improvement of London Bridge, as well as many plans for the creation or improvement of canals, mills and waterwheels.

Cambridge University Press has long been a pioneer in the reissuing of out-of-print titles from its own backlist, producing digital reprints of books that are still sought after by scholars and students but could not be reprinted economically using traditional technology. The Cambridge Library Collection extends this activity to a wider range of books which are still of importance to researchers and professionals, either for the source material they contain, or as landmarks in the history of their academic discipline.

Drawing from the world-renowned collections in the Cambridge University Library and other partner libraries, and guided by the advice of experts in each subject area, Cambridge University Press is using state-of-the-art scanning machines in its own Printing House to capture the content of each book selected for inclusion. The files are processed to give a consistently clear, crisp image, and the books finished to the high quality standard for which the Press is recognised around the world. The latest print-on-demand technology ensures that the books will remain available indefinitely, and that orders for single or multiple copies can quickly be supplied.

The Cambridge Library Collection brings back to life books of enduring scholarly value (including out-of-copyright works originally issued by other publishers) across a wide range of disciplines in the humanities and social sciences and in science and technology.

Reports of the Late
John Smeaton

Made on Various Occasions,
in the Course of his Employment as a Civil Engineer

VOLUME 2

JOHN SMEATON

CAMBRIDGE
UNIVERSITY PRESS

CAMBRIDGE
UNIVERSITY PRESS

University Printing House, Cambridge, CB2 8BS, United Kingdom

Cambridge University Press is part of the University of Cambridge.
It furthers the University's mission by disseminating knowledge in the pursuit of
education, learning and research at the highest international levels of excellence.

www.cambridge.org
Information on this title: www.cambridge.org/9781108069786

© in this compilation Cambridge University Press 2014

This edition first published 1812
This digitally printed version 2014

ISBN 978-1-108-06978-6 Paperback

REPORTS

OF THE LATE

JOHN SMEATON, F.R.S.

VOL. II.

REPORTS

OF THE LATE

JOHN SMEATON, F.R.S.

MADE ON

VARIOUS OCCASIONS,

IN THE COURSE OF HIS EMPLOYMENT

AS

A CIVIL ENGINEER.

IN THREE VOLUMES.

VOL. II.

LONDON:

PRINTED FOR LONGMAN, HURST, REES, ORME, AND BROWN,
PATERNOSTER-ROW.

1812.

CONTENTS.

―――――◆―――――

London Bridge.

Forth and Clyde.

Comparative

CONTENTS.

HEWICK

CONTENTS.

CONTENTS.

Coquett

CONTENTS.

SEACROFT

CONTENTS.

MILL

CONTENTS.

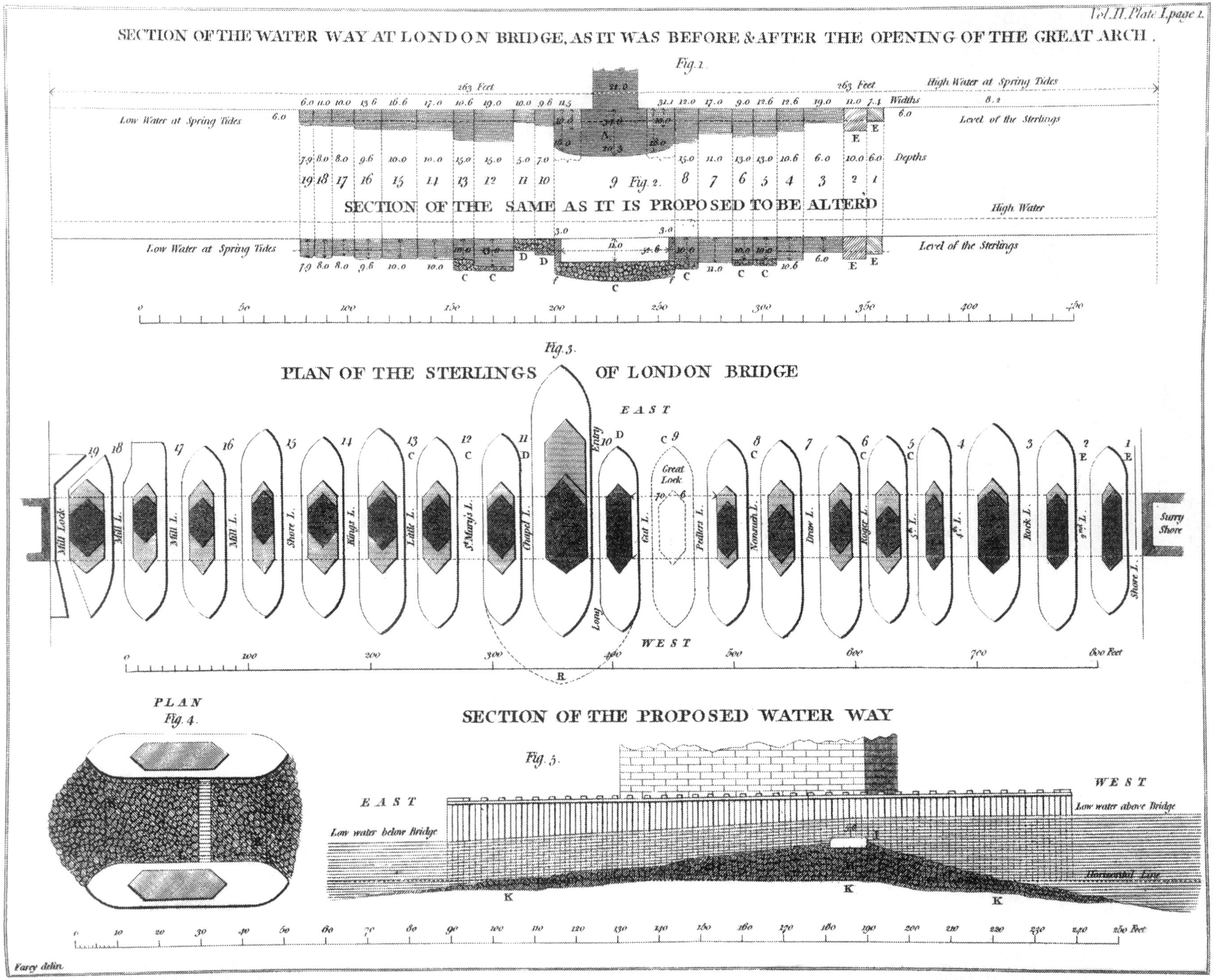

SECTION OF THE WATER WAY AT LONDON BRIDGE, AS IT WAS BEFORE & AFTER THE OPENING OF THE GREAT ARCH.

Fig. 1.

High Water at Spring Tides

Level of the Sterlings

High Water

Widths

Depths

Low Water at Spring Tides

SECTION OF THE SAME AS IT IS PROPOSED TO BE ALTER'D

Fig. 2.

Level of the Sterlings

High Water

Low Water at Spring Tides

PLAN OF THE STERLINGS OF LONDON BRIDGE

Fig. 3.

EAST

WEST

Surry Shore

1st L. — 2nd L. — Rock L. — 4th L. — 5th L. — Roger L. — Draw L. — Normch L. — Pedler L. — Great Lock — 6th L. — Chapel L. — St Marys L. — Lock L. — Kings L. — Short L. — 3dd L. — 3dd L. — Mill L.

Shore L.

Entry

Long

Mill Lock

SECTION OF THE PROPOSED WATER WAY

Fig. 5.

WEST

Low water above Bridge

Horizontal Line

EAST

Low water below Bridge

PLAN
Fig. 4.

REPORTS, *&c.*

LONDON BRIDGE.

The REPORT of JOHN SMEATON, upon the QUESTIONS proposed to him by the Committee for improving, widening, and enlarging London Bridge.

Queſtion 1ſt. IN what degree, at a monthly average, is the natural power of the water, and effect thereof, upon the wheels of London Bridge water-works diminiſhed, by removal of the pier, and opening of the great arch?

Anſwer.—By calculation it appears, that the natural power of the water, as the cir-cumſtances of the great arch were in the beginning of February, 1763, was diminiſhed at the monthly average as 2000 to 1277; and that the effect thereof upon the wheels, as appears by the company's regiſter, for above ſix years paſt, was diminiſhed in propor-tion of 2000 to 1300.   The proceſs of which inveſtigation is hereto annexed.

Queſtion 2d.—What are the methods, whereby the effect of theſe works may be re-ſtored to the ſame quantity, with reſpect to raiſing water, as ſubſiſted before the open-ing of the great arch?

Anſwer.—The methods that can be put in practice, for reſtoring the effect of the works, are reducible to two, viz: by improving and enlarging the preſent works, or

VOL. II.   B   by by penning up the fame head of water as before the alteration. With refpect to the pre-fent wheels, though they are all capable of improvement, yet they cannot all be improved in the fame degree in which they are now deficient, nor can they be fo much improved taken collectively; it will, therefore, be neceffary, not only to improve fome of the prefent wheels, but to add a new one; but as the old wheels cannot be much improved without rebuilding, and *one* new one cannot be depended upon to make good the defi-ciency, without rebuilding fome of the old ones; as the power, while acting, muft be increafed, in order to do the fame bufinefs in a lefs time, and as the prefent mains are but barely adequate to the prefent wheels, (which appears from their refufing a part of the water thrown up by the engines during the action of the ftrong fpring tide, as has happened during my examination of the works), it follows, that the addition of wheels will require an addition of mains, which, upon the whole, will create a great expenfe, and of a very complex nature. On the other hand, the very fame expedient that will be neceffary for fecuring the great arch, will alfo contribute confiderably towards the relief of the water-works; I, therefore, think it more eligible for the committee not to engage in building or improving the water-works, but to proceed to fuch method as will reftore the head of water; which may be done, as I conceive, at a lefs expenfe, and not only without prejudice to the navigation, but in fome refpects be an aid thereto, and that is by raifing the bed of the great lock, and fome others, that are unneceffarily deep, and by ftopping the water-way below the ftarling of two others, which appear to be of little or no ufe or confequence to navigation.

Queftion 3d.—If by ftopping up arches, or raifing the bed of the river under the arches, which arches will be beft to be ftopped up, or the beds under them raifed; and what will be the expenfe thereof?

Anfwer.—The addition of water-way is altogether made under the great arch; and if the reduction were to take place there alone, it would totally defeat the purpofes of the committee in doing fervice to the navigation, by the opening thereof; it is there-fore I propofe to do only a part here, fuch as is confiftent with thofe views, and the reft under fuch other arches as will admit thereof without prejudice.

For the particular manner of executing this fcheme, I muft refer to a drawing (fee plate I.), containing two fections of the water-way under London Bridge, the firft figure fhewing the water-way as it was before the alteration; and afterwards, as it was at the beginning of February, 1763: figure 2d fhews the water-way as it is intended to be after the alteration now propofed.

The

The abstract of the first section, figure 1. is as follows:

Feet, superficial.

The area of the water-way below the starlings, before the alteration, was - - - 2073

The increase of water-way by removing the solid under the great arch, and by the subsequent increase of depth, - - - - - - - 861

The diminution of water-way by stopping up the two locks next the south shore, - 154

The difference is the increase of water-way, - - - - - 707

The sum is the water-way after the alteration, the beginning of February, 1763, - - 2780

It therefore appears, that in order to restore the water-way, and of consequence the water-works, to what they were before the alteration of the great arch, the area of the water-way below the starlings, such as it was the beginning of February, 1763, must be reduced by 707 superficial feet; and in order to do this without prejudice to the navigation, I would propose 1st, to raise the bed of the river under the great arch, so as to be eleven feet at its highest part under the top of the starlings, in which case it will have the same depth of water as the draw lock, which was accounted the most useful for large craft before the alteration; and which, as well as the great lock when the water-way is restored, will never have a less depth of water than five feet at low water.

2dly. By advancing the starlings into the great lock, so as to double the fence, as proposed to the committee by Mr. PHILLIPS, a further reduction will be effected, and this being done three feet on each side, will still leave a water-way under the great arch in breadth fifty-two feet, which will be very sufficient for the craft.

3dly. I find there are four locks, whose depths exceed that of the draw lock, all of which are, on account of their widths, unfit for the passage of the larger craft. I therefore propose to raise the beds of those locks one foot higher than the level of the draw lock and the great lock. St. Mary's lock was justly accounted the best for large craft before the alteration, being the widest (viz. nineteen feet in the clear) and one of the deepest, viz. fifteen feet below the starling, that is four feet deeper than the draw lock; this depth of fifteen feet I apprehend to be totally unnecessary, and rather tending towards the insecurity of the bridge; but as the preserving one lock deeper than the rest, may possibly on some occasion be useful, I propose raising the bed of St. Mary's only two feet, in which case it will be thirteen feet below the starlings, that is two feet deeper than what is proposed for the great lock.

4th.

4th. As Long Entry and Chappel locks, which lie between *St. Mary*'s and the great arch, appear to be of lefs ufe than any of the reft, being both narrow and fhallow, and by their indraft tend to bring the craft foul upon the intermediate ftarlings, without a poffibility of their paffage, I would therefore propofe thofe to be ftopped up to ftarlings' height, the fame as the firft and fecond locks on the fouth fide now are; and that the craft may not be entangled with thofe locks or their ftarlings, but always find a paffage either through the great lock or *St. Mary*'s, I would propofe fender piles to be drove at about three feet diftance in a circular manner, from the upftream point of the ftarling of the great arch, to the upftream point of the ftarling of *St. Mary*'s, as reprefented in the draft fig. 3, at R.; thofe piles to be well braced, and their heads capped over about two feet above high-water mark.

Thefe things thus done, the account will ftand as follows:

PROPOSED REDUCTIONS of the area of the water-way below the ftarlings.

	Feet, superficial.
By raising the bed of the great arch,	314
By extending the starlings three feet on each side,	66
By raising the bed of little lock,	$52\frac{1}{2}$
By ditto of *St. Mary's*,	38
By shutting up Chappel lock,	50
By ditto Long Entry,	$66\frac{1}{2}$
By raising the bed of Non-such lock,	60
By ditto of Roger lock,	27
By ditto of the fifth lock,	$37\frac{1}{2}$
Water-way to be reduced,	$711\frac{1}{2}$
Increase of water-way as before stated,	707
Overplus,	$4\frac{1}{2}$

Now there will ftill remain an increafe of water-way above the ftarlings by the removal of the fhaft of the middle pier; but as the effect of this can only take place while the water is above the ftarlings, when the water-way is at the greateft: the effect upon the water wheels thence arifing does not appear by computation to be above $\frac{1}{160}$ part of the whole, which will be much more than compenfated by the above overplus of $4\frac{1}{2}$ feet below the ftarlings.

It is poſſible, notwithſtanding the attention I have paid to this matter, that ſome circumſtances may intervene, which have not occurred to my conſideration ; but I apprehend that none can, which may not be compenſated by raiſing the bed of *St. Mary*'s lock to the ſame height as the great lock : to which I can ſee no objection, if experience ſhould ſhew the ſame to be neceſſary.

The eaſieſt, cheapeſt, and moſt effectual method of raiſing the beds, not only of the great lock, but of the others, is, by throwing in rough ſtones, ſo as to form a ſlope both ways, the interſtices of which in time will fill up with gravel and the ſullage of the river, and become as compact and durable as a rock. The heavieſt, rougheſt, and largeſt ſtuff, is beſt adapted for this purpoſe, and none better than large *Kentiſh* rubble ; but for want of a ſufficient quantity thereof, *Portland* or *Purbeck* rubble may be uſed, provided it is more large in proportion, as it is leſs ſharp and heavy. For the particular application thereof to the great arch, I refer to the drafts, figs. 4, and 5.

N. B. If the following method ſhould be thought more eligible ; inſtead of raiſing the bed of *St. Mary*'s lock, the firſt open lock next the water-works may be ſtopped up, by two pair of gates pointed weſtward ; forming a navigable pen lock between them ; which would enable ſuch craft as can row or ſail againſt tide to paſs the bridge upward on tide of ebb, without waiting for tide of flood. And in this caſe the ſtopping up of Chappel and Long Entry locks may be diſpenſed with. In what extent this ſcheme may be of advantage to the navigation, I am not enabled to judge, but am of opinion, that if not of material uſe, what was firſt laid down is the moſt eligible, as the gates would need frequent repairs.

Queſtion 4th.—If by improving the old wheels, or adding new ones, which will it be beſt to improve, or where add the new ones ; and at what expenſe will this be done ?

Anſwer.—I apprehend, as was mentioned before, that all the wheels may be improved, but the wheels in the firſt and fourth arches from the north ſhore are the moſt ſuſceptible thereof, and the moſt want it ; if any wheel were to be added, it ſhould undoubtedly be in the fifth arch, which, though a capacious lock, is not much uſed by craft going down on tide of ebb, on account of the ſhipping commonly lying before it, but as I look upon it to be more eligible to reſtore the effect of the water-works by the more certain, and perhaps leſs controvertible method above deſcribed ; I have not made eſtimates for theſe erections and improvements, to do which, with any tolerable degree of exactneſs, (being very compounded) would take up ſome time.

Queſtion

Queſtion 5th.—If arches are ſtopped up, or beds raiſed, what effect will this have upon the channel of the river?

Anſwer.—If done in the manner before propoſed, no ſenſible effect of any kind whatever.

Queſtion 6th.—What effect upon the channel of the river has been produced by opening of the great arch, and by what means are the inconveniencies, if any, to be removed, and particularly with reſpect to what is alledged in the memorial of the Waterman's Company?

Anſwer.—Upon opening the great arch, the water finding a more free and open paſſage there than in any other part of the bridge, the main current would, of courſe, be drawn thither, and in conſequence, acting more ſtrongly upon the bed of the river than before, ſome thouſand tons of matter appear to have been moved by the action of the tides ſetting in each way, above, between, and below the ſtarlings; the moſt groſs and heavy parts ſubſiding as ſoon as the ſtrength of the ſtream is ſpent, form ſhoals or banks of matter both above and below bridge; yet ſuch banks, though encreaſed by the freſh action of the great arch, appear, in a great meaſure, to have been formed before, by the rapidity of the current ſetting through the arches; and I remember to have ſeen the ſhoal above bridge many times dry before the alteration took place, and if I am not miſtaken that below likewiſe. The preſent channel below bridge, ſtands quite fair to the great arch, and has about $4\frac{1}{2}$ feet navigable water over it at low ſtill water; but above bridge the caſe has been more untoward, for the body of the former ſhoal happening to lie ſomewhat to the ſouth of the direction of the great lock, the new-formed matter has, in a great meaſure, lodged itſelf in addition to the north ſide thereof, which has of conſequence thrown the main channel of the river conſiderably more to the north ſide, ſo that the current, on tide of ebb, making a ſudden turn near the bridge, has rendered it difficult for the craft to make the great arch, notwithſtanding the great enlargement of its width. The effect of this grievance will, in ſome meaſure, be remedied by the alterations propoſed in the bridge itſelf, becauſe the ſet of the tide being leſs ſtrong to the great arch, as well as more equally all over the bridge, the direction of the current will be leſs diverted towards any particular place: towards this end the ſtoppage of Long Entry, and Chappel locks will alſo contribute; for being ſituated between two of the principal arches, the balance will be preſerved, and the current leſs diverted in its paſſage through either of them. This, together with the fender piles before theſe two locks, will direct the craft, ſo that they cannot fail of hitting

one

one or the other, when there is water enough over the ſhoal above ; but at or near low water, I don't apprehend that any thing leſs than taking the whole ſhoal away, down to the depth of the main channel, can procure a true ſet of the tide ; however, it may in a great meaſure be remedied, by making a paſſage or channel through the ſhoal, which channel, to prevent as much as poſſible the effect of the ſuperior indraught of the main channel from acting upon it, ſhould be exactly in the direction of the great arch, but its weſt end declining a little to the ſouthward ; this new channel ſhould be at leaſt four feet deep at low water, and as wide as the great arch, and I know no better method of performing it, than by the ballaſt lighters ; it will alſo be adviſable to widen and deepen the preſent channel through the ſhoal below bridge, ſo as to make two and a half feet more water over the ſame but this will be a much leſs work than the former.

I come now to conſider the matter of the memorial of the Watermen's Company.

I apprehend, when all is finiſhed, the great lock will be found moſt uſeful to ſmall as well as large craft, as no boat can catch upon the ſtarlings there, being wide enough for moſt veſſels to go through ſideways.

I apprehend, the ſmall locks moſt uſed by ſmall boats ſhould be planked or lined, which will be of ſervice to the ſtarlings, and at the ſame time prevent boats from catching upon the irregularities of the piling, in which, I apprehend, conſiſts the greateſt danger in paſſing London Bridge on tide of ebb, with the water below the ſtarlings.

I am informed there are ſeveral ſtumps of piles in ſeveral of the locks ; but while I was making my obſervation, there was always too much water to ſee them ; it would be right to ſaw or cut them off with a large chiſel made for the purpoſe, but not to draw them or drive them down.

Reſpecting the bank oppoſite Fiſhmonger's-Hall ; the methods foregoing tend to procure a true ſet of the tide through the great arch ; but I know of no bank, either about or below bridge, that is of material conſequence to the water-works ; the current ſets over from the water-works towards the great arch, not becauſe it is impeded by a ſhoal but becauſe it finds a readier paſſage that way.

REFERENCE

REFERENCE to the plans annexed.

Plate I.

Fig. 1.—A, The fhaded fpace fhews the pier, ftarling, and bed of the river, which, before the alteration, were folid, but which, at the beginning of February, 1763, were converted into water-way.

N. B. The light fhaded fpaces fhew the dimenfions of the water-way in each lock, taken at fuch part where the bed of the lock is higheft. The figures above the fpaces are the widths of the refpective locks, (the piers and ftarlings between them being omitted), the figures beneath are the depths taken from the furfaces of the ftarlings at the place where the depths are leaft.

E E are two locks—the firft and fecond from the *Surry* fhore, which, before the alteration, were open, but which were clofed up by dams of piles, in the beginning of February, 1763.

Fig. 2.—The light fhaded fpaces fhew the intended dimenfions of the water-ways of the refpective locks, as they will be after the propofed alteration, the widths, except the great lock, are the fame as fig. 1. and the depths are fpecified beneath.

C C C C C C—Solids, added in order to raife the depths of the refpective locks.

D D—Two locks to be clofed up with dams of piles to the height of the ftarlings.

E E—Two locks already clofed, which are to remain fo.

f f—Solids, to be added to ftrengthen the ftarlings within the great arch, three feet. thick on each fide.

N. B. Both thefe fections fuppofe the water-ways laid together, both above and below the ftarlings, and all the folids removed, except fuch as are the fubject of the altera-tions.

The meafures differ fomething from thofe before taken, by reafon of their being taken at different places.

Fig. 3.

Fig. 3.—R fhews the difpofition of the fender piles, extending from the great lock to *St. Mary*'s, to prevent the craft getting foul of the two intermediate locks.

D D—Locks which are to be clofed up as before ftated in fig. 2.

E E—The fame as fig. 2, and C, the fix locks, to be raifed as in fig. 2.

The names of the different locks in this figure, and the numbers, which are the fame in fig. 1 and 2, will point out the names in the latter figures.

Fig. 4.—Plan, and fig. 5. fection of the propofed alterations under the great arch.

H—A body of *Kentish* rubble for lining the bottom of the river under the great arch, to refift the action of the water, and for raifing the bed of the lock for the fervice of the water-works, leaving five feet depth of water over the crown of the dam at low water, for the fervice of the craft.

I—Blocks of *Portland* ftone, eight feet long up and down ftream, and two feet fix inches thick for capping the top of the rubble.

K K K—Ground pier of larger rubble ftones than the reft, for fupporting the fmaller until a compleat bed is formed.

L L—Fig. 5. dotted lines, fhewing in what manner the ftarling of the great arch may be enlarged three feet on each fide.

APPENDIX.

INVESTIGATION of the anfwer to queftion 1ft, viz.

1ft. IN what degree, at a monthly average, is the natural power of the water, and effect thereof, upon the wheels of London Bridge water-works diminifhed, by the removal of the pier, and opening of the great arch?

An exact folution of this queftion is, perhaps, impoffible, on account of the very great variety of circumftances that affect it; yet from the great number of obfervations that I have made upon the bridge and water-works, relative thereto, by attending to the moft material and leading circumftances, and by repeating fuch as are of fmall account, or whofe effects are likely to balance each other; I flatter myfelf that what is here offered, will be found fufficiently near the truth to fulfil the purpofes for which the queftion was propofed. And here I muft premife this principle, that where water moves through confined paffages, which occafion a fenfible pen or difference of eleva-tion of their furfaces, above and below the paffage, the water moves through the fame as one column, that is, nearly the fame at bottom as at top, abftracted from friction; and therefore, that an increafe of depth is of equal confequence with an in-creafe of width. This appears from reafon to be the cafe, fince every part of the effluent column is put in motion by a column of the fame height, viz. one whofe height is equal to the difference of the two furfaces; this is alfo confirmed by expe-rience, and in the prefent cafe is remarkable, fince it was the action of the effluent column upon the bottom of the river, that has removed fuch a large quantity of matter from under the great arch.

On this account I have taken the fection of all the water-ways below the ftarlings afrefh, and as I did not confine myfelf to any particular part of the locks, but where I found the fhalloweft part, there I took the width between the ftarlings, this occafions fome difference between thefe and former meafures.

Plate 1. Fig. 1.—Is a fection of the water-way at London Bridge, in which the dark fhaded place A. reprefents the folid of the pier, ftarling, and ground upon which

they

they ftood, before the alteration; which folid being converted into an opening, and the firft and fecond arches from the fouth fhore being clofed up to ftarling's height, fhows the fection of the water-way the beginning of February, 1763, from whence it appears, that when the furface of the water is at the heights hereunder expreffed, the fum of all the water-ways, taking width and height together, was

	3½ Feet above the Starlings.	At Starling's Height.	3½ Feet below the Starlings.
Before the alteration, - - -	4365	2524	1656
Beginning of February, 1763,	5245	3231	2244

Within thefe limits the principal part of the bufinefs of the water-works is on an average performed. The mean proportion therefore of the increafe of water-way, above and below the ftarlings, will ftand thus:

At 1 foot 9 inches above the ftarlings, as 3444 to 4238.
At 1 foot 9 inches below the ftarlings, as 2090 to 2737.

Let us now confider what alteration will be produced in the motion of the water through the openings enlarged in the proportion above fpecified.—London Bridge may, in fome meafure, be confidered as a fluice, through the openings whereof the water of one part of the river is difcharged into the other, moving one way or other according as each furface is refpectively higher or lower.—Now, if the water's paffage is enlarged, the water will be difcharged fooner from the higher to the lower; but not fo much fooner as in proportion to the increafe of opening, becaufe that would require the fame velocity to be continued, and confequently the fame difference of elevation of furfaces to be preferved; which cannot be, unlefs a greater quantity poured into the pool below in the fame time, could run off without fwelling to a greater height, which is impoffible. The difference of furfaces therefore being leffened, the velocity of the water will be diminifhed, which depends thereupon, and the time proportionably lengthened. On the other hand, fince the quantity of water, fooner or later, muft pafs through every fection of the fame river, if any fection is enlarged, the velocity will be diminifhed, and diminifhed in the fame proportion as the fection is increafed, where the water has a free paffage and goes off as it comes, without being pent up; but when, being pent up, a part of the obftacle is removed, the velocity of the iffuing

water

water is diminifhed on the account before mentioned, but not fo much diminifhed as the fection is increafed, becaufe, were that the cafe, the head of accumulated water would be as long in running off through a larger opening as through a fmaller, which is contrary to our firft pofition, and to matter of fact. The enlargement of the water-way therefore acts two ways; that is, partly by diminifhing the time in which a given head is run off, and partly by diminifhing the velocity of the water fo iffuing; and fince each of thefe principles has an equal claim, they will be feparately and inverfely, as the leffer water-way to a mean proportional between that and the greater.

Now, if the effects of thefe two principles upon the water-works, followed the fame proportions, they might be both confidered together as one; and the diminution of effect of the water-works would be fimply as the increafe of water-way; but this they do not, for the diminution of effect on account of time will be as the diminution of the time fimply; but the diminution of effect on account of velocity, will be as the fquare of the water's velocity; for the wheels will not only diminifh their velocity in the fame proportion as the velocity of the impelling fluid diminifhes, but their velocity will be further diminifhed in proportion as the impulfe is weaker thereupon, and the impulfe being weaker in proportion as the diminution of velocity, the velocity of the wheels acting againft the fame load, will be diminifhed in duplicate proportion, that is, as the fquares of the velocity of the water, as I have had formerly an opportunity of proving by many experiments.

The effect therefore of the enlargement of water-way, upon the water-works, being compounded partly of a fimple, partly of a duplicate proportion of the enlargement of the water-way, will be in a proportion between the two, that is, the proportion which is called fefquialtera.

Now, taking the fefquialtera proportion of the above-mentioned enlargements, viz.

	Direct proportion.	Sesquialtera.
Above the ftarlings -	3444 : 4238	3444 : 4702 i..e. 1 : 1.365
Below the ftarlings -	2090 : 2737	2090 : 3120 1 : 1.500

the diminution of the effect of the wheels will therefore be,

At a medium above the ftarlings, as 1.365 : 1

At a ditto below ditto   -   -   1.500 : 1

But the general proportion upon the whole will be compounded of the above proportions, and of the times of the continuance under the above proportions refpectively.

At

At a neap tide, February 8th, the wheels at a medium worked five hours, and at a spring tide, February 15th, 6½ hours. The respective times above and below starlings, were as follows:

					Above the starlings.	Below the starlings.
Neap tides	-	-	-	-	1¾ hours	3¼ hours
Spring tides	-	-	-	3 ditto	3½ ditto
				Sum 4¾	6¾

Now, as it is the proportion, and not the absolute quantity of time, above and below the starlings, that comes under the present consideration, it is of no material consequence, though the length of time during which the water wheels continue at work each tide should differ sometimes from those observed, provided the proportion of the time above the starlings to that below is nearly the same, which we may reasonably presume: Multiplying therefore the number 1.365, expressing the proportional difference of the effect of the wheels while the water is above the starlings, by the sum of the times 4¾, we shall have 6.48375; and in like manner, multiplying 1.5, the number expressing the proportional difference of effect below the starlings, by its correspondent sum 6¾, we shall have 10.125.

Now 4¾   6.48375
6¾   10.125

Sum of the times 11½, the sum of the products 16.60875; and dividing the sum of the products by the sum of the times, we shall have 1.444, which will express the mean proportion of effect above and below the starlings, the proportion of time considered and included; which proportion of 1.444 to 1, or 1444 to 1000, would be the proportion of the power of the water before and since the alteration, did not some other cause intervene, which, though of less consideration, may not inconveniently be brought to account.

The water-way of the great arch being much more open and free than the rest, it is reasonable to suppose that the velocity of the water is less impeded by friction than in the others, which will increase the proportional discharge by the great arch, and in consequence increase the above proportion. I found by experiment, that when the water was at its medium workable height below the starlings, its velocity was ⅕ quicker through the great arch, than through the rest at a medium; but as the difference when the water is above the starling will be considerably less than as above specified,

and

and yet it will be fomething, I therefore take $\frac{2}{3}$ of the above quantity, that is, $\frac{2}{15}$ for a medium of the whole. Now, $\frac{2}{15}$ of 1444 is 193, which, being added, make 1637 : 1000, for the proportion once corrected.

2dly. On account of the increafe of water-way at London Bridge, I find that the tides rife at a medium, four inches higher above bridge than they did before ; and that at a medium the rife is from low water above bridge, about eleven feet; fo that the body of water is increafed in perpendicular height, and at leaft in quantity, by $\frac{1}{33}$ part of the whole. Now this additional quantity will act upon the wheels in the fame manner as the main body, that is, partly by increafing the velocity, and partly by lengthening the time ; its effects will, therefore, be in a fefquialterate proportion of the quantity, that is, nearly $\frac{1}{23}$ part of the whole, and of courfe will leffen the number 1637 before ftated ; therefore $\frac{1}{23}$ part of 1637, which is 71, being taken away therefrom, the remainder will be 1566 ; fo that 1566 to 1000, will exprefs the proportion in which the natural power and effect of the water upon the water-wheels is diminifhed by opening the great arch, exclufive of fuch caufes and irregularities as cannot be brought into the account.

Ever fince the year 1756, a daily regifter has been kept at the water-works, containing the time of high and low water, the height to which the tides rofe above bridge, with the number of ftrokes of flood and ebb that was made by each pump of the wefternmoft or firft wheel in the fourth arch. To this book having, by permiffion of the company, had free accefs, I have endeavoured to extract therefrom fuch materials as might afcertain their average performance, before and fince the alteration of the great arch : this I did by extracting three adjoining of the beft fpring, and three adjoining of the worft neap tides, during the courfe of a month, at each quarter of the year ; the months were March, June, September, and December, which take in the equinoctial and folftitial tides, and the fummer waters will be balanced againft the winter ; fo that the average tide for each year was collected from the medium of forty-eight tides, critically chofen at fuch feafons as were moft favourable, and moft unfavourable for the performance of the engines. An abftract of which is as follows :

	Number of strokes.
Average tides for the latter half of the year 1756	2803
for the year - - 1757	2888
do. - - - 1758	3029
Carried over	8720

	Brought over	8720
Average tides for the year - - 1759	3215
do. - - - 1760	3099
do. - - - 1761	3142
for the firſt half of the year 1762	2998
	Sum	21174
	Mean	3025

		Number of strokes.
Average tide for July - - - - 1762	2846
Auguſt - - - do.	2448
September - - do.	2193
October - - do.	2121
November - - do.	2090
December - - do.	1689
February - - 1763	2248

From the above it appears, that for ſix years preceding the alteration, the average performance of this engine was 3025 ſtrokes per tide.

That from the time of beginning to take away the ſtarling, the performance gradually leſſened, till December, when it fell remarkably ſhort, and as remarkably increaſed in the month of February; which irregularities were undoubtedly owing to the remarkable drought in the month of December, preceding the froſt, and at a ſeaſon when the tides at ſea naturally run ſhort, and the other to the remarkable tides and floods that ſucceeded the froſt; ſo that balancing thoſe two months againſt each other, the average will be 1968; but if we take the four months of October, November, December, and February, ſince the alteration, the average will be 2037. So that, comparing the average performance of thoſe engines, before and ſince the alteration,

they will be as $\begin{cases} 3025 \text{ to } 1968, \text{ that is as } 2000 \text{ to } 1300. \\ 3025 \text{ to } 2037 \qquad\qquad 2000 \text{ to } 1346. \end{cases}$

but by calculation as 1566 to 1000   2000 to 1277.

Hence

Hence it appears that the deficiency, as obtained by the company's regifter, being at moft 700 in 2000, is not fo great as refults from calculation, which amounts to 723, whereas from confidering the great proportional increafe of friction in the works, when acted upon by a fmaller power inftead of a larger, one would have expected the difference to have been on the other fide.

But this I apprehend to be balanced, and perhaps more than balanced, by the following circumftance:

It has been already proved, that the wheels work flower than before in a duplicate proportion of the difference of the water's velocity; fuppofe this, for argument's fake, to be $\frac{1}{4}$; now if the wheels work at a medium $\frac{1}{4}$ flower than before, the water will move $\frac{1}{4}$ flower through the pipes, which amounts (refpecting the engine) to the fame thing as if the area of the pipes were made $\frac{1}{4}$ larger: in confequence whereof the water moving to its place of deftination with greater facility, will not require fo tall a column to impel it, fo that the water will not rife to fo great a height in the ftand pipes, and in confequence the engine acting againft a lefs column of water will tend to revolve fafter, and in fo doing will, I apprehend, in the prefent cafe, produce an equivalent to the greater proportion of friction.

1763.   J. SMEATON.

ESTIMATE for the works fpecified in anfwer to queftion the 3d.

To 3000 ton of rubble stone for raising the bed of the river under the great arch, and securing the same, at 10 shillings per ton, - - - - - - - £1500 0 0
To 320 feet running of additional wharfing to the starling of the great lock, at £4 - 1280 0 0
To 1040 cube feet of *Portland* blocks, for capping the crown of the dam of rubble under the great arch, laid in place, at 3s. 6d. - - - - - 182 0 0
To 240 feet superficial of plank piling, for stopping up Chappel and Long Entry locks, at 5s. 60 0 0
To 2000 ton of rubble, for raising the beds of the five other locks, at 10s. - - 1000 0 0
To 4200 cube feet of fir timber, to make the fender piles before Long Entry and Chappel locks, including brace piles and braces, every thing fixed in place, iron work included, at 4s. per foot, - - - - - - 840 0 0

4862 0 0
To contingencies upon the above accounts, at 10 per cent. - - - 486 0 0

Total £5348 0 0

N. B.

N. B. It will take 2000 tons of rubble to line the bed of the river under the great arch, for the security thereof. The second article is chiefly calculated for the same purpose; and the last article, being wholly for the service of the navigation, which may be applied or not after the other works are performed, as the use or necessity of the thing shall direct. These articles being deducted from the above estimate, there will remain £1742 to be expended on account of the water-works; which, with its proportion of contingence, comes to £1916. The necessary repair of the starlings, on account of the damage done by the ice, is not included in the above estimate.

The REPORT of John Smeaton, engineer, concerning the state of the great arch of London Bridge.

HAVING carefully viewed the piers and ftarlings, and taken foundings of the depth of water, both underneath and for fome fpace both above and below the great arch of London Bridge, before, at, and after low water, on Friday, the 25th July, 1766, I have the pleafure to affure the committee, it does not appear to me, that either of the piers, or the arch dependent thereon, has given way in the fmalleft degree, nor does there appear to be any crack or fettlement therein, by failure of the foundation, or otherwife, that I obferved ; and though a part of the out-works of the fouth pier has fuffered fome derangement by the continued action of a very rapid and powerful current, yet, by the feafonable application of 200 tons of rubble ftones, it appears to be fecured for the prefent, fo that I don't find the ftructure in immediate danger. Having faid thus much for the fatisfaction of thofe who may have fuppofed otherwife, I will now proceed to report the refult of my own obfervations, as to the caufes of the prefent failure, and what appears neceffary for further fecurity : but firft I muft premife, that

When I was called upon, in the beginning of the year 1763, I found the depth of water under the great arch, at low ftill water, to be twenty-two feet, the current making hourly depredations upon the ftarlings, the fouth-weft fhoulder of the north pier undermined fix feet, and the original piles, upon which the old works had been built, laid bare to the action of the water, and feveral of them loofened. In this perilous ftate, when a fettlement of that pier muft neceffarily have taken place in a few days, I propofed the only remedy I knew of, that was likely to be attended with fuccefs in circumftances fo preffing, viz. that of fecuring the bed of the river with a body of rubble ftone, upon which the faid angle was under-pinned, and fo far fecured, that it has never yet fhewn any defect. It was alfo thought proper to raife this body of rubble fo high as to pen up the water for reftoring London Bridge water-works to the fame power as before the alteration, and to defend the ftarlings, next the great opening, by an *additional* out-work of three feet broad, on each fide, furrounding the former cafing; which matters were fully fet forth in my former report, dated 18th March, 1763.

On founding at low ftill water, on Friday 25th, I found no where more than eleven feet water, and in general not above eight feet, immediately under the great arch, but above and below, in a line, and at the points of the ftarling, I found twenty-four and
twenty-five

twenty-five feet water below, and from twenty-two to twenty-six feet above bridge, I also found in going downward, in the direction of the stream, at the distance of about thirty-five or forty feet below the points of the starlings, the water to deepen to thirty-three feet; and at seventy or eighty feet, which was the deepest place, thirty-eight feet. On the upstream side of the bridge, about fifty feet above the point of the starlings, I found twenty-eight feet water.

From the above soundings it appears evident, that from the continual wear of the very powerful and rapid current which sets through this arch, that the bed of the river, above and below the points of the starlings, has been worn away to a much more considerable depth than it was in the year 1763; in consequence of which the bed of the river, from the points of the starlings upward and downwards, forming a slope too great for the rubble to lie upon, when impelled by so strong a current, the skirts of the rubble will naturally slide into the cavities; this, in consequence, has impoverished the body of rubble immediately under the arch, which it did on the south side so as to loosen the piles for about twenty feet in length, of the additional casing that had been driven into, and had their fixing in the body of rubble; but which, by a timely supply of same material, are now made firm, and this part of the body of the bed of rubble is as high as any of the rest, so that no immediate danger is hence to be expected.

The reasons why the failure of the outward casing happened on the south side, I apprehend to be the following:

1st. That as it was the north pier that complained in the year 1763, the principal attention would naturally be directed to that side, and, in consequence, the best and largest rubble be deposited there.

2d. From the natural set of the current of the river, which for the space of about 100 yards above the bridge sets obliquely over from the north side to the south, and will, in consequence, strike against the north face of the south starling of the great arch, this cross set of the tide is rendered still more oblique by the particular situation of the arches that are stopped up. When my report of 1763 was made, I found the two locks next the *Surry* shore stopped up, and advised the two small locks next the great arch, on the north side, called Long Entry and Chappel locks, to be stopped up likewise; this, with the proposed alterations under the great arch, and the raising the beds with rubble to a certain height, of five other locks, which I found unnecessarily deep, (all which were specified in a draught attending my report), appeared, in my opinion, sufficient to restore
the

the water-works to their former power. Upon my view, on Friday the 25th, I obferved that the two locks next the *Surry* fide were open, and was informed that none of the beds of the five locks had been raifed, but that in lieu thereof, the fifth lock from the north fhore, had been and remains ftopped up. Now, if all the locks on one fide of the great arch were ftopped up, the current would be obliged to make its way towards the open fide; the ftopping up, therefore, of the fifth lock, on the north fide, and the opening of two upon the fouth fide, muft neceffarily conduce to carry the current more obliquely towards the fouth fide than it would naturally do; whereas the ftopping up of the two arches propofed by me on the north fide, would have been no more than a counter balance to thofe I found ftopped on the fouth fide. I did, indeed, fuggeft in my former report, that in cafe it fhould be thought eligible for navigation, two pair of pointed gates, fo as to form a navigable pen lock, fhould be placed in the fifth arch from the north fide: in which cafe, the raifing of the bed of *St. Mary*'s lock, and the ftopping up of Chappel lock, or Long Entry, might be difpenfed with; but this I did not recommend as preferable to my firft fcheme, unlefs more eligible on account of navigation. Upon the whole, it is to be wifhed, for the fake of fecurity to the bridge, as well as navigation, that fome equivalent could be formed to the water-works, fo that all the arches might be unftopped; the great arch remaining as it is.

I now come to advife what is further neceffary, in my opinion, to be done for the fecurity of the bridge, which principally confifts in guarding againft the derangements that probably will arife from the deepening and wearing away of the ground above and below bridge; and for this purpofe I would advife, that, with all poffible difpatch, a further quantity of *Kentish* rubble be depofited between and at the four points of the ftarlings of the great arch, fo as to leffen the too great depth and flope of the bed of the river and of the rubble, to the end that the rubble fo depofited may not work down into the pools; it would be advifeable fo far to leffen the depths thereof, by dropping in rubble, as to make a firm footing for the rubble that lies immediately more contiguous to the piers and ftarlings; further than this would be rather detrimental than ufeful.

After all, it will be neceffary that the condition of the rubble bed fhould be frequently examined, and to have always in readinefs a quantity of rubble to fupply fuch deficiencies as, from time to time, may happen; for, though this method will, like the ftarlings and the reft of the ground-works of this bridge, need occafional temporary fupplies, yet, being duly attended to, there is fufficient reafon to fuppofe that it will prove a lafting fupport; and, if imperfect, yet deferves the greateft attention, becaufe I do

not

not know any practical method by which the great arch of this bridge, so founded and so circumstanced, can be maintained and supported without incurring much greater expenses with less certainty; and when, by the disposition of rubble, the cavities above and below are hindered from pooling, and the foot of the rubble bed, within the starlings, supported thereby, and the whole by time consolidated, it may be expected that the repairs of this part will be very inconsiderable.

J. SMEATON.

London, 28th July, 1766.

To

To the right honourable the Lord Mayor, Aldermen, and Commons, in Common Council assembled.

MY LORD AND GENTLEMEN.

IN anfwer to the order of common council, of the 13th day of March, in which I am defired to *take into consideration the state of London Bridge ; of the navigation under the same, and of the London Bridge water-works ; and also the proposed alterations suggested in the committee's report, and to give my opinion upon the same ;* I muft beg leave to refer the honourable court to my former report, of the 5th of February, wherein is contained the moft material part of what I can obferve upon the fubject.

I, therefore, confider what I now further offer as fupplemental thereto.

Refpecting the ftate of London Bridge, I look upon its greateft weaknefs to be in its foundations, being chiefly built upon piles, fawn off above low water mark ; and this conftruction confidered, I am of opinion, that the lefs the fall and velocity of the water is, in paffing the bridge, the lefs fubject thofe piles, together with the ftarlings and other works for defending the piers, are to get out of order, and into difrepair, and confequently the *less fall,* the greater fecurity to the bridge.

The alterations fuggefted to the committee (as appear in my paper above referred to), directly tend to diminifh the fall and velocity of the water ; and, therefore, fo far as that difference goes, it is in favour of the foundation of London Bridge.

I apprehend this honourable court is, by the act of parliament for the alterations of the bridge, obliged to keep up a head of water for working the engines as effectually as they were worked before the alterations ; and, if fo, I am clearly of opinion, that the fall and velocity muft be maintained greater than they will be, in cafe the alterations *suggested in the committee's* report, are carried into execution ; and, therefore, that confiftently with the maintenance of power to the London Bridge water-works, the fall will be reduced, by the propofed alterations, as much as it can be, and for this reduction that the grant of the further arches may, in fome meafure, be confidered as a compenfation for the lofs of power that they would otherwife fuftain by the other parts of the propofal.

With

With refpect to the navigation immediately under or through the bridge, as nothing can be more clear than that whatever tends to diminish the fall and velocity of the water there, muft be of benefit to the navigation, the only queftion that can remain on this point is, whether, by diverting the navigation from the firft and fecond arches on the *Surry* fide, it will lofe any thing that is material.

The particular ufes of thefe arches to navigation is matter of experience and practice which I have not had ; but, in my own opinion, I apprehended them to be in this refpect of very little benefit, and am the more confirmed in this opinion, by having found them both ftopped up, (and the only ones that then were fo), for the affiftance of the water-works, when I was called upon in this affair, in the year 1763, and the reafon then given was, that thofe two arches were looked upon as the leaft ufeful of any in the bridge, for the purpofe of navigation ; the difference, therefore, in point of navigation, immediately under the bridge, feems, upon the whole, in favour of the alteration propofed ; but, if we confider the navigation of the *Thames* above bridge, I am of opinion, that were the fall at the bridge confiderably reduced, *by any means whatever*, the navigation of that part of the river would be *materially* affected.

It is difficult at this time to determine, whether the bed of the river *Thames* was as high above bridge before London Bridge was firft erected, as now it is ; and whether the ftoppage of the water at the bridge, acting as a dam, was an *expedient* to retain more water in the river at low water over thofe fhoals for the fake of navigation ; or whether this ftoppage arifing from the conftruction has, in its *effects*, occafioned the fullage of the river gradually to gather, and the bed to rife nearly in the fame proportion as the water's furface at low water is kept higher ? however this might be, 'tis certain the bed of the river above bridge, is now, in *proportion*, *higher* than it is below bridge.

If London Bridge were, therefore, to be taken away, the river would become fo fhallow above bridge, at low water, that the navigation would be greatly impeded for hours each tide.

If this difference of bed was *original*, we muft expect it to remain ; but if an *effect*, the caufe being taken away, the river would gradually reftore itfelf ; but as this might probably take up feven or eight hundred years, (the time it has probably been gathering), the work of reftitution would go on far too flowly to anfwer the demands of the prefent generation.

That

That a ftoppage at London Bridge, in the prefent ftate of the bed of the river above bridge, is neceffary to the prefent navigation thereof; and that it cannot greatly be reduced, without detriment, is more than fpeculation. The ill effects were experienced during the time that the bottom of the river was gulled underneath the great arch to the depth of above twenty feet at low water, when, by means of fo great a waterway, the tide-water was difcharged fo much more fpeedily, as not only to prove greatly detrimental to the power of London Bridge water-works, but alfo to the navigation of the river above bridge.

If, therefore, the keeping up of a certain head of water be neceffary to the navigation above bridge, and to enable this honourable court to acquit themfelves of what is enacted refpecting the water-works; if the water-works will, by agreement, (in confequence of availing themfelves of the additional arches), be contented with *such a head of water* as will fubfift after the dam propofed to be removed, and the other alteration fuggefted made; then will this honourable court avail itfelf of all the reduction that the cafe, in all its circumftances, will admit of; and as it appears that fome head of water muft be kept up and maintained, not only for the water-works, but for the navigation, then the more *useful purposes* this head of water is applied to, the more beneficial it will be to the community.

The petitioners fet forth, *that wheels under the four arches in their possession, would not act with the same velocity, at any time, as they did before the alteration of the bridge, when two arches were laid into one :* if this allegation is true, then it will follow that the head of water has never been fo great, fince the alteration, as before. And on this occafion I muft, in juftice, declare, that the feveral matters contained in my report of 1763, on purpofe to reftore the head of water, have never been fully put in execution; and that, except what has been done under the great arch, no part of the directions therein contained has been purfued, except the ftopping up of Long Entry and Chappel locks; the doing whereof has fince been complained of, as making an eddy; and that, in lieu of other alterations by me fuggefted, the fifth arch was ftopped up as it now remains, and with which, I fuppofe, the company were contented; but which, in my opinion, was not equivalent to the head of water that would have been gained, had the feveral alterations propofed by me taken place. Hence it moft evidently appears, that if the honourable court fhould not confirm the propofed agreement between the committee and the company, inftead of reducing the head of water ftill further than it now is, by the further propofed alterations, the court will be obliged to fupport a greater head of water than has ever fubfifted fince the re-

moval

moval of the old pier under the great arch; which, as has already been said, is what, by all means, ought to be avoided, both on account of the bridge and of the navigation.

As the Long Entry and Chappel locks were stopped up in consequence of the directions in my report of 1763, it may not be an unseasonable digression to say a word or two on that subject. It appeared to me from computation, that in order to restore the head-water, as prescribed by the act, it was necessary to stop up some of the locks, and raise the bed of others. In order to specify which locks should be stopped-up, I enquired which of them were least useful to navigation; and, among others, Long Entry and Chappel Locks were mentioned, as being both too narrow and too shallow In order to destroy, as much as possible, the effect that might be produced by stopping up of locks on one side of the great arch, more than of the other, I proposed to have stopped one of the locks next to the great arch, on the Surry side, which, together with the two locks that I found stopped, (and proposed to have remain so), next the Surry shore, would have made three stopped arches on the Surry side, and two on the London side; but, instead of this, the dams on the Surry side were removed, and the fifth arch from the London side stopped up, which made three stopped locks on the north side, and none on the other, by which means the balance of current on each side of the great arch not being preserved, the current of the great arch must, necessarily, tend to that side where it meets the least resistance; that is, towards the two adjacent stopped arches, and form a greater eddy there than it would have done, had its tendency been equal both ways.

In my last report, I have declared it as my opinion, " that three locks being stopped " up on the north side of the great arch, and none on the south, is the greatest artificial " cause of the eddy complained of, and that the changes proposed are the most likely " and easy means to remedy the same."

It must not, however, be expected, that the eddy of the great arch will be wholly removed by taking away the dams from the locks above-mentioned; for so far as it depends upon a superior strength or column of water, penetrating into the body of more still water below, it must be expected always to remain.

Upon the whole matter, I am clearly of opinion, that considering the advantages to the security of the bridge, arising from an easement of the fall below, and the being clear of the pipes above, to the navigation through the bridge, and also to the navigation of the Thames above bridge, this honourable court cannot do better than con-firm the propofed agreement, which the committee have formed with the proprietors of the water-works; it being, I fuppofe, clearly underftood that the faid proprietors give up all claim to further alteration of the bridge, from its prefent ftate, on account of the head of water directed by the act of parliament to be maintained.

I am, my lord, and gentlemen,

your moft obedient,

and moft humble fervant,

J. SMEATON.

London, 23d June, 1767.

INSTRUCTIONS

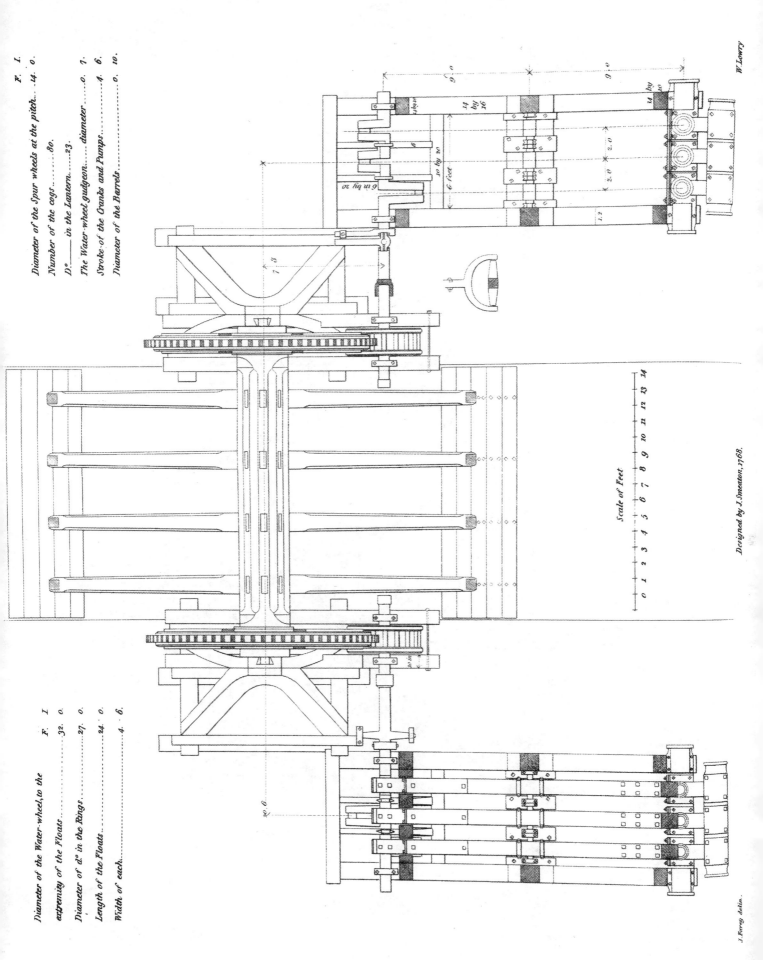

W. Lowry.

Diameter of the Spur-wheels at the pitch... 14. 0.

Number of the cogs.........80.

D°._____ in the Lantern.......23.

The Water-wheel gudgeon.... diameter.....0. 7.

Stroke-of the Cranks and Pumps.......4. 6.

Diameter of the Barrels.........0. 10.

Diameter of the Water-wheel, to the   F.   I.

extremity of the Floats.........32. 0.

Diameter of d° in the Rings.......27. 0.

Length of the Floats.........24. 0.

Width of each.........4. 6.

Scale of Feet

0 1 2 3 4 5 6 7 8 9 10 11 12 13 14

Designed by J. Smeaton, 1768.

J. Farey delin.

ELEVATION of the GREAT ENGINE, for the FIFTH ARCH of LONDON BRIDGE.

Spur Wheel, 80 cogs.
Lantern......23 dº
Pumps, 10 inches diameter.
Length of strokes, 4 fᵗ 6.

Scale of Feet.

0 1 2 3 4 5 6 7 8 9 10 11 12 13 14 15 16 17

Designed by J.Smeaton,1768.

Level of the Sterlings top.

INSTRUCTIONS for Mr. NICHOLLS to proceed with the water-wheel and spur-wheels for the great engine, to be erected under the fifth arch of London Bridge.

	Feet.	Inches.
1st. The diameter of the water-wheel, to the extremity of the floats, - -	32	0
2d. The diameter of ditto in the rings, out and out, - -	27	0
3d. The width of the wheel, or length of the floats, to be as great as a sufficient clearance of the lock will allow ; I suppose - - - - - -	15	6
4th. The number of the floats to be twenty-four,		
5th. The width of each to be - - - - - - -	4	6
6th. The diameter of the spur-wheels to be, in the pitch circle, - - -	14	0
7th. The number of cogs eighty,		
8th. The number of ditto in the lanthorns, twenty-three,		
9th. The water-wheel gudgeons to be in diameter, - - - -	0	7
10th. Length of the cylinder part, - - -	0	7

REMARKS.

I suppose the water-wheel to stand as near the arch as possible, to leave a sufficient clearance, and to go as near the bottom as possible to be safe.

2dly, I propose the tails of the starts to reach through the rings, so as to support the inside floats, and to make up the whole breadth four feet six inches; the rings to be clasped by the boards, half-way from one side, and half-way from the other.

3d. I suppose that four rings will be sufficient, at four feet distance, middle and middle ; more rings than necessary, on account of strength, being prejudicial.

4th. I suppose the gudgeons to be fixed on crosses, seven inches in breadth on the face, and $7\frac{1}{8}$ breadth on the inside, and to be $3\frac{1}{2}$ inches thick. One hoop at each end will be sufficient, and should be five inches broad, and at least $\frac{3}{4}$, (but better of one inch), thick ; it should have eight holes for strong short spikes or dowells, two in each quarter, and about $1\frac{1}{2}$ or two inches from the inside ; the hoops should be adapted to the crosses, but to bring all home tight, a grouve in each end should be left for iron wedges, as described in the elevation plate II. As the intersection of the cross will form a base

larger

larger than the gudgeon, this is to be filled up by a fillet of about ½ an inch projection, according to the length of the axis, which will not only ftrengthen the joining of the gudgeon to the crofs, but make a regular fhoulder, which will keep the end of the axis clear of the head ftock. The different breadths between the faces and back of the arms of the crofs forming a dovetail, will, when wedged in, be alone fufficient to keep them faft, but not depending altogether on the wood, four fcrews, of about one inch diameter, are put in about four inches within the ends of the crofs, as defcribed in the elevation plate 2. which faften with four nuts, let fideways into the diftance of about eight inches behind the infide of the hoops, which done, the holes are plugged up again in the manner of bed fcrews. The heads may be cylinders of about two inches diameter, and one inch in length, to be funk with a fquare fhoulder, flufh with the arms. N. B. The hoops muft be as large as the wood will allow.

5th. I fuppofe the conftruction of the arms, rings, ftarts, and floats, to be according to the method of fixing which has been found, by experience, to anfwer beft, as alfo the fpur-wheels are propofed much larger than what has been ufed before; that in cafe compafs arms cannot eafily be procured, in order more equally to fupport the circumference, the fquare formed by the clafp-arms may be one foot on a fide, if neceffary, larger than the axis, which, being filled with blocks, and wedged, may be fixed as tight as if the axis were of that fize.

J. SMEATON.

Newcastle, 17th July, 1767.

To

To the Committee of London Bridge Water-Works.

GENTLEMEN,

HAVING carefully examined and confidered the conftruction of your fire engine, I am of opinion as follows : That as in the prefent fituation of the pipe for fupplying the well, the engine cannot have a proper quantity of water at low water, I entirely approve the propofition of laying an horizontal fuction pipe, in order that the engine may take the water immediately out of the Thames.

This will, however, lay fome additional burthen upon the engine, but as at prefent it works by a fecondary power, that is, the power of the atmofphere lifts a weight, which weight, in defcending, raifes the column of water; the addition of burthen would be more than compenfated, were the engine made to raife its column immediately by the power of the atmofphere.

To produce this effect, it will be neceffary to change the working barrel, with the valves and machinery dependent thereon; and at the fame time that the working piece is changed, a new one may be introduced, whofe proportion will be more advantageous.

At prefent, the whole of the water raifed by the fire engine is raifed to the top of the tower, 120 feet high, as I am informed, though it moft commonly happens that one half, two thirds, or three-fourths of that height, would be fufficient. I would, therefore, propofe it as a very material improvement, not only to fave fuel, but to raife more water, to unite the main from the fire engine with thofe from the water engines, by which means the fire engines will, like the water engines, never be burthened with a greater column than is fufficient for the fervice then on: to this I am fenfible there is an objection, viz. that the column to be lifted by the fire engine being then variable, the fire engine will not work with the requifite degree of fteadinefs, which would certainly be the cafe in the ordinary way of applying the injection, but as I have found out and experienced a method of applying the injection, whereby the engine-keeper is enabled, extempore, while the engine is working, to vary the quantity proportionable to the column to be lifted; hereby the ill effects arifing from a variation of the column will be altogether prevented, and a proportionable faving will be made in the fuel.

As

As the engine at prefent, when lifting the whole column to the top of the tower, would bear a greater load, and would confequently be confiderably under loaded when the column was lowered as above propofed, inftead of a twelve inch, as now, I would propofe to put in a thirteen inch working barrel, by which means the engine will every ftroke raife one fifth more water, and by virtue of the changes above mentioned, I expect the coals will be reduced from three bufhels per hour, as at prefent, to two at an average.

Some other fmaller alterations may alfo be made with benefit: and the whole machine wants a repair, particularly the arch heads and chains, and the beam will want to be ftrengthened.

The boiler is too fmall for the cylinder, and not of the very beft proportion; but as I underftand it is in good condition, and may ferve fome years, I do not at prefent recommend any alteration in that part.

J. SMEATON.

London, March 8, 1771.

P. S. I do not mean that the engine will lift the water to the top of the tower with two bufhels of coals an hour; but I expect it will (befides the additional load of fuction) work a thirteen inch working barrel in lieu of a twelve, and without any additional quantity of coals, in confequence of altering its mode of lifting.

J. S.

11th March.

The

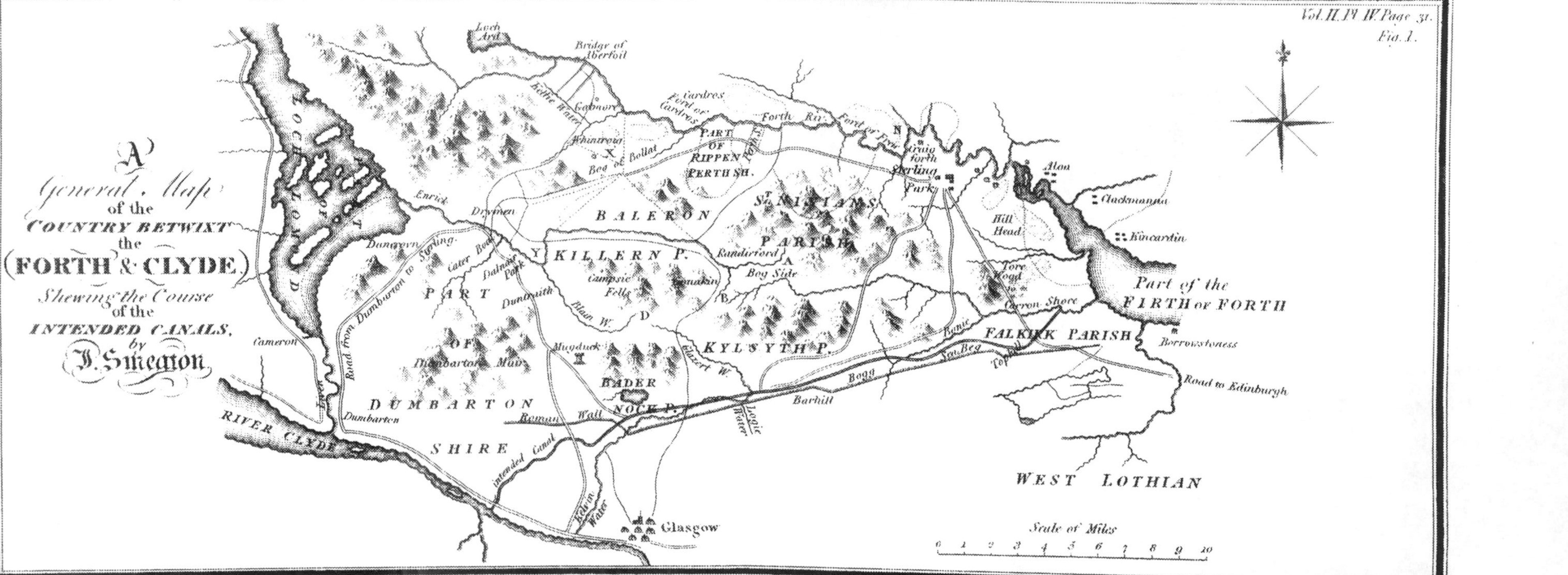

A General Map of the COUNTRY BETWIXT the (FORTH & CLYDE) Shewing the Course of the INTENDED CANALS, by J. Smeaton.

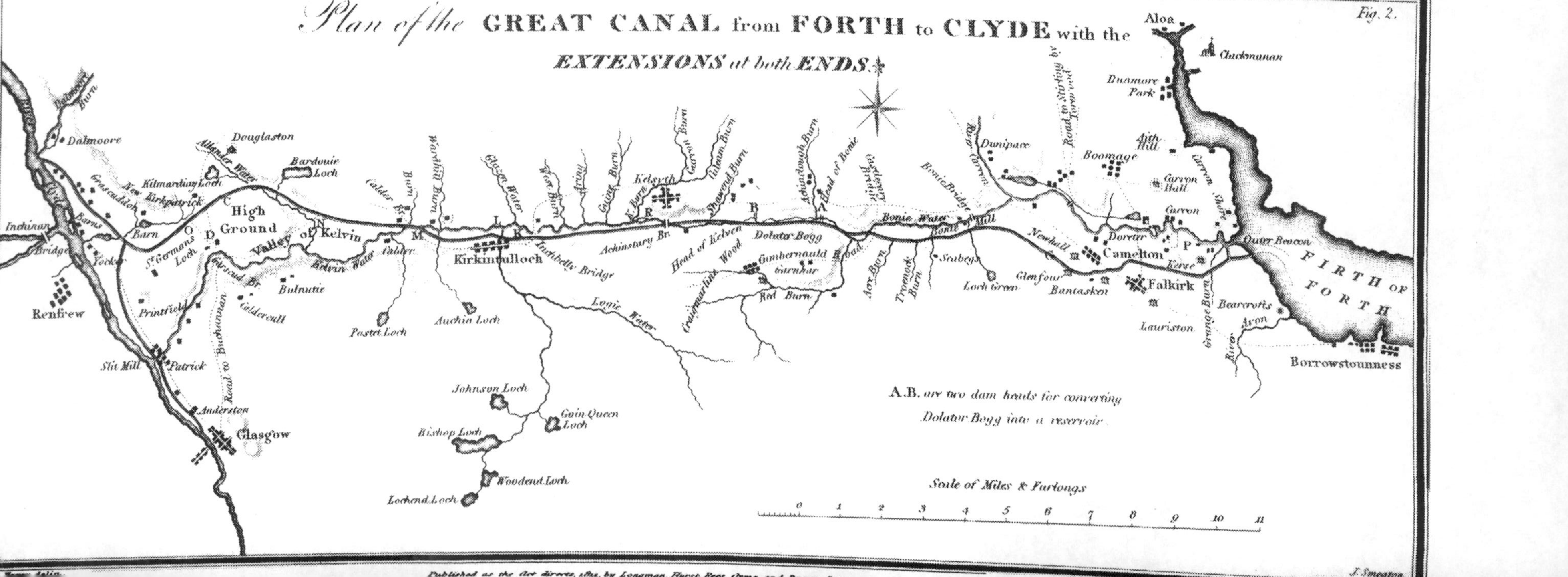

Plan of the GREAT CANAL from FORTH to CLYDE, with the EXTENSIONS at both ENDS.

Fig. 2.

A.B. or two dam heads for converting Dolator Bogg into a reservoir.

Scale of Miles & Furlongs

The FORTH and CLYDE NAVIGATION.

To the Honourable the Trustees for improving Fisheries and Manufactures in Scotland.

The REPORT of JOHN SMEATON, Engineer and F. R. S. concerning the practicability and expense of joining the rivers Forth and Clyde, by a navigable canal, and thereby to join the east sea and the west.

THE great utility of a navigable communication between the east sea and the west, has given occasion to several projects for this purpose, in different parts of the kingdom of Great Britain. The principal of which are the Thames and Severn, the Trent and Severn, the Trent and Weaver, the Calder and Mersey, and the Forth and Clyde. But from what I have seen and heard of these matters, I am well convinced, that by far and most easy to be accomplished, not only in point of distance, but in point of perpendicular height, is the last, and it is somewhat remarkable, that notwithstanding the country at this place lies in general as high, if not higher, than any of the rest ; yet, that through this high ground, there happen to be two different passages, both lower than any of the others, and so much appearance of equality in point of practicability, that upon ocular surveys, it has remained a doubt which of the two ought to be preferred.

One of these passages is from the river Carron, by way of the water of Bonnie, through the bog of Dolater into the Kelvin, and from thence into the Clyde by way of the Yocker Burn. See the plan, pl. IV. fig. 2.

The other is by following the river Forth some miles above Stirling, and then crossing over through the bog of Bollat into the water of Enrick, down to Loch Lomond, and from thence by the river Leven into the Clyde, at Dumbarton (fee the map, plate IV. fig. 1.), and as I have, by order of the honorable the trustees for improving fisheries and manufactures in Scotland, been at some pains in taking proper observations in order to determine the preference of the two passages, I shall first endeavour to settle this question, and then proceed more minutely to explain the several leading points that occur for laying out the design for the preferable tract.

GENERAL

GENERAL DESCRIPTION of the CARRON PASSAGE.

See the plan, fig. 2. and the black line, fig. 1.

From the Forth to about three miles up the river Carron, to a place called Carron Shore, there is at prefent a good navigation for fea-veffels, drawing from nine to ten feet water, at fpring tides, and from feven to eight at neap tides, which, from the cuts propofed to be made acrofs the Loops, as I am informed, is likely to be improved; and fo far I confider the prefent fea-navigation to extend.

From Carron Shore to Top-hill, a little below Camelton Bridge, the ground is gently rifing, and adapted as well as poffible for a canal. From thence to the point above Camelton, between Newhall and Glenfour, the ground rifes near fixty feet in half a mile, and is more uneven, but yet attended with no confiderable difficulty; from thence the canal may be managed upon the decline of the high ground fo as to run upon a dead level to Caftle Cary Bridge, being about four Englifh miles; this tract of ground is interfected with haughs and burns, which occur very frequently, and will therefore require a confiderable expenfe in making banks, aqueducts, and other extra works, yet nothing occurs but what may be conquered at a reafonable charge. The principal fingle work will be an aqueduct bridge in croffing Bonnie mill Burn, near Bonnie Bridge, over which the canal muft be carried upon arches. Another, but lefs work, of the fame kind, will be required in croffing the river Bonnie, at Caftle Cary Bridge. From hence to the bog of Dolater, which begins a little above Wineford; the paffage is fufficiently eafy by a gentle afcent, and through the faid bog, (extending almoft to Craigmarline wood), there is fcarcely four feet defcent either way, in the compafs of two miles; about the middle of which is the fummit between the two feas, and is elevated above high water neap tides, at Carron Shore, about 147 feet, being diftant therefrom about nine Englifh miles. On each fide of Dolater bog the country rifes high, which affords a number of fprings and rills, which brings down confiderable quantities of water on the leaft rain; but the principal, which is conftant all the year, is the burn of Achinclough, which difcharges itfelf into the bog, and which, though it now runs into the waters of Bonnie, might, with very little trouble, be turned, fo as to run the other way into the Kelvin; for a drain that now paffes lengthways through the bog, has, for half a mile in the middle thereof, no fenfible run at all; fo that the fame trench is the head both of the Bonnie and Kelvin. This bog, as it is called, is properly a peat mofs or morafs, and is in general near half a mile broad, but contracted

at

at each end by the high land at A and B, to about 120 yards; fo that it is adapted as well as poffible for the formation of refervoirs, canals, and paffages for water: indeed the whole bog might eafily be put under water, by a fmall dam between the high grounds at each of the contracted places. From the fummit, or point of partition, weftward, the Kelvin runs in an open valley for about thirteen miles, upon a very regular and almoft infenfible decline as far as Garfcud Bridge, where it begins to be more rapid, and runs through a very deep channel, quite confined by rocks and precipices, in fome places almoft perpendicular, and down to Partick, where it falls into Clyde. Of this diftrict, all that can be faid, is, that it is not impoffible to be made navigable, but it would coft a great fum to make a bad navigation; for this reafon, thofe who have turned their thoughts towards this project, have fought out for a new paffage into Clyde; and Nature has kindly furnifhed two that are practicable, one by the valley of the Allander, leaving Douglaftoun, Kilmardoney, and new Kirkpatrick, all on the right, and cutting through a rifing ground between the two laft places at C, we are conducted into a valley leading down to Grafcaddon. The other paffage leads out of the Kelvin's valley, juft above Garfcud Bridge, and going by way of Canny's Burn, traverfes a rifing ground at D, and falls into St. German's Loch, and from thence, by the fame valley as the former, to Grafcuddon. Neither of thefe paffages is without difficulty, on account of the rifing ground to be cut through, being above thirty feet perpendicular above the general level, yet both are greatly preferable to the paffage by the Kelvin, below Garfcud Bridge, which, befides the incumbrances already mentioned, is filled with mills of various kinds, which occupy the ground where any cut could eafily be made. At Grafcuddon the ground falls confiderably in a little fpace, and then the paffage is very eafy and regular down to Yocker Burn, which difcharges itfelf into the Clyde, almoft oppofite Renfield, and a little above a place called Barns of Clyde; from hence downward the Clyde is navigable at all tides with any veffels that can be expected to navigate upon the intended canal, which I would propofe to be fuch as could occafionally go from Port Glafgow or Greenock, to Leith. From the point of partition of the waters, in Dolater bog, to Barns of Clyde, by the courfe of the canal, will meafure about eighteen Englifh miles (to which meafure I all along confine myfelf), fo that the whole length will be about twenty-feven Englifh miles.

GENERAL DESCRIPTION of the Paffage, by way of Loch Lomond.

See the Map, pl. IV. fig. 1.

BEGINNING from Clyde to Dumbarton Caftle, we enter into the river Leven, which, though but $6\frac{1}{4}$ miles long, meafuring by the links or turnings thereof, is yet a large river, and brings down a confiderable quantity of frefh water at all times of the year; for having the great lake, called Loch Lomond, for a refervoir at its head, which receives the water of feveral confiderable rivers, it always affords a confiderable currency in the dryeft feafons, and is never fubject to any great or fudden floods on account of the great capacity of the aforefaid lake. This river is, in a ftate of nature, navigable for flat-bottomed veffels from the Clyde to Loch Lomond, in winter, and wet feafons in fummer; but, in dry feafons, the navigation is in a manner ftopt; for though it affords a good deal of frefh water at all times, yet as it is broad, and runs in a fhallow channel (except about a mile above Dumbarton,) and has about twenty feet fall from Loch Lomond to high water at neap tides, which are fenfible at a place called Dalquurn Bleachfield, about a mile and a half above Dumbarton, confequently there is twenty feet fall in lefs than five miles, which makes the water fomewhat rapid, and, in dry feafons, leaves not above eighteen inches in depth over the fhoals, which are pretty numerous; and though it might not be difficult to improve this navigation, fo as better to anfwer its prefent purpofes, yet it is not capable of any remedy adequate to the purpofe of a general navigation, otherways than by locks and dams, to pen up the water from one to another; for as the river checks in with the high ground, firft on one fide, then on the other, the valley is but ill adapted to a canal the whole length.

From the head of Leven, we traverfe about feven miles of Loch Lomond, in which there is no impediment till we come near the mouth of the river Enrick, which having its fource in a high country, comes down very rapidly, and brings a great quantity of fand from the mountains, which is lodged near its mouth in the lock, and there, as it were, forms a bar; over which, in the fummer ftate of the loch, there is not above eighteen inches water, while it ferves as a ford in moft feafons. To attempt making a paffage through this fand, by lifting the matter, would be endlefs; for, being fituated in a bay which lies open to the weft wind, the fand is brought back upon the fhore by the action of the wind and waves. I have fince been informed that at the fouth corner of this bay, near a place called Aber, the fhore is clear of fand; but being there rocky, to make a paffage through the fame, of fufficient depth below the furface of the loch

in

in its loweft ftate, for the paffage of loaded veffels, would probably be more expenfive than the removal of the bar itfelf, which, I apprehend, might be done by running out two jetties about ¾ of a mile into the loch, (for, according to my information, 'tis that diftance from fhore, in dry feafons, before we fhould find any navigable water), in which cafe the current of the river, being confined between the jetties, would fcour out the fand, and keep the paffage clear.

From the mouth of Enrick, the river runs in an open valley, in a deep channel, and almoft upon a dead level, as far as Cater Boat, oppofite Drummond; which, in a right line, is about 3½ miles, but by the courfe of the river, which is very winding, cannot be lefs than fix. From hence to the fouth-weft corner of Dalnair Park, the river meafures 2½ miles; the perpendicular rife is almoft eleven feet. The valley fo far continues open, but as the channel is very winding, and checks in with the high ground on both fides, feveral times, the beft way of making it navigable will be by locks and dams, as the Leven.

A little above this, that is, a little above where the water of Blain falls into Enrick, the valley is quite contracted, and the river falls over a fet of rocks, forming a re-markable cataraĉt, called the Pot of Gartnefs, marked Y, to pafs which, in the channel of the river, would be very difficult, but this may be avoided by cutting acrofs the neck of land from the fouth-weft corner of Dalnair Park into the river above Gartnefs Mill, the length will be about a quarter of a mile, and the perpendicular rife about thirty-feven feet. In this paffage we meet a hill, whofe fummit rifes above forty feet higher than the level of the river, at the aforefaid cataraĉt; but as the length of the hill is not great, it may be pierced with equal eafe, and by the fame means as thofe mentioned in croffing from the Kelvin to Grafcuddon, in the other paffage. This paffage, by the way of Enrick, in the general map, fig. 1. is fhewn by a line of ftrong dots.

Having regained the river Enrick, above Gartnefs, there feems no way but to follow the courfe thereof, 1¾ mile, in which fpace there is a rife of twenty-five feet. This paffage is fufficiently rugged, and the valley confined by fteep brays and rocks; nor do I fee any adequate means of avoiding it, as the country hereabout is very fteep and uneven; however, it may be managed by locks and dams, upon the fame principle as thofe below. From hence, it is propofed to pafs by canal to the bog of Bollat, the fummit of which is diftant a little more than a mile, and is elevated above the aforefaid point in the river, 129 feet, viz. from the river's furface to the top of the

bray

bray thirty-five feet, and from thence to the summit of the bog 94 feet; so that the summit of the bog of Bollat is elevated above the surface of Loch Lomond 202 feet: and above high water neap tides in the Clyde 222 feet, the whole distance being twenty-five miles.

The bog of Bollat is about half a mile long, and about one-eighth of a mile in breadth; from the middle or summit, the ground falls away pretty quick both ways, and the valley widens: no constant running waters discharge themselves into this bog, except one small rill that comes in on the north side; there is, however, a burn, called the Auld Hurr, marked X, which falls into the decline of the valley towards the Forth, which might easily be conducted to the summit; but this, at the time I saw it, after a remarkably rainy season, was far short of supplying a navigation. There is also another burn to the west of Bollat, affording much about the same quantity, which falls into Enrick, the upper part of which might be intercepted and carried to the summit, but still there would remain a great deficiency. I should not, however, despair, did the point of difficulty rest here, of bringing a sufficiency of water to the bog of Bollat, and making proper reservoirs there; for it appears, from the face of the country, practicable to lift the water of Enrick, four or five miles above, so as to bring any quantity required to the point of partition, or even to bring a part of the Keltie water upon a sufficient elevation round the hills to the same place.

From the point of partition in the bog of Bollat, a valley gradually forms itself to the north, into which the waters from the neighbouring hills collect themselves, and form a burn; which, passing by Achintroig, and a place called Offrings, falls into the river Forth, a little below the mouth of the Keltie. From the summit of Bollat to the river Forth is above four miles, the last mile of which being through the valley of the Forth, has but little descent, so that the greatest part of the whole descent is in the three former miles.

From the mouth of Keltie to Stirling, the river Forth runs in a remarkably winding and deep channel, and with a slow current; which is, however, interrupted at three principal places, viz. at the foot of Cardross, where there is a fall of about two feet; at the ford of Frew, where there is the like fall, and at Craigforth mill, at N. about 1¼ mile above Stirling, where there is a fall of four feet.

The falls at Cardross and Frew are occasioned by gravel shoals formed there in the river, and brought down from the hills by the burns that fall into the Forth, near those places;

places; but that at Craigforth mill, N. is occasioned by a natural rock, which runs acrofs the river, the defects of which, as a dam, is made good by art, in order to produce a fufficient pen of the water, to work the mills, and to catch falmon. The fpring tides come up to Craigforth mill, and the river is then navigable for lighters or barges to this place, and, according to my information, at neap tides, within half a mile below this place.

Above Craigforth mill, a navigation is at prefent carried on upon the Forth in fmall boats as high up as Gartmore, except in very dry feafons; thefe boats bring down lime-ftone from the rocks near that place, as I was told. Two locks and one dam would ᵗmake an open navigation from Gartmore to the Firth of Forth, at all feafons of the year; and were there any trade of confequence up this extenfive valley, would be worth the while, independent of a navigable communication between the two feas. One lock ought to be placed oppofite Craigforth mill, and the other lock and dam at the ford of Frew; this, with a little clearance of the fhoal at Cardrofs, would make a navigable paffage over the fame.

From the mouth of the Keltie to Stirling, is about twelve miles, in a ftraight line, but by the courfe of the river, appears to be more than twenty. It may feem that this might be helped, by cutting through fome of the moft remarkable loops, but the furface of the river lies fo much within foil, that thefe cuts, though fhort, would be required to be fo deep, that the advantage gained would be by no means adequate to the expenfe, nor would the making an entire new canal be fo eafy a matter as the feeming flatnefs of the valley would indicate, for, being interfperfed with moffes and broad flades, where the burns come down, which are numerous, the furface of the mofs is elevated feveral feet above, and the flades depreffed as many below the mean furface; fo that what with extra cutting in fome places, and extra banking in others, together with the works neceffary in croffing the burns to get rid of their waters in rainy feafons, this work would be made very expenfive.

From Stirling to Alloa, there is at prefent an open navigation, but through fo winding a paffage as is almoft without a parallel, it being accounted four miles by land, and twenty-four by water: fome of thefe loops, I apprehend, it would be worth while to cut, for the improvement of the navigation.

According to this defcription, it meafures by the courfe of the rivers, &c. from the fummit of the bog of Bollat to Alloa, at leaft forty-eight miles, and from Dumbarton

to

to the fummit of the bog twenty-five miles, fo that the whole courfe, from Dumbarton to Alloa, by the loops of the rivers, will be feventy-three miles; and it further appears, that by the moft direct paffage that can be made, it will be at leaft forty miles.

COMPARISON OF THE TWO PASSAGES.

FROM what has been already obferved, it is manifeft that the Loch Lomond paffage is confiderably further about than the Carron paffage, and that it is at leaft equally embarraffed with difficulties; the only point, therefore, in which a competition can be fuppofed, muft be on account of expenfe. In order to bring this matter into a narrow compafs, I will fuppofe the Loch Lomond paffage to be done at the leaft poffible expenfe, in which cafe the diftance, as before mentioned, will be feventy-three miles. The impediments being fuppofed equal, the difference of expenfe will lie in two articles, *viz.* the difference of length of the artificial part of the navigation, and the difference of the lockage. To reduce the difference of expenfe arifing from the difference of length of the artificial part, to the moft fimple form, I will fuppofe in favor of the Loch Lomond paffage, that the parts of the river to be made navigable, by means of dams, &c. to be the fame expenfe as the fame length of canal, though in general, this kind of work turns out more expenfive in fuch embarraffed fituations.

	Miles.
The length of the river Leven, in a right line,	4
The river Enrick from Loch Lomond, to the S. W. corner of Dalnair Park, in a right line,	5
From hence to the point of departure from the Enrick,	2
From thence by canal to the summit of the bog of Bollat,	1
	12
From the summit of the bog of Bollat to the nearest part of the river Forth, near Offrings,	4
The cuts and dams upon the Forth being reckoned at one mile, will be greatly undervalued,	1
Total length of the artificial part in the Loch Lomond passage,	17
The Carron passage, measures	27
Difference of length of artificial navigation, in favor of Loch Lomond,	10

The

	Feet	Inches
The perpendicular height of the summit of the bog of Bollat, above the neap tide high water surface of the Clyde, at Dumbarton, - - - -	222	0
The height of the summit of Dolater bog, above high water neap tides, at Carron shore, -	147	0
Difference of perpendicular height, in favor of Carron shore, - - -	75	
But this difference of perpendicular height in point of lockage will be double, - -	150	

	£.
I estimate the cutting at £1250 per mile, extra banking, bridges, tunnels, towing-paths, and contingent works, therefore ten miles will come to - - -	12,500
The lockage I estimate at £133 per foot, rise or fall, the difference being 150 feet, comes to	15,000
The difference of expense in favor of Carron, - - - - -	£2,500

So that the Loch Lomond paffage will coft £2500 more than the Carron paffage, is further about by almoft forty miles, and will be attended with an additional lofs of time in paffing, and expenfe in keeping between thirty and forty locks in repair.

I prefume I have now faid enough on this head.

FURTHER PARTICULARS relating to the Carron Paffage.

HAVING already given a general defcription of the path by which the canal may be conducted from Carron fhore to Barns of Clyde, it remains that I now enlarge upon fome matters relative to this fcheme, not yet fufficiently explained.

Firft, As to the method of fupplying the canal with water. The firft and grand principle is to bring a fufficiency of water to the point of partition, which is here in the middle of Dolater bog, between A and B fig. 2. not only to anfwer the expenfe of water in filling the locks on the paffage of veffels, and of the conftant leakage that will always be, more or lefs, through the gates thereof, but alfo to make good any wafte that may happen by foakage through the banks, and into the ground where the canal is carried, above foil, or upon an elevated part of the country, or that may arife from the exhalations from the furface, by the fun and winds in dry feafons; and it is very evident that the fupplies for all thefe purpofes muft depend upon the water brought to the point of partition till other fupplies can be brought into the canal at a lower level in aid thereof. Now, the fpace of canal that I look upon will chiefly depend upon the point of partition for its fupply, will be from the falling in of Redburn, a little above Caftle Cary Bridge, eaft, to Inch Belly Bridge, upon the Kel-

vin,

vin, weft, which includes a fpace of $7\frac{1}{2}$ miles; but before we can determine the quantity of water neceffary to fupply this fpace, we muft firft determine the kind of veffels
propofed to navigate this canal, and the tonnage of goods that may be fuppofed to
be carried thereon.

With this view I examined the ga boats, which ply upon the Clyde, and are capable of navigating that Firth in all common feafons, and which I apprehend by the
fame rule would navigate the Firth of Forth, between the canal and the port of Leith,
if found requifite fo to do. I found that a middling ga boat of fifty-fix feet long,
ftem and ftern, feventeen and a half feet wide, and drawing four feet water, will carry at
leaft forty ton, and this I look upon to be the largeft fize that will be convenient for
an artificial navigation. Now, fuppofe twenty of thefe boats pafs per day, at an
average, in the dryeft feafons, that is, ten each way, and fuppofing them to go full
loaded from the Clyde to the Forth, and half loaded from the Forth to the Clyde,
they will carry 600 tons of goods per day, and 4200 tons per week; in which I reckon
feven days: for though the boats fhould not work on Sundays, yet, as the water will be
amaffing, the capacity of carriage, in point of water, will not be diminifhed, which
will alfo be the cafe, though they do not go regularly. But cafting off the 200 tons
for accidents and difappointments, and reckoning upon 4000 tons per week, this will
amount to 208,000 tons per year, exclufive of what may be further done, when
the fupply of water is unlimited. The lock duty, exclufive of freight, would, I apprehend, very well bear five fhillings per ton; but at two fhillings and fix-pence,
208,000 tons, would amount to £26,000 per annum; from hence I would infer, that
if we could furnifh water in dry feafons for twenty boats per day, there will be water
enough at the ioweft tonnage that can be fuppofed, to raife a much greater annual fum
than can be wanted for repairs, and to difcharge the intereft of the capital to be expended thereon, and in all probability, more water than the trade can poffibly want;
but I would not have it inferred from hence, that becaufe the canal can carry
208,000 tons per year, that there will be 208,000 tons per year to carry; confequently, the price of tonnage muft be fixed upon the probability of carriage, which
I leave to the decifion of thofe who are better fkilled than I am both in the general
trade of this kingdom, and the particular trade of thofe parts.

Now it is certain that every veffel in its paffage cannot require more than two locks-
full of water, out of the canal of partition; one at its entry, the other at its departure,
which locks-full of water, will either furnifh the fame veffel, or fome other, with a
paffage through all the reft; fuppofing them of no greater perpendicular rife: I fay,

cannot

cannot require *more*; becaufe, if two veffels meet at a lock, one going down, the other up, they may both pals with one lock-full; but fuppofing the worft, that is, that every veffel take two locks-full from the canal of partition, the fize of the veffels being as before mentioned, will require the locks to be fixty-four feet long, and eighteen feet wide; and fuppofing the rife at each lock to be three feet from the canal of partition, to Redburn one way, and to Inch Belly Bridge the other, the water neceffary for the paffage of a veffel will be 3,456 cube feet at each lock, at the extremities of the canal of partition, at A and B, and at both 6912, and for twenty boats 138,290 cube feet per day.—A well made lock will not leak above its own capacity in twenty-four hours, but as things cannot always be equally in order, we will allow two locks-full per day leakage, and as they may be mifufed by the carelefsnefs common to watermen, we will allow two more on this account, amounting to 13,824 cube feet, which, added to the lockage, 138,240 cube feet, makes 152,064 cube feet of water, to be expended in twenty-four hours. Now, within the limits before-mentioned, this will be the whole; for, in regard to foakage, as the canal will be carried in the fink of the valley, and dug out of the folid, it will not be above the common drainage of the country, and therefore will be more likely to acquire water by cutting of fprings than to lofe by foakage. In regard to exhalations, I have obferved in refpect to ponds, canals, and other ftagnant waters of a competent depth, that had no fprings, but yet were water tight, that in the greateft extreme of the late dry fummer, the lofs never amounted to above ten inches in their perpendicular heighth before they were replenifhed by cafual rains: now I propofe that every part of the canal fhould be dug one foot deeper than the draft of the water of the boats, by which means the boats will have fufficient freedom to their motion; and the canal will contain within itfelf a refervoir or magazine for fupplying the exhalations.

The next point is, to enquire how this quantity of 152,064 cube feet of water is in dry feafons to be collected, and brought to the fummit of Dolater bog every twenty-four hours; and the firft thing that prefents itfelf is the burn of Achinclough, which difcharges itfelf into the bog as already mentioned.—This burn, according to information I received from the miller at Achinclough, is capable of furnifhing water in the dryeft feafons fufficient to work his mill $4\frac{1}{2}$ hours per diem; now, from the quantity I found neceffary to work the mill, this will amount to 89,343 cube feet per diem.

The burn of Kylfith, by the fame kind of examination, at Kylfith mill, I find will deliver 87,003 cube feet in the fame time; the fum is 176,346 cube feet that thefe two burns will furnifh in twenty-four hours in the dryeft feafons; that is, more by 24,281 cube feet, than as before required, which affords a furplufage of feven locks-full per diem, to anfwer contingencies, befides fome other fupplies that fhall be mentioned.

The burn of Kylfith falls into the Kelvin a little below Achinvole, marked R. at a place about twenty feet below the fummit of the bog, but as this burn comes down from the high ground, it can be intercepted and carried to the point of partition. The moft eafy and convenient way of doing this, will be to take it up at the tail of Kylfith mill, and carrying it eaftward upon the declining ground, it will crofs the ridge of hills to the north-eaftward of a place called Craigftones, and fall into the Shawend burn; with which it can eafily be carried to the place of deftination: from an obfervation not very minute, it appears to me, that there is a fufficiency of elevation to carry Kylfith burn over into Shawend burn, from the tail of the mill; but fhould there, upon a more accurate level, prove fcarcely enough, yet the purpofes hereby intended would not be defeated, as there is a fall of between thirty and forty feet juft above the mill, fo that by raifing the mill wheel, an additional fall of twenty feet might be gained from the tail thereof, without detriment to the mill.—Shawend burn is faid to afford fome water in the dryeft feafons, and in common feafons a confiderable quantity —At the eaft end of the bog, at Wineford, a coal-pit drain difcharges itfelf, which affords a conftant fupply of water, amounting to fome locks-full per day: I have not, however, brought thefe fupplies to account, becaufe, it appears from the preceding calculation, that the two burns of Achinclough and Kylfith are of themfelves fufficient: I have taken notice of thofe others therefore, as a means of fupplying deficiency, if any fuch fhould happen to arife from unforefeen caufes. Nor are thofe the only ones than can be fo applied, becaufe, if it fhould prove neceffary, Redburn, which, together with Achinclough, forms the river Bonnie, a little above Caftle Cary Bridge, may eafily be brought to the point of partition; for the level of the furface of Redburn, at the point of confluence, is not more than twenty-three feet below the fummit of the bog, and diftant about a mile, and as Redburn is very rapid, it may be lifted fo as to run to the point of partition, without going any confiderable diftance up the burn. This burn I look upon from infpection, to be at leaft as good as either Achinclough or Kylfith burns, fo that, together with the refervoirs for water that may be formed both in the bog, and in the hollows of the contiguous hills from whence thefe burns proceed, it does not admit of a doubt but that nature has furnifhed the means of bringing as much water to the point of partition as can poffibly be wanted; and that by methods fufficiently fimple and eafy.

I have

I have mentioned that the space depending upon the waters to be brought to the point of partition, is that between Redburn, east, and Inch Belly Bridge, west; for at all those places the accession of waters, by taking an occasional supply from Redburn into the canal on one hand, and on the other by taking in the Kelvin at Inch Belly Bridge, will not only contervail any deficiency by leakage or soakage, between their junction with the canal and its termination each way respectively, but also to allow of one foot of greater height in the locks below those points, to avoid too great a multiplicity thereof.—By taking the Kelvin into the canal at Inch Belly Bridge, the whole collection of little rills and springs that fall into the Kelvin below the level of Kylsith mill, (of which there are several constant ones) will be brought to account, and the communication with the river Loggie, in crossing the same at Kirkintilloch, will be an absolute security against any defect in that branch below the same; so that there is all possible certainty that the canal will be amply supplied from end to end.

Secondly. Having now supplied the canal with water, it remains with me to show how this is to be done without injury to the mills and other works upon the rivers Carron and Kelvin; which, below the mouth of the Bonnie on the Carron, and below Garscud Bridge on the Kelvin, are not only valuable, but said to be in want of water in dry seasons.

In order to put this matter in a clear light, let us examine for how much water we have, by the preceding scheme, taken credit of the aforesaid rivers; and it will appear that we have taken the burns of Achinclough and Kylsith, with so much from the others as shall be wanting to make up the deficiencies of the former, in supplying the lower parts of the canal with water.

	Cubic feet
For supplying the locks of three feet rise between Redburn and Inch Belly Bridge, we shall want as much as before estimated per day, - - - -	138,240
Now, if the locks were proposed to be of three feet all the way, the same water would serve them quite through, but being proposed of four feet below those points, they will require an addition of one-third of the foregoing quantity, which will amount to - -	46,080
The leakage of the locks were before estimated at - - - -	13,824
Total required -	198,144
The Achinclough and Kylsith burns supply - - - -	176,346
Deficiency to be taken from Redburn, Kelvin, and Loggie - - -	21,798

which

which deficiency is nearly one-eighth of the whole quantity, fupplied by the two burns ; that is, as thofe two burns are faid to turn their refpective mills, 4½ and four hours, which is together 8½ hours, as much water muft be taken from the other rivulets as will turn a common burn mill, with an overfhot wheel, a little more than an hour, in order to make good the furplufage above the two burns.

I have not brought the exhalations, and foakage through the banks, into the above ac-count; for, with regard to exhalations, as the whole is propofed to be dug to an extra depth, in order to take in a furplus quantity of water in time of rains (at which times the mills have water enough to fpare), this may be laid out of the account : and in regard to foakage, though a fupply for that purpofe will be wanted from the burns, yet as that will return again into the rivers, it is not loft to the mills ; and though fome part of the foakage may fall into different valleys, yet as it muft be expected, that the rills and fprings, which will be interfected by the canal, that ufed to run into thofe valleys, will counter-balance the foakage of thefe parts of the canal, there will ftill be no water loft to the mills on account of foakage ; the quantity then that we are to reftore, is the value of the two burns, one equal to Achinclough, and the other to Kylfith, and one eighth part more of each refpectively ; and this may be done by intercepting and turning fome waters that fall into other rivers, into the Carron and Kelvin.

Having, for this purpofe, examined the country, I find that a part of the river En-rick, near its fource, may be intercepted a little above a place called Randeford, which, being carried to the weft of the ruins of St. John the Graham's caftle, and paffing by a place called Bog Side, through a morafs, lies nearly upon a level between the two rivers will fall into the Carron ; the place is marked in the general map, fig. 1. with a line A.   The waters of Enrick, at the point where it is propofed to be turned, appear to be more than double thofe of the burn of Achinlough ; but if unforefeen occafion fhould happen to require a further fupply, it may be procured by turning the upper part of the burn of Gonakin into Carron, which rifes from fprings in Campfie Fells, and now falls into Enrick ; this is marked with a dotted line B.   With refpect to Kel-vin, the water of Glazert, which falls into Kelvin, and the water of Blain, which falls into Enrick, have their fource near the fame place, there being a continued flat valley between them, like the Kelvin and Bonnie : near the point of partition of the waters, is a remarkable cafcade, or cataract, called the fpout of Ballagin, marked D, which, falling from the fouth fide of Campfie Fells, immediately forms a confiderable burn, whofe general courfe is into the waters of Blain, but is fo critically fituated, that in time of floods, a part of its waters is fometimes difcharged into the Glazert, and with a

very

very little trouble would be made to do fo conftantly; this is marked in the general map by D. This burn appears equal to the burn at Kylfith mills, and, therefore, may be fubftituted in its ftead; and the eighth part thereof ftill wanting may be made good by intercepting a number of fprings which iffue out of the hills on both fides at a fufficient elevation to be carried over the point of partition into the Glazert, which now falls into the waters of Blain. Thofe fpings may be intercepted on the north fide, if occafion fhould require, almoft as far as Duntraith.

To this method of fupply it may poffibly be objected, that in order to re-pay the mills on Carron and Kelvin, I have robbed thofe on the Enrick and Blain : but here I muft obferve, that where there is more work than water, to take away a part thereof, though fmall, is a proportionable lofs ; but to take away a fmall quantity of water, where there is much more water than work, is no lofs at all.

Upon the water of Enrick and Blain, there are no mills of any confequence ; the moft confiderable are thofe at the Pot of Gartnefs, Y, which, as I am infomed, are far from being fully worked, and, if they were, as a confiderable part of the fall there is unemployed, their power might be greatly increafed, though the quantity of water were a little dimi-nifhed; and it muft be remembered, that the whole quantity wanted hence, is only an equivalent to Achinclough burn, and one-eighth more, which burn turns an overfhot mill $4\frac{1}{2}$ hours per day ; whereas the river Enrick, at the Pot of Gartnefs, is a confiderable river, and has a fall there, as before obferved, of between thirty and forty feet. However, to cut off all altercation refpecting mills, the truftees to be appointed for the execution of this fcheme may fafely engage to give the mills on the Carron and Kelvin as much water in dry feafons as they take away, and to make good all damages that may accrue to the others. I fhall now more particularly defcribe fome things, which, though hinted at in the general defcription of the Carron paffage, have not been fully explained.

Thirdly. The fize of the canal I propofe to be twenty-four feet in the bottom, and the fides to be floped at a medium, in the proportion of five to three ; that is, for every three feet depth to widen five feet on each fide, or ten feet on both fides ; to be five feet deep of water, and at a medium feven feet deep within foil, fo that its mean width at the furface of the water will be forty feet eight inches, and to be made wider in convenient places, and at proper diftances, for veffels to pafs eachother. To make the canal lefs for boats of $17\frac{1}{2}$ feet wide, drawing four feet water, would make them draw hard, and to make it larger, would induce an unneceffary expenfe. The locks I would propofe not to exceed four feet rife at each, in order to fave water in working them, and if made with lefs rifes, would be too troublefome on account of

their

their number; only between Redburn and Inch Belly Bridge, as water may there be more fcarce, I have propofed them of three feet rife each.

The feveral hollows that the canal will interfect in its paffage from Camelton to Caftle Cary Bridge, I have propofed to pafs by banking acrofs, fo as to raife the water in the fame up to the level of the canal: the deepeft of thefe is the firft to the weft of Camelon, marked G in the plan of the canal, fig. 2. which is about eighteen feet deep in the deepeft part, and feventy yards wide at the top, the others are lefs confiderable, except at Bonnie mill, which I have propofed to crofs on arches, this will be about thirty feet deep, and feventy yards wide at top, but the arches need not extend above thirty yards, the reft may be done by wharfing and banking. The aqueduct bridge that will be wanted in croffing the river Bonnie juft below Caftle Cary Bridge, will be neceffary, not on account of the depth of the valley, but to give the river a free paffage under the canal; four fmall arches of ten feet each, will here be fufficient; but if it be thought more eligible in the execution, the croffing the river may be effected by forming a dam or weir acrofs the river Bonnie, but this will deftroy the fall of Caftle Cary mill. From hence to Dolater bog there are no difficulties.

The canal of partition extending about 1½ mile between the two locks at its extremes, marked A and B in the plan, I would propofe to be dug out fifty feet wide at bottom, with flopes as before, which will form a refervoir for water, capable of holding three days fupply upon a foot in depth: this is as much as can be wanted, efpecially at firft, till the canal has got a large trade upon it, and if any thing further fhould appear neceffary, efpecially after the Redburn is brought to the point of partition, I would advife it to be done by banking in a part of the bog, by which means it will be formed fufficiently high to yield its contents into the canal of partition by a fluice, as occafion may require; there will be here included a fpace of twenty Scots acres, which, at four feet deep, will contain a fupply of twenty-five days.

From hence to Inch Belly Bridge there is no difficulty. Below this bridge I propofe to put a dam at K, in order to force the water at a proper elevation into the canal K L, which is intended to drop into the Loggie above the mill-dam at L; which dam will again force the water into the canal L M, and which, by proper locks, conducts us into the river at Calder Bridge. A little below this bridge, in like manner, I propofe a dam or weir M, to throw a proper body of water into the canal M N, by which we are conducted to the Allander, where the two paffages divide, by either of which we may pafs upon the fame level into the valley leading down to Grafcuddon, one up the Allander,

the

the other by Canny's burn into St. German's Loch; a little below which, both paffages would unite in the fame point at O. It remains, therefore, to give a reafon, why, of the two paffages, I prefer the latter.

For this purpofe, having taken a level from the furface of the water of the Kelvin, juft below Calder Bridge, to the fummit of each high ground, I found them to differ but inconfiderably in their perpendicular height, the fummit of the high ground by the Allander paffage being thirty-two, and that by the Canny's burn paffage being thirty-three feet above the faid water: I likewife upon examination found the bafe of the hill in the Allander paffage, to be a mile and a half, or twelve furlongs, that would require an extra depth of cut; fix furlongs of which would be above fifteen feet deep: whereas, in the Canny's burn paffage, the whole length that would require extra cutting is not above $2\frac{1}{2}$ furlongs; and there is not quite one furlong that would be above fifteen feet deep. If nature had furnifhed no other paffage than that of the Allander, fomething might be done towards leffening the expenfe by retrograde locks, but as this would caufe hindrance of time, as well as an additional coft in the article of lockage, there appears to me no way by which it can be done either fo well, or fo cheaply as by the Canny s burn paffage.

Where the depth exceeds fifteen feet, which will not be above two hundred yards, I would propofe the paffage to be by a vault under ground; and though the matter of the hill feems to be a loofe gravel, yet I apprehend the means are not very difficult by which it might be perforated. I would begin at an end and open the ground from the top, the fame width I intended it at bottom, that is, eighteen feet, like the locks, with an allowance for the thicknefs of the walls; and cutting down the fides perpendicular, would fhore the fame from fide to fide with boards, beams, and braces, of ordinary Scotch fir, thefe keeping up the matter till the arch is built, the fhores that are above the arch may be taken away, and replaced as the work advances, and the crofs braces that interfere with the walls of the arch may be walled in, and afterwards cut out when the whole is completed.

From St. German's Loch, I would carry the canal upon the decline of the hill upon fuch a level, as, with a little extra cutting, to pafs the high ground that lies on the fouth of the mill-dam of Grafcuddon; this I propofe, in cafe it fhould not be agreeable to the lord of Grafcuddon to carry the navigation through this dam; which is an ornamental piece of water, with planting and walks round it; otherwife this part of the canal will be more eafily done by paffing through this piece of water.—From Grafcuddon

cuddon mill-dam there is between twenty and thirty feet fall in a little fpace, but from thence to the Clyde the ground lies upon a gradual defcent, in which there is no difficulty.

At entering the Clyde, a jetty, formed of rubble ftone, may be neceffary to keep clear the tail of the cut.

Fourthly. With refpect to the number of locks : this depends upon the perpendicular height of the whole, and the perpendicular rife propofed to each lock.—The perpendicular rife of the whole I make to be 147 feet, from the high water furface at a neap tide at Carron fhore, to the fummit of Dolater bog. But as I would propofe to keep the furface of the canal fomewhat below the furface of the land there, we will call the perpendicular afcent 145 feet. In this I cannot pretend to be exact, for the violent fqualls of wind and rain that occured the whole time of the furvey, made it exceedingly difficult and troublefome to manage any inftrument for this purpofe : it is poffible I may have erred fome feet in the whole perpendicular, but as from the direction in which the wind generally blew, I am inclined to believe that what error there may be, is by exceeding the truth, if it fhould be found, upon re-examination, to be five feet lefs than I have made it, (or even double that quantity,) it will make no alteration in the general defign, nor produce any other effect than the faving fo much in the lockage of that perpendicular; and hence it appears, that an exact knowledge of the perpendicular is not at prefent neceffary to determine the practicability of the fcheme.

It may likewife be queftioned, whether the perpendicular elevation of Dolater bog be the fame above the Firth of Clyde, as it is above the Firth of Forth; and it is poffible it may not; but as water undifturbed naturally places itfelf upon a level, and as the difturbing caufes, viz. the wind and tides, frequently act in oppofite directions, we may be affured that the mean height of the two feas is not fo different as to make any material difference in the fcheme propofed. Accounting therefore the perpendicular height both ways 145 feet, we fhall have the number of locks as follows :

	No. of locks
From Carron shore to the intended level of the canal above Camelton, I make to be 108 feet, which, at four feet each, requires - - - - -	27
From thence to Castle Cary Bridge, the canal being upon a level, will require none, but from Castle Cary Bridge to Redburn, there is a rise of sixteen feet, which, at four feet each, will take - - - - - - - -	4
From Redburn into the canal of partition, being twenty-one feet, at three feet each, will require - - - - - - -	7
No. of locks from Carron shore to the point of partition - - -	38

From

	No. of Locks.
Brought over -	38

From the canal of partition to Inch Belly Bridge, I estimate the fall at thirty-six feet, this, at three feet each, will take - - - - - - - 12

From thence to the Clyde there will remain 109 feet, which, at four feet each, will take - 27

Number of locks in the whole - - - - - 77

N. B. The above is the greateſt number that can be required, for if in the execution it ſhould appear, that the ſupply of water ſhould turn out more liberal, or a leſs number of veſſels expected to navigate, than I have ſuppoſed, then ſome expenſe may be ſaved by making the riſes of the locks greater in general ; and without preſerving a ſtrict equality, by which means they might be made more readily to ſuit the ground : it muſt, however, be obſerved, that the higheſt lock determines the quantity of water to be uſed : ifor were there but one lock of eight feet in the whole collection, this would occaſion as great a conſumption of water, as if they were all of the ſame height ; unleſs an extra ſupply could be brought to this lock, to ſet againſt the extra quantity.

Fifthly. Wherever the canal is by banking held up above ſoil, that is, above the natural ſurface of the land, it is propoſed to cut back drains behind ſuch banks, to receive any leakage that otherwiſe might hurt the adjacent ſoil, and to conduct the ſame to ſome common water-courſe ; alſo to lay tunnels under the bottom of the cut to communicate ſuch water-courſes acroſs the ſame as require to be preſerved, or whoſe elevation may not ſuit to be brought into the canal. Alſo to make bridges in proper places over the canal, for communication between ſuch properties as ſhall be ſevered by the ſame, and to make gates upon the towing-paths between every fence, and double gates, if required, for the diviſions between different properties, with over falls in proper places for diſcharging the overplus waters, and whatever elſe may appear neceſſary for preſerving each perſon's property as nearly as may be in the ſame ſtate as before the execution of this project.

Sixthly. It is very difficult, if not impoſſible, upon one ſurvey, to take in every view of a ſubject ſo complicated ; I therefore conſider what I have reported as ſhewing the general practicability of the propoſed ſcheme, with one method of executing the ſame ;

but as it is moſt probable that many improvements may be made whenever it is re-conſidered for execution, the truſtees appointed ſhould be inveſted with ſufficient power for doing whatever might appear neceſſary.

ESTIMATE of the EXPENSE.

	£
The canal being twenty-four feet mean width in the bottom, and seven feet mean depth, with slopes as five to three, the width at top will be forty-eight feet, and mean width thirty-six feet, will contain twenty-eight cube yards in each yard running, and in a mile 49,280 cube yards of digging, which, at 3d. per yard, come to £616 per mile, and for twenty-seven miles in length, to - - - - - - - -	16,632
To extra digging in the canal of partition, to make it fifty feet bottom, 53,308 yards, at 3d.	667
For extra work in making several passing places, and some additional measure in turning the angles, allow upon the whole the value of one mile, - - - -	616
The canal being forty-eight feet top, allow the same width for the banks on each side, that is, forty-eight yards in breadth over all, which makes in one mile 84,480 square yards, which is 15¼ Scotch acres nearly; and for twenty-seven miles, 418½ acres, which, supposing them to be purchased at a medium at £20 an acre, come to - - - -	8,370

> N. B. The greatest part of the tract is corn ground, but as there is a considerable quantity of pasture and moor grounds of small value, and the banks being possessed by the owners of the respective lands, are generally supposed to be of half value, I suppose the above price, at a medium, may be sufficient, if otherwise, the extra value must be added.

	£
Besides what is immediately occupied by the canal and its banks, land will be wanted for conveniences, trenches, reservoirs, &c. which in the whole may amount to twenty acres,	400
The number of locks being seventy-seven, £400 each, amount to - - -	30,800
To extra digging in passing a narrow gripe, between two rising grounds a little to the south-westward of Mungull's house, - - - - - -	100
To extra banking across six haughs, or hollows between that place and Bonnie mills, at an average £100 each - - - - - - -	600
To an aqueduct bridge in passing Bonnie mill burn, and extra banking up the same, with an over-fall for discharging the overplus water of the canal, - - - -	1000
To extra work in passing Seabeg's wood, Trannock burn, Acre burn, and some other haughs and risings, from thence to Castle Cary Bridge, - - - -	400
To an aqueduct bridge for passing the river Bonnie near Castle Cary Bridge, and extra banking there, - - - - - - - -	800
	£60,385

To

Brought up   £60,385

To a shuttle for taking in water, as occasion may require, from Redburn, and an over-fall for
discharging the overplus water - - - - - - -   20

To two miles of trenching, in bringing Kylsith burn, together with Shawen burn, into the canal
of partition, which, at one shilling per yard running, comes to £176; and allowing seventy-
four pound for extra cutting, with such small tunnels and bridges as may be wanted under
and over the same, together with a shuttle and over-fall for taking in the water, and discharging the superfluous ditto, this work will come to - - - -   250

To building a dam or weir across the Kelvin below Inch Belly Bridge, with proper shuttles for
drawing the water off occasionally - - - - - -   500

To building a ditto below Calder Bridge - - - - - -   500

To an aqueduct bridge for crossing the river Allander, and extra banking there - -   300

To extra work in piercing and vaulting a passage through the high ground, between Canny's
burn and St. German's Loch - - - - - - -   1200

To extra work in passing Grascuddon - - - - - -   100

To extra work in making a jetty to defend the mouth of the canal, and clearing the passage
into Clyde - - - - - - - -   150

To five public road bridges, viz at the road from Falkirk to Carron, at Camelton, at Castle
Cary, at Canny's burn, and a Yocker, at £100 each - - -   500

To twenty-one bridges, where the lesser roads intersect the canal, at seventy-five pounds each   1575

To private bridges that may be wanting for preserving the communications between the lands,
No. 28, making, with the public road bridges, at an average, two in a mile, at fifty pound
each - - - - - , - -   1400

To thirteen large tunnels for communicating the lesser brooks under the canal, at forty pounds
each - - - - - - - -   520

To sixty-eight small tunnels for preserving water-courses, making, together with the large
ones, at an average, three in a mile, at ten pound each - - -   680

To making towing-paths, back drains, gates, towing-bridges, &c. at per mile, twenty pounds   540

To bringing the water of Enrick from a little above Randeford, into Carron - -   150

To bringing the burn of Ballagin, with some springs rising out of the hills to the west of the
spout of Ballagin, into Kelvin - - - - - -   100

To temporary damages to lands and mills, impediments and works, unforeseen accidents,
engines, utensils, and supervising - - - - - -   10,000

Total   £78,970

APPENDIX.

APPENDIX.

AS the nobleſt work of the kind that ever has been executed, viz. The Canal Royal of Languedoc, has been generally eſteemed not to anſwer the expenſe, this will undoubtedly be made an argument againſt the preſent propoſition, I ſhall therefore oppoſe ſome matters of fact relating to the French canal by way of parallel.

Canal of Languedoc.

Length of the canal between Port de Cette, on the Mediterranean, and Thoulouſe, is 152 Engliſh miles, beſides a river navigation from Thoulouſe to Bourdeaux, which is above 100 miles more, ſo that the whole length of inland navigation is above 250 Engliſh miles between the ſea ports.

Suppoſe the veſſels make way, at an average, 1½ miles an hour, the paſſage will take fourteen days of twelve hours each.

The perpendicular height of the point of partition is 639 Engliſh feet above the two ſeas.

The canal is navigated by 100 locks, of above eight feet riſe each.

The expenſe of this undertaking has been £612,500 ſterling.

Canal propoſed between the Forth and Clyde.

Length of the canal between the Carron ſhore and Barns of Clyde, is twenty-ſeven Engliſh miles. Sea veſſels going up to Carron ſhore, and to Dumbarton, the diſtance of which by the canal will be thirty-three miles, but reckoning from port Glaſgow to Leith, the diſtance will not exceed ſeventy Engliſh miles.

Suppoſe the veſſels to make way, at an average, 1½ miles per hour, the paſſage will be but four days of twelve hours each.

The perpendicular height of the point of partition in the propoſed canal, is 145 feet.

This canal is to be navigated by ſeventy-ſeven locks, of between three and four feet riſe each.

The expenſe of this undertaking is eſtimated at £78,970.

This

This work is ftill incomplete, by reafon of fhoals in the river Garonne, below Thouloufe, which, in dry feafons, greatly interrupts veffels, to carry the canal below which, as the only remedy, is eftimated to coft £43,750.

This canal drops into the rivers Carron and Clyde, in the tides way, in places where there will be no obftruction to the veffels propofed to navigate the canals: in the dryeft feafons, at high water, to proceed upon their voyage to their refpective fea ports.

Now, if it may be admitted, that as great a trade is likely to be carried on between the Forth and Clyde, where the fea navigation is long and dangerous, the inland fhort and eafy, as between the Gulf of Lyons and the Bay of Bifcay, where the fea navigation is open at all times, though long; the inland navigation tedious, and, at fome times, almoft impaffable for want of water over the fhoals in the river; the inference is very plain; that the fame tolls which will hardly keep the French canal in repair, will make this a very beneficial undertaking to the Britifh adventurers.

POSTSCRIPT.

Firft. On further confidering the matter of the oregoing report, I am of opinion, it will be eligible to carry Redburn into the canal of partition, not as a fubfidiary expedient in cafe of need, but to be applied in the firft conftruction; the charge thereof will be amply recompenfed by the omiffion of five locks between Redburn and Inch Belly Bridge; for the three feet locks propofed in that diftrict may then be four feet, the fame as all the reft; and it will appear from what has been faid above, that the water expended on the whole, will be precifely the fame: fo that then the number of locks will be thirty-fix each way, and in the whole feventy-two.

Secondly. Since the foregoing report, with the preceding article, by way of poftfcript, was delivered to the honorable the truftees for fifheries and manufactures in Scotland, Mr. Smeaton has difcovered that, notwithftanding the care and pains he took to be correct, he has committed an error, in fuppofing the Scotch chain, with which the meafures of the length of the tract for the land were taken, to confift of feventy feet each, whereas, in reality, it confifts of feventy-four, this difference will not, however, in any refpect affect the general principles upon which the preceding fcheme is built, nor the eftimate in any articles, except where the length of the canal is concerned. This length now appears to be $28\frac{1}{4}$ miles inftead of 27, as before fuppofed, and this addition

of

of 1¼ mile, will affect the eftimate in the whole £1322. 10s. but as by the foregoing article of poftfcript, there will be a faving of at leaft £1600, the eftimate upon the whole will be rather lefs than greater, in which there is ample allowance for contingencies.

N. B. The true meafure of the canal is 2010 Scots chains, which, at feventy feet per chain, comes to twenty-fix miles, five furlongs, feven poles, 4½, this, for a round number, was called twenty-feven miles, but at feventy-four feet to the chain, makes almoft 28¼ Englifh miles.

J. SMEATON.

Comparative ESTIMATE between a canal from Forth to Clyde, for veffels of forty tons, and for thofe propofed in the contracted fcheme.

	£	s.	d.
Estimate, as it stands in Mr. Smeaton's report - - - -	79,970	0	0
To this add the expense upon additional length, as mentioned in the postscript -	1,322	10	0
Sum total	82,292	10	0

Deduct as follows:

	£	s.	d.
Saving in making of thirty-six locks, of eight feet high, instead of seventy-two of four feet - - - - - - - - - -	9,200	0	0
Saving by half a foot depth of canal - - - - -	1,678	0	0
Ditto, by extra digging, in widening the canal of partition, which, as there will be plenty of water in proportion to the work, will be unnecessary - - -	667	0	0
	11,545	0	0
	68,647	10	0

	£	s.	d.
Saving, by making thirty-six locks of ten feet wide instead of eighteen, at £75 each	2,700	0	0
Ditto, by making the canal twelve feet narrower, and one foot shallower, than supposed in my estimate, by which means it will be equally accommodated to boats of nine and a half, as I have supposed for seventeen and a half - - - -	8,076	0	0
	10,776	0	0

Saving

	£	s.	d.
Brought over -	10,776	0	0
Saving, by purchase of lands, the whole width being thirty-two feet instead of forty-eight, one-third must be deducted, amounting to - - - -	2,918	10	0
Ditto, by bridges, tunnels, and other extra articles, amounting in the whole to £9,095, one-sixth thereof I suppose will be saved - - - - -	1,516	0	0
	15,211	0	0
The reduction of several articles, as before - - - -	15,211	0	0
Reduction of £10,000 allowed for contingencies, in proportion to the above reduction	2,594	0	0
	17,805	0	0
The first estimate reduced on account of water - - - -	68,647	10	0
Difference in the value of a canal, according to the dimensions proposed for a little one	£50,842	10	0

This sum is £100 too little, but was so delivered in.

QUERIES

QUERIES propofed by the Right Honorable Lord CATHCART.

1. Is not the beft way to render the Leven navigable, the removal of a bar where it iffues out of Loch Lomond, which would lower the lake ten feet perpendicular, and reduce the fall into the Leven ten feet, which would facilitate the making two locks and and dams to complete the navigation?

2. Is it not probable that the grounds gained off the lake in this manner would do more than anfwer the expenfe of removing the bar? and is not £4,125, a proper charge for ten feet of lockage, and two and a half miles of artificial navigation in the Leven?

3. Is it not probable, that were the channel of the Forth ftraightened from Alloa to Offring, the tide would rife to that place, as it is but eight feet above the level of the prefent neap tide above Stirling Bridge, and the tide which comes from Leith to Alloa, in three quarters of an hour, takes the fame length of time to go up to Stirling Pier, which is not four miles in a ftraight line, and a quarter of an hour from thence to the Bridge which is but a few yards when ftraightened? And would not the ground gained by this operation in all probability anfwer the expenfe of it?

Thefe rivers, made navigable either by the method propofed, or fome better expedient, it is to be confidered, how the junction of the two feas can beft be made; whether, by the Carron canal, or joining the Forth to the Enrick and Leven, or by carrying goods over land from Forth to Enrick?

Is it not better to cut a canal from the upper part of Leven to Oatter along the ground, which will be uncovered by lowering the lake, than to navigate the lake, which is tempeftuous, and the Enrick, which will be difficult to enter at the mouth, after the lake is lowered?

4 Does it appear impoffible to run a level drift from that part of the Enrick where he propofed to infert his canals under the fummit of Bollat, and iffuing out of the hill on the fide next the Forth, and pointing towards Cardrofs, the faid drift not exceeding two miles in length, and gaining one hundred and twenty-nine feet of perpendicular height?

The

The reafon for believing this poffible is the defcription of the ground in Mr. Smeaton's report, which fays, the diftance from the Enrick to the fummit of Bollat is one mile, the height one hundred and twenty-nine feet; that from the fummit the ground falls towards the Forth, and it is apprehended in a proportion not unlike that in which it rifes from Enrick; in that cafe the drift would be the bafe of a triangle, the fides of which would be two miles, and therefore would itfelf be lefs than two miles.

5. Might not two hundred and two feet of height be faved by a drift of four miles from the level of Oatter, and coming out fomewhere in the channel of Keltie burn?

6. What ought the dimenfions of fuch drifts to be, in order to carry the canal through the hill at one or the other of thefe points, and what, at an average, might be the expenfe?

The rubbifh will certainly be drawn up to the furface through pits made for that purpofe, which will not be deep; and, therefore, it is apprehended, the length of the drift ought not to increafe the proportion of the expenfe; it is alfo imagined, in fuch a drift, a towing-path would be unneceffary.

7. Would it not be a fhortening of the navigation, both in the Bollat and Oatter paffages, to join the Forth at Cardrofs, rather than Offring?

8. Suppofing that the public, or that private adventurers were to undertake the Carron canal, how long would it probably be before the faid canal could be navigated, and what would be the tolls, or rather, what would be the fum of the toll neceffary for re-imburfement and repairs?

9. Suppofing the Oatter paffage to be undertaken, how foon would it be practifed?

10. If the Bollat or Keltie paffage were to be made choice of, the Oatter paffage would be a part of either; but though the Carron paffage were pitched upon, would the temporary advantage, and the eventual refource of the Oatter paffage, in cafe of accidents in the Carron canal, together with the toll which would arife from it before the Carron canal could be finifhed, be proper and fafe motives for laying out £12,000 upon it?

11. How many men and horfes are required to draw a ga boat of forty tons?

ANSWER to the queries propofed by the Right Honorable the Earl of CATHCART to JOHN SMEATON, engineer.

TO the firft. I apprehend it would be a very difficult and expenfive piece of work to deepen the channel of the Leven, fo as to reduce the water of Loch Lomond ten feet, becaufe the water of the loch is not held up by any one fingle fhoal, but by a fuc-ceffion of fhoals from the mouth of the loch, to within a mile of Dumbarton, all of which I apprehend muft be deepened in different degrees, in order to produce the defired effe&t. There is indeed a bar or fhoal in the mouth of the lake, which would pen the water, in cafe the fhoal next below the boat of Balloch was removed, which now pens the water over the fhoal at the entrance; and from the extenfion of width in that place, the water is nearly ftagnant : but fuppofe both thefe fhoals removed, the water would ftill remain penned up by the natural bottom of the river, which is in general fhallow from thence quite to the tides way; and fuppofing the whole cleared away to ten feet below the uppermoft fhoal, unlefs the fe&tion of the river were very much increafed, the natural declivity of the furface, requifite for difcharging fo great a quantity of water, would ftill produce a pen upon the water of the loch.

The deepening of rivers for any extent I have found a very tedious and expenfive piece of work; and, in fhort, I fee no means of reducing the water of the loch at the fame, or, indeed, any thing near the fame expenfe, as making it navigable by locks and dams.

Secondly. How far it might anfwer with refpe&t to the drainage of the bordering lands, I cannot be a judge, unlefs I know the quantity and value of the lands that would be uncovered by a certain number of feet redu&tion, having never feen any part of the coaft, except about the mouths of Leven and Enrick; but it is probable that if a quantity of land were drained hereby, the land-owners would, with great difficulty, be brought to pay for the improvements : £4,125 is the average expenfe of ten feet of lockage for $2\frac{1}{2}$ miles of artificial navigation; but I am of opinion, double that fum would not draw off ten feet water from the furface of the loch, nor would this lockage be faved, if the navigation were carried further, for what is fubtra&ted below muft be added above.

Thirdly But I apprehend the gained lands would by no means defray the expenfe of making the cuts; becaufe it would be fome years before the old loops would be-come

come land, and a good deal of prefent land would be converted into water ; befides, the great depth of cutting required, would make the work turn out a greater expenfe than the making a new canal for navigation only from Offrings to Alloa.

This may appear at firft fight a paradoxical pofition, but if it be confidered that the furface of the water in the channel of the Forth, above Craigforth mill, is in general ten feet within foil, and I fuppofe in general twenty feet depth of water ; the new cuts would in general be twenty feet deep to make them equivalent with the old river, which I take from view to be eighty feet wide at a medium, and as I fuppofe at leaft one-third muft be new cut to make it tolerably ftraight, the expenfe would be much greater, width and depth confidered, together with the drainage of the water for what lies below the river's furface, than digging a new canal fufficient in width and depth for the purpofes of navigation only. It perhaps may be imagined that digging the cuts of fmall dimenfions, and down to the water's furface at firft, may fuffice, and that the current will wear out the cores below water, and widen the channels; but though this might be the cafe between Alloa and Stirling, where the tide runs flow, yet above it would be fo checked by thofe narrow obftructions, that the water being obliged to follow chiefly the old courfe, would be fpent in the meanderings, and the propofed effect deftroyed, and as the reflow of the waters from the country would in like manner take that courfe which upon the whole would be eafieft ; it would probably be many years before the new cuts would be wholly converted into a current river, unlefs the old loops were entirely fhut up by dams, which would ftill greatly increafe the expenfe.

4th, 5th, 6th, and 7thly. It is difficult to fay what cannot be conquered by art, induftry, length of time, and unlimited expenfe, except fuch things as are impoffible in nature, of which kind I do not reckon the piercing the hill of Bollat, or that between Lomond and Keltie, with a navigable paffage, but this kind of work generally turns out fo very expenfive, and fo very uncertain in its expenfe, that I think it fhould not be undertaken but in cafes of abfolute neceffity. That it will be a great and expenfive work to make a navigable canal, which muft be at leaft twenty feet wide, and at leaft two miles under ground, we are very fure, and the moft eligible cafe of it is where there is a free ftone rock fufficiently foft to make it work eafy, and fufficiently compact to make it ftand without fupporting ; in cafe it is fofter, fo as to make it neceffary to arch it, it greatly adds to the expenfe, but in cafe it fhould prove gravel, or any other loofe kind of matter, it will be very difficult to get it to ftand, till it can be fupported by an arch ; and if a running fand, fo far as I am acquainted, impoffible, without opening it from the top, which, though not impracticable at a great depth, is exceffively

expenfive;

expenſive: on the other hand, if it ſhould turn out a hard whinſtone rock, ſo as not to be worked without blaſting, the work would not only go on very ſlow, but vaſtly expenſive, one furlong of which, or a running ſand, might overſet the undertaking after the reſt had been ſuccefsfully executed.—Drains for coal-pits or mines are much more eaſily managed, becauſe, being of ſmall dimenſions, if of ſoft matter, they are more eaſily made to ſtand till ſupported, and if of middling or hard, there is leſs matter to cut.

In the canal of Languedoc, a hill was pierced for the navigation to paſs, and made wide enough for a horſe-path by the ſide of it. The paſſage was pierced through a ſoft rock, thought at firſt ſufficiently firm to ſtand by itſelf, but has ſince been obliged to be ſupported by an arch, as I have been informed; what might be the coſt of it I know not, but this is certain, that though the extent in our meaſure be not above 256 yards, yet it is accounted one of the prodigies of that very expenſive undertaking. The Duke of Bridgwater has lately made the drain of his colliery navigable, and it is carried upon the ſame level as his canal; but the ſubterraneous navigation is only 5½ feet in width; and they were near two years in advancing it half a mile, though no part of it needed blowing: yet many parts were obliged to be ſupported by walls, and an arch of brick: what the particular expenſe has been, I have not been informed, but imagine it not a trifle, nor is it in any degree to be compared with one near four times the width.—To pretend to give an eſtimate for a work of this kind without any data of quality of ſoil, which it is not eaſy to know before hand, would be ſo very vague, that I can hardly tell how to ſet pen to paper in the way of calculation. It may be imagined perhaps that a hole, pierced the ſize of the Duke of Bridgwater's, might anſwer the end, but the inconvenience of twice altering the bulk, with the expenſes contingent thereon, will be conſidered under another head.

Eighthly. I apprehend that ſuppoſing as many men could be ſet to work as could conveniently be properly ſuperintended, the Carron paſſage might be completed in ſix or in moſt ſeven years, but the difficulty of ſupporting and maintaining the neceſſary number of men (ſuppoſe 250) upon the works, in the parts of the country where any of the paſſages are practicable, is an article of conſideration; for if they were obliged to be lodged in barracks, the charge of building and removing them, will be a further article of expenſe.

As to the money to be raiſed by way of toll, £4000 per annum is the intereſt of £80,000 at five per cent. but if undertaken by private adventurers, it ſhould be by no

means

means limited to that sum, as has been done in most of the late Navigation Acts passed for England; but that if the tolls will bear it, to give the adventurers the prospect of making ten per cent.; this not to be exceeded: but I apprehend absolutely requisite, in order to draw the necessary supplies out of private pockets. In this light, it will never be practicable to reimburse the principal, because if it affords less than five per cent. the means are wanting, and if more * unjust, nor indeed is there any reason for it, because the increase of tolls to be taken in order to produce the reimbursement, must be an additional load laid upon the present generation, who are supposed to contribute the whole, in order to ease the succeeding, who will have the benefit of this great work done to their hands, and only have to raise the interest to maintain and support it, which the benefit their trade will receive will insensibly do.—But if done at the public expense, then I suppose it will be thought equitable that the public money should be reimbursed, and in order that it should be done in some reasonable time, the tolls should clear at least £8000 per annum.

With respect to repairs, they would greatly depend upon the first execution, but suppose this done reasonably well, as I have supposed in my estimate, I look upon it, after every thing is completed, that two gangs of ten men each, with an overseer to each gang, and a general surveyor of the works over them, will keep the works in order as to common repairs.

This I have fully stated in the enclosed copy of an estimate I have drawn out for the service of the Board of Trustees, which amounts at the utmost to £1600 per annum, supposing that provision is made from the first, for the great repairs that in time must be wanted, but if the current expenses only are raised at first, and the other committed to the increase of trade, that must in a manner necessarily take place, then the sum of £1136 per annum is the sum required.

Ninthly. I suppose the necessaries for the Oater passage could not be executed in less than three summers, as the first is commonly consumed in getting the works in motion.

Tenthly. I think from what has been already advanced upon the former heads, that it would be by no means worth the while to complete the Oater passage, with a view to serve as a temporary expedient till the Carron passage is perfected, for before the

trade

* If I expose my money to loss upon the uncertain prospect of gain, 'tis apprehended to be unjust to pay off the winning stake, as the tolls may be lowered in proportion.

trade could get into a fettled tract at Oater paffage, the Carron paffage would be completed, and if expected, would hinder the merchants from turning their attention fo effectually upon the benefits that might accrue from the other; for though it might be very eligible, upon a feparate plan, to make both the rivers Leven and Forth navigable, yet to effect a communication over land, two additional large warehoufes, with additional clerks for the receipts and deliveries, muft be planted at the head of each, and all the apparatus neceffary for land carriage muft be provided, all which in a great meafure would be ufelefs when the Carron paffage was completed: and as to ferving in cafe of accident, there is hardly any accident that can happen in the common courfe of things, but would be fooner repaired than the hands and apparatus neceffary in the other paffage could be got in motion: for fuppofing a lock wanted rebuilding, which is the greateft ftoppage that can be expected, the work is not begun till every thing is ready, and then veffels being planted on both fides, the goods are delivered over from one to another, which is certainly much lefs trouble than to employ two warehoufes, two clerks, and a land carriage of five or fix miles between; nay, it is much to be queftioned whether, with fuch an interruption of the water carriage, goods would not be carried cheaper by land the whole way; for fuppofe the Carron paffage interrupted by a fpace of five or fix miles in the middle, which could not be made navigable, I apprehend that goods could be carried cheaper by land the whole length, than by this compound carriage; for the altering the bulk is not only a great expenfe and lofs of time, but by interrupting the certainty of the delivery, much lefs eligible to merchants: for, befides the expenfe of two additional warehoufes as before ftated, there will be the loading and unloading of a veffel, and the loading and unloading of the land-carriage extra in this method of conveyance, and for want of carriages and veffels keeping exact time with each other, neglect of clerks, and the like, if a merchant puts a quantity of goods on board at one end of the canal together, he never could be certain that they would be delivered together at the other end, or in due time.

An experiment was tried in our river this winter: the town of Brighoufe, at the prefent head of the navigation, fetch their coals by land-carriage five miles, from a place within half a mile of the river: fome cargoes of coals were brought up by water, and though the tolls upon coals are but one quarter of merchant's goods, yet they could not deliver them cheaper than by land-carriage, for the loading and carriage down to the river fide, and the loading and carriage from the river to people's houfes, together with freight and tolls, made the coals rather dearer than if fetched by

land

and carriage from the pits, and yet the tolls were no more than one half penny per ton per mile.

Eleventh. I apprehend two men and a boy, two horses and a driver, will be neceſſary to navigate a ga boat of forty tons.

———————

ESTIMATE of the expenſe attending the maintaining and preſerving of the canal from Forth to Clyde by way of Carron water, and alſo for collecting the tolls thereof.

	£	s.	d.
To sixteen labourers, at one shilling per day each, (or their equivalent in men whose labour is reckoned at a different rate), their wages per annum will be fifteen pounds, twelve shillings each, and for sixteen, - - - - -	249	12	0
To two masons and two carpenters, at one shilling and sixpence per day each, (or their equivalent), which comes to twenty-three pounds, eight shillings, per year each, and for the four, - - - - - - -	93	12	0
To two overseers, at forty pounds per year each, to take account of the time, and oversee to the works of the above men, supposed to be divided into two gangs, -	80	0	0
To a surveyor per annum, for directing the repairs of the whole, - -	80	0	0
To workmen's wages, - - - - - -	503	4	0
The tolls and materials consumed by the working hands, I suppose will be equivalent to their wages, the two first articles will therefore be - - - -	343	4	0
Expense of common annual repair, - - - -	846	8	0
To the clerk for collecting the tolls from the toll-gatherers, and paying the same to the treasurer once a month, and keeping the accounts relative thereto, per annum, -	80	0	0
To two toll-gatherers, one at each end of the canal, at forty pounds per year each, -	80	0	0
To one ditto at the point of partition, - - - - -	30	0	0
To six men stationed as lock-keepers on different parts of the canal, to be a check upon the bargemen from doing damage to the works, by running against the lock-gates, leaving the clough running, so as to let off the water, &c. at three shillings and sixpence per week, - - - - - - -	54	12	0
To books, paper, &c. per annum, - - - - -	15	0	0
To letters, messengers, and other contingent expenses, - - -	30	0	0
To clerks, toll-gatherers, lock-keepers, &c. - - - -	289	12	0

To

	£	s.	d.
Brought over -	289	12	0
To common annual expenses, - - - - - -	846	8	0
Common annual expense, - - - - -	1136	0	0
Supposed to be laid by every year, to answer the above, and for the purposes after mentioned, - - - - - - - -	1600	0	0
Then the annual overplus will be - - - - -	464	0	0
This, in twenty years, will amount to - - - -	9280	5	0

I suppose in twenty years time many of the locks will want new gates, all which will gradually fail in a few years after, I therefore suppose them all made at the end of twenty years, and therefore seventy-two locks at sixty pounds per lock, - 4320 0 0

The bridges, and other works of timber, will likewise want repair, which I suppose upon the whole to amount to one-sixth of the locks, - - - - 720 0 0

Sum of these repairs, - - - - - 5040 0 0

This, taken from the above accumulated sum of £9280, leaves in hand at the end of twenty years, - - - - - - - - 4240 0 0

The interest of this sum for twenty years more, at three per cent, - - 2544 0 0

The accumulation of the overplus sum of £464, will, in the second twenty years, amount to, as before, - - - - - - - 9280 0 0

Money in hand at the end of forty years, - - - - 16,064 0 0

The second set of lock-gates will, at the end of twenty years more, want renewing as before, and as the thresholds will want renewing also, I estimate the repairs at seventy pounds per lock; this, for seventy-two locks, will be - - - 5040 0 0

And as the second repair of other work may be in the same proportion, that is, one-sixth of the locks, this will be - - - - - - 840 0 0

Repairs wanted at the end of forty years, which deduct from the above sum on hand, - - - - - - - - 5880 0 0

Remains in hand at the end of forty years, - - - 10,184 0 0

The interest of this sum, at three per cent, for twenty years, - - 6110 8 0

The accumulation of the overplus sum of £464, will, in the third twenty years, amount to, as before, - - - - - - - 9280 0 0

Money in hand at the end of sixty years, - - - 25,574 8 0

At

	£	s.	d.
Brought over, -	25,574	8	0

At the end of sixty years I suppose the locks in general may need re-building, but as the greater part of the stone, and some other of the materials may be of service, and the excavation ready made, I suppose they may be as good as at first, for £300 each, at an average, therefore, will cost - - - - - 21,600 0 0

And if we suppose the other works to follow in the same proportion as before, they will cost, to make all good as at first, - - - - - 3,600 0 0

To making all the works as good as at first, at the end of sixty years, which deduct from the money in hand, - - - - - - 25,200 0 0

There remains an overplus in hand, at the end of sixty years, after every thing is made as good as at first, - - - - - - - 374 8 0

N. B. In the preceding estimate I have endeavoured to shew what sum of money applied from the beginning, will preserve the work to perpetuity; but I apprehend this to be altogether a needless supposition; for if the work will defray the common expenses and repairs for the first twenty years, viz. £1136 per annum, there is no doubt but that the increase of trade naturally following the use of these undertakings, will answer the greater repairs that must afterwards follow.

J. SMEATON.

Austhorpe, 22d December, 1764.

To the honorable the board of trustees for fisheries, manufactures, and improvements, in Scotland.

MINUTES concerning Forth and Clyde.

WHEN Mr. SMEATON was applied to to furvey the canal from Forth to Clyde, no eftimate was given him of any quantity of tonnage that might be fuppofed yearly to pafs from one Firth to the other; and being generally fuppofed, that the great dif-ficulty would be to provide water for the neceffary tonnage, the great object of his report was to fhew how large a tonnage of goods might be navigated with the water that was clearly and certainly producible; and that to work the largeft fize boats, that, in his opinion, would be convenient for an artificial navigation: upon that propofition, he formed his eftimate, obferving, at the fame time, "that if in the execution it fhould appear, that the fupply of water fhould turn out more liberal, or a lefs number of veffels be expected to navigate than he had fuppofed, (that is, fuppofing a leffer tonnage to pafs than computed upon), then fome expenfe may be faved by making the locks greater in general (that is higher) and without preferving a ftrict equality." The higheft lock determines the quantity of water to be ufed.

The quantity of tonnage that this canal was eftimated to be capable of navigating in the dryeft feafons, was 208,000 tons a year; but the quantity expected, it feems, is not above 25,000, which is fhort of one-eighth part of the quantity computed upon; fo that hence, it appears, the locks may be of any height that convenience may fuggeft. I am very far from thinking that 25,000 will be the whole annual tonnage of the canal, if executed, to carry fuch veffels as I have propofed; but fuppofing four times as much, that is, 100,000 tons per year; then by the fame rules, the locks may, inftead of four feet, be made eight feet high, this would reduce their number to thirty-fix inftead of feventy-two; and as an eight foot lock may be built as well for £600, as a four foot lock can be built for £400, the whole expenfe in this article only will be reduced to £9200.

It is mentioned in the report, that the canal is propofed to be dug one foot deeper than the draft of water which the barges will draw, in order that the canal may be a refervoir within itfelf, to anfwer the exhalations in dry feafons; but it appearing from the reduction of the fuppofition of tonnage, that there will be a great deal of water to fpare, the extra depth on this account will be unneceffary, and as fix inches beyond their draft of water is very fufficient to allow the veffels freedom of motion, the whole canal may be reduced fix inches in depth, which will produce a faving of £1678, the article of

widening

widening the canal of partition, will alfo become unneceffary, amounting to £667 Therefore, after adding £1322. 10s. the fum by which the eftimate is affected on account of an increafe of length, (fee poftfcript, page 54) and deducting the above articles, the eftimate will then become * £68,647. 10s. 6d. to execute the plan firft propofed, that is, to carry ga boats of forty tons.

In order to come at the merits of the queftion, let us now fee what reduction can be made by a fuppofition of ufing veffels eight feet narrower, and drawing fix inches lefs water than according to the former fuppofitions; and deducting the value of thofe favings, according to the former eftimate, we fhall have the true comparative value of the two plans; for to compare, according to the eftimate of different perfons, working from different dimenfions, and fuppofition of prices, can never be a true comparifon at all, for whatever may be faved by working cheaper, is equally applicable to the plan propofed by Mr. Smeaton; as by the new propofed canal, and therefore leaves the difference nearly the fame.

Suppofing then the locks to be of the fame height and number as above propofed, viz. thirty-fix of eight feet high each, the fame in the fmaller canal as in the larger, (which is a fuppofition in favour of the fmall canal, as the multiplying of locks can never leffen the expenfe) as the walls of the fmaller locks muft be of the fame height and thicknefs, and nearly of the fame length, they muft be of the fame ftrength; there will therefore be nothing faved but in the width of the gates, the floor and the excavation; the whole of which, as nearly as I can judge, will be about £75 per lock, which for thirty-fix, amounts to £2700.

The width of the canal may with propriety be about twelve feet lefs for a $9\frac{1}{2}$ feet boat than for a $17\frac{1}{2}$ feet boat, and drawing three feet fix inches water inftead of four feet; it may be half a foot fhallower: the difference of thefe dimenfions will at the fame prices produce a faving of £8076. 10s.

The ground occupied will be lefs in breadth by twelve yards, which, upon the whole canal, at the fame price per acre, faves £2918. 10s.

The difference in the bridges, tunnels, and other extra articles, affected by a difference in width, I eftimate at £1516.

The

* This fum is wrong, fee page 55.

The sum total of these four articles of saving is £15,211; but as it may be supposed that the round sum of £10,000 at the tail of my estimate, for contingent expenses, will be proportionably less upon a lesser work, the proportional reduction on this account will be £2594, which, with the saving before, amounts to £17,805 difference between the two plans. They will then stand thus:

	£	s.
To execute the greater canal to carry ga boats 17½ feet wide, and drawing four feet water, of forty tons, as first proposed, having plenty of water, - - . -	68,647	10
To execute the less canal for boats of 9½ feet wide, and three feet six inches draft of water - - - - - - - -	50,842	10
Difference -	17,805	0

which is a saving of little more than one-quarter of the sum necessary for the greater canal.

But if, instead of reducing the depth of the greater canal six inches, we suppose the vessels to load down to four feet six inches, then as there will be no fear of wanting water, they will carry twelve tons more, and occasionally may draw four feet nine inches, which will enable them to carry eighteen tons more.

As I was unwilling to over-rate any thing in my report, I have called the tonnage of the vessel therein proposed forty tons, but in reality they ought to carry at four feet draft of water forty-eight or rather fifty tons neat weight; but as the water carriage is often lumped, and more than neat weight carried, that nothing might fall short in practice, I called them in round numbers forty tons; reckoning therefore, according to neat weight, four feet will carry forty-eight tons; four feet six inches, sixty tons; and four feet nine inches, sixty-six tons.

By a like computation, a vessel fifty-five feet long, 9½ feet wide, and drawing three feet six inches water, may (lighter built) carry in smooth water, upon the canal only, thirty-three tons neat weight; but if built so as to sail upon the Firths, cannot carry above 26½ tons: a vessel also of the same width and depth, but thirty-six feet long, may (lighter built) in the canal, carry 20½ tons, but built for the Firths, not above 16¼ tons neat weight.

Supposing

Suppofing the canal to be of the depth originally computed upon, in order from the furplufage of water to take an advantage of a fuperior tonnage of veffels; the article of reduction arifing from fix inches depth of water, muft then again be added; the account will then ftand thus :—

	£	s.
To making a canal from Forth to Clyde, for vessels 17½ feet wide, drawing four feet six inches water; and navigating the Firths equally with the canals, and occasionally the coasts in summer, of fifty tons burthen - - - - -	70,335	10
To making a canal from Forth to Clyde, for vessels of 9½ feet wide, and drawing three feet six inches water, and navigating the Firths, (I apprehend with much less safety, and the coasts not at all) of 26½ tons - - - - - -	50,842	10
Difference of expense -	19,493	0

The merits of the two plans I leave to others to determine.

It is to be obferved that the charge of carriage of goods does not depend upon the toll of the canal, but of that with freight, and other charges conjointly.

It now remains to explain a paffage in my report:

" I found that a middling ga boat of fifty-fix feet length, ftem and ftern, 17½ wide,
" and drawing four feet water, will carry at leaft forty tons; and this I look upon to be
" the largeft fize that will be convenient for an artificial navigation."

I never meant by thefe words to limit the fize of veffels fit for any artificial navigation; nor do I in fact determine any thing, but my own opinion was, that this was as large a fize as would be convenient for an artificial navigation fo circumftanced in regard to water, trade, and profpect of funds to execute.

That this was the largeft fized veffels that could be carried upon an artificial navigation, would have been very abfurd in me to fay, having very frequently afferted the poffibility in point of art, to carry firft rate men of war acrofs this tract; but in this fcheme I paid more regard to what I thought was fit and likely to be carried into execution, than what was poffible in itfelf; and this is not determined any way otherwife than as a matter of opinion at that time. So that if any body afterwards fhould be able to prove the expediency of larger veffels, I have no hefitation of the poffibility of performance.

performance. As to the increafe now mentioned, it is not a new thought; but what I faw as clearly at that time as now; and imagine every body, (being nothing more than a deduction of common fenfe) would fee as clearly as myfelf.

In fine, though after the lapfe of nearly four years, my ideas may be prefumed to be extended; yet I have no great reafon to vary my opinion much on this point; for I conceive, that after having provided a paffage for fuch veffels as can navigate both the Firths, very well and fafely, and coaft occafionally; that to extend the canal to fuch dimenfions as to navigate veffels capable of going foreign voyages, would fo greatly enhance the expenfe, that confidering the quantity of goods coming under this denomination only, the additional convenience betwixt going through and reloading, would no way anfwer the intereft of the additional capital; and if not, I confider it to be no matter whether the money goes from public or private hands, for if the former, the fum had better be applied to cheapen thofe means that are naturally adapted to do the bufinefs the cheapeft way.—After all, this being a matter not within my profeffion as an engineer to determine, whenever the contrary is clearly evinced to the fatisfaction of thofe whom it may concern, I readily fubmit.

ESTIMATE of the probable expenfe of executing a canal through the tract formerly propofed by Mr. Smeaton, from Carron fhore to Barns of Clyde, for veffels drawing eight feet water, twenty feet wide and fixty feet long, and upon a fuppofition that at a medium, each lock full of water paffes fifty tons of water from weft to eaft, and twenty-five tons from eaft to weft, and to navigate at this rate 100,000 tons per year.—

£

For vessels of twenty feet wide, I suppose the canal to be twenty-seven feet bottom, and $8\frac{1}{4}$ deep of water, and at an average $10\frac{1}{4}$ feet deep, within soil; this, with slopes, as in the former estimate, will make the width at top sixty-two feet; this, at 5d. a cube yard at an average, will come to £1906. 13s. 4d. per mile, and for $28\frac{1}{4}$ miles - - **53,864**

The width of the canal at top being sixty-two feet, and as the matter cannot be heaped up so high in proportion to the depth of the canal, as when it is shallower, instead of allowing twice the breadth for the canal for depositing the matter, I think it necessary to allow three times, so that the whole width over all will be eighty-three yards; this will contain $30\frac{2}{10}$ acres per mile, and for $28\frac{1}{4}$ miles, $853\frac{2}{10}$ acres, which, if purchased at a medium of £20 per acre, as before supposed, comes to - - - - - **16,064**

Carried forward **69,928**

Supposing

		£
Brought over		69,928

Supposing the vessels to have their bowsprits to turn up, or unship, then the locks need not be above seventy-two feet long, and being 20½ feet wide, they will pass (in vessels at a medium capable of carrying fifty tons, as before mentioned) 100,000 tons annually, and the locks may be of eight feet rise; but being of greater depth and strength than before required, they will probably cost £1000 each, and being thirty-six in number, will come to   36,000

The extra cutting, banking, bridges, tunnels, and other works, including contingencies in the former estimate, comes to £23,178, but for this work I estimate them at double, which will therefore be - - - - - - - - - 46,356

             152,284

N. B. If small locks are made by the side of the former for lighters, they will cost about half as much as the great ones, that is, an addition of £18,000; and if the vessels are made with fixed bowsprits, the locks must be made considerably longer, and will therefore cost more money, and navigate less shipping, unless the locks are also increased in number.

INSTRUCTIONS for Mr. MACKELL relative to the proposed Canal from Forth to Clyde.

Mr. MACKELL is desired by Mr. SMEATON to examine afresh all the circumstances relative to the tract of the intended canal through Dolater bog, and of the resources for supplying the same with water; and if the time will allow to get the level formerly taken by Mr. SMEATON verified, also to examine the different tracts or passages, whereby the canal proposed by Mr. SMEATON, to terminate in the Carron shore and barns of Clyde, may be extended to the Forth of Clyde into deeper water; and to examine the soundings of the two Firths, till every impediment is incontestably out of the way of all such vessels as are proposed to navigate upon the said canals. To examine the matter or soil of the fords or shallows where obstructions are likely to happen, and particularly such as lie below the mouth of Dalmoore burn, upon the Clyde, as also Erskin sands, and Dumbuck ford, with the length of space that each impediment continues.

Also, to examine wherever any difficulty shall appear likely to arise, in the practicability of cutting the ground, to the average of ten or eleven feet below the soil, whether

ther from rock, bog, mud, or running fand, to afcertain the fame by boring, or fuch method as may appear from circumftances the moft proper in each particular cafe : alfo particularly to note the moft proper tract for carrying the canal from Grafcuddon to Dalmoore burn, how far the canal can be carried below the fame, and what difference of depth of water can be obtained by fuch further extenfion, and particularly, if the canal can be carried into Clyde below Erfkin fands.

To examine likewife the tracts and levels of the grounds moft proper for carrying off the branch canals to Glafgow and Borrowftonefs, noting the length of bafe and fection of fuch grounds as are likely to be above the level of the canal ; and alfo the breadth and fection of fuch hollows as muft neceffarily be filled up or banked, in order to fupport the canal acrofs the fame, with all fuch circumftances as appear to Mr. Mackell proper to be enquired into, in order to afcertain the method of execution of the propofed fcheme, and the true cofts that may be likely to attend the fame.

The above is defired to be done the firft week in July, when, or as foon after as pof-fible, Mr. SMEATON expects to be in the country, to examine the premifes.

J. SMEATON.

London, 16th of May.

The

The Second REPORT of JOHN SMEATON, engineer, and F. R. S. touching the practicability and expense of making a navigable canal from the river Forth to the river Clyde, and thereby joining the east sea to the west, for vessels of greater burthen and draft of water than those which were the subject of his first Report.

HAVING, in my former report, endeavoured to explain as clearly as possible the general situation of this tract of country, together with the principles upon which a work of this kind must proceed; having also explained my reasons for preferring the Carron Passage to that by way of Loch Lomond, and it further appearing to me, that every argument against the Loch Lomond Passage holds the stronger, as the size of the canal is supposed larger;—I shall now confine myself to the Carron Passage, and to such points as the difference of the present design and circumstances likely to attend the execution, make necessary to be more particularly explained or enlarged upon. Supposing, therefore, my former Report, signed the 1st of March, 1764, in the hands of every one concerned, I shall consider this as supplemental thereto.

I must here beg leave to observe, that in the forming of the scheme that was the subject of my former report, I had no data to proceed upon, no estimate of quantity of tonnage, dimensions, or sorts of vessels proper to be used: my instructions from the honorable board of trustees, which were verbal, being, as I understood them, to give them a design for the sort of canal, that in my judgment would best suit the country, when taken in every point of view.

As I considered the making a canal across the kingdom from sea to sea, (though here the most practicable of any place I had seen), as a work of very considerable expense; to have made a design upon so large a scale, as to have stood no chance of being executed, seemed to me in no respect fulfilling the general purport of my orders. I had therefore no idea of designing a canal for sea-built vessels, proper for going upon foreign voyages, no such canal having yet appeared in any part of the world, and this being in a manner, the first attempt toward an artificial navigation by locks in Scotland.

My business, therefore, seemed to be to look out for the least sort of vessel that would safely navigate the two Firths, and occasionally go coastways. Of those I assigned the dimensions and tonnage, and those " I looked upon to be the largest size, that would be convenient for an artificial navigation," for to make larger veffels and works, without making them proper for fea veffels, feemed to me to be likely to be attended with additional expenfes, to which the difference of utility would be no ways adequate. This way of reafoning produced the above expreffion of limitation, not any appearance of natural impracticability: but this matter having been taken up by gentlemen verfed in trade, who are of opinion, that a paffage for fea veffels would be of great public utility, and that fentiment being efpoufed by the principal noblemen and gentlemen of the country, it now becomes probable, that fufficient funds, which before feemed the principal obftacle, may be raifed for the execution thereof.

The propofition therefore now ftands thus:

To afcertain the practicability, general defign, and expenfe of a canal capable of navigating veffels drawing eight feet water, and of fuch as are fixty feet long, or twenty feet wide, with a fupply of water for the paffage of 100,000 tons of merchandize yearly, upon a fuppofition that two-thirds of the above quantity pafs from weft to eaft, and one-third from eaft to weft, and that among different forts of veffels, they carry, when loaded, fifty tons at a medium.

It is alfo propofed to carry off a branch to Glafgow of equal dimenfions to the main canal, and to extend the canal at the weft end as far down the river Clyde as Dalmoore's burn foot, at the leaft; alfo to extend it at the eaft end as far down as a place called the Hewk Farm, near the mouth of the river Carron.

I am alfo defired to confider what will be the difference of expenfe, in cafe the canal is made of feven feet, or of ten feet deep. Thefe are the propofitions or data upon which I am now to proceed.

As a neceffary fupply of water is the principal requifite, I fhall begin with this point firft.

If two-thirds of 100,000 tons of fhipping pafs one way, and the veffels return fome loaded, fome half loaded, and fome empty, fo as upon the whole to carry back half loads, or one-third of 100,000 tons, it will be the fame thing in point of water, as if they carried two-thirds of 100,000 tons each way: now, two-thirds of 100,000 is $66,666\frac{2}{3}$ tons, which, divided by fifty tons, the average quantity of each veffel, gives $1333\frac{1}{3}$ for the number of trips made by the veffels altogether in a year;

and

and as each veſſel will take four locks-full of water for her paſſage backwards and forwards, that is, one at each end of the canal of partition going, and the ſame in returning with her half load, the number of locks-full expended yearly will be 5333⅓.

The tonnage and number of veſſels being conſiderably leſs than what was ſtated in my former report, the locks may be admitted of greater capacity, and perpendicular riſe, as mentioned page 49, which will render the navigation more ſimple, and the ſtoppages leſs frequent. Suppoſing them now to be of eight feet riſe each, to take in veſſels of ſixty feet long, and twenty feet wide, they ſhould be ſeventy-two feet long, and 20½ feet wide, which locks will contain 11,808 cube feet of water to fill them, this, multiplied by 5333⅓ the number of locks-full in a year, gives the annual lockage of water equal to 62,976,000 cube feet, which, divided by 365, gives the expenſe per day 172,482 cube feet.

If we allow the leakage per day equal to four locks-full, as per former report, (page 41), then the leakage will be 47,232 cube feet per day.

With reſpect to exhalations, as the whole canal was before propoſed to be dug one foot deeper than the draft of water propoſed for the boats, and as alſo obſerved (page 41), the loſs by exhalations does not exceed ten inches, the canal would have contained in itſelf a reſervoir of ſupply thereof, but as we now propoſe, in order to ſave expenſe in extra digging, to make every advantage of depth of water, the exhalations muſt be otherways accounted for, and compenſated. When deep ſtanding waters loſe to the amount of ten inches perpendicular, it muſt be in a long drought of three or four months continuance; now, if we ſuppoſe ten inches to be loſt in 100 days, this will be at an average of one-tenth of an inch per day, which very well agrees with ſuch obſervations as I have made ſince that time.

The whole extent of the canal will be about thirty-ſeven miles, and about 55½ feet wide at the water-line, but to take in paſſing places, turning places, and other extra widths, we will take the average width at ſixty feet; the whole ſuperficies then will be 11,721,600 feet, which, at one-tenth of an inch deep, will contain 97,680 cube feet for the mean daily evaporation.

The ſoakage is a matter that it is impoſſible to make any eſtimate of, as it depends upon circumſtances that cannot be allowed for; and though in ſandy and gravelly ground it will at firſt be conſiderable, yet experience ſhews that in a moderate ſpace of time,

time, thefe foils, when fully faturated and wrecked up with the fediments of the water, become as tight as other foils; and as in the courfe of the canal many fprings muft be expected to be cut, which will afford a fupply of water, it is probable that the lofs by foakage after a year or two will be next to nothing*. Yet, that we may not be deceived, I will fuppofe the foakage equal to the exhalations, which I look upon to be an ample allowance, the account will then ftand thus :

			Cube feet.
Water expended by lockage per day,	-	-	172,482
by leakage,	-	-	47,232
by exhalations,	-	-	97,680
by soakage,	-	-	97,680
Total expense of water per day,	-	-	415,074

When I made my former report, it appeared, that the fupply of water, even in the dryeft feafons, was fo fuperior to what was likely to be wanted for the locks then propofed, that I did not fee it neceffary to enter into the computations of fome fupplies that were but barely mentioned; but as we now want a fupply of 415,000 cube feet water per day, inftead of 152,000, as before fet forth, page 41; and as the affair of water has been moft in doubt, I fhall endeavour to fet forth this article more at large, and for this purpofe, I have not only re-examined the fupplies formerly mentioned, but fhall bring fome to account that I was not then informed of.

The burns or brooks that more immediately offer themfelves for fupplying the canal of partition, are the following:

1ft. Achinclough Burn. 2d. Wineford coal-pit drain, or level. 3d. Redburn. 4th. The Garron Burn, which turns Kilfyth Mills, and in my former report, called Kilfyth Burn. 5th. Culeam Burn, not noticed before; this falling into Garron Burn, fome diftance below the mill, forms the Burn of Kilfyth. 6th. Shawend Burn.

Firft. The Achinclough burn, in regard to difcharge of water, I find, as before reprefented, viz. that is, to produce 89,343 cube feet per day, in the dryeft feafons.

Achinclough

* It is to be obferved, on the article of soakage, that in soils subject thereto, thofe parts of the canal will be filled in rainy seafons, and kept supplied with muddy waters; fo that it is to be reckoned upon, that all the most confiderable leakages will be stopped before the canal is opened for ufe.

Achinclough mill has indeed loſt half its water by its being diverted to another mill; but as this mill's water falls into Dolater bog, the whole quantity reckoned upon Achinclough Burn will remain the ſame.

2d. Wineford coal-pit drain is ſaid to be conſtant the whole year; I gaged its water the 17th of Auguſt laſt, and it then amounted to 25,000 cube feet per day.

3d. Redburn has no mill upon it, ſo that it is not eaſy to ſay what its diſcharge may be in the dryeſt times; in common it is as big as Achinclough, which led me to ſuppoſe it the ſame at others, but on my laſt view I was informed that in dry weather it is far leſs, yet always has running water, and never leſs than Wineford coal-pit drain.

N. B. Caſtle Cary mill takes in all the preceding, but its dam head is ſo leaky that no concluſion can be formed therefrom in dry times.

4th. The Garron Burn is ſet down in my former report at 87,003 cube feet in dry times, but from information and meaſures taken on my laſt view, it comes out only 65,365. The miller laſt informed me that the late drought was one of the greateſt he ever knew, yet it continued only to affect them with a ſcarcity of water betwixt ſeven and eight weeks; and that in the dryeſt part of this time they could work the mills three hours per day. From this the above computation is taken, whereas, in 1763, I was told this mill could go four hours, as in all ordinary dry times I ſuppoſe it can, which ſufficiently accounts for the difference of the above reſults.

5th. Culeam Burn was not before pointed out to me; it has no mill upon it, but all accounts agree that it is at leaſt one-third of Garron Burn at all times; it will be carried along with the Garron Burn into Shawend Burn.

6th. Shawend Burn turns a lint mill near a place called Townhead, about a mile from Dolater bog: this mill is ſometimes aſſiſted by a part of the water of Culeam Burn, ſometimes not, but excluſive of any ſuch aſſiſtance, they can, in the dryeſt times, by the water of Shawend Burn only, work two hours a day, and if they get the whole of Culeam Burn, the miller ſuppoſes they could work four hours. Having gaged their water ways, I find this mill in two hours will ſpend 27,432 cube feet, excluſive of a conſtant waſte for want of tightneſs at the dam head, &c.

Beſides

Befides the above, by extending an aqueduct trench to the weftward from the Kilfyth mills, a number of fprings may be intercepted, which form fmall burns that now run down into the Kelvin; among the number of which are the weft burn, Arnig, and Guine Burn, and by like means on the eaftward of Redburn, are Acre Burn, Tranock Burn, and feveral others.

Previoufly however to any computation, I muft take notice, that if 100,000 tons of goods are to be navigated yearly, it is not neceffary to fuppofe that the fame proportion thereof is to pafs in the dryeft time of the whole year as when water is plenty; there are very few navigations but what are fometimes fcarce of water; if therefore a navigation be capable of carrying at the rate of 100,000 tons a year in the very dryeft times, it is capable of carrying a much greater quantity upon the whole, and that without fenfible inconvenience.

I take it for granted that generally the mills upon the burns have a full fupply of water, fo as to keep them going the whole twenty-four hours for full feven months in the year, and that they are fupplied with water to work twelve hours in twenty four for at leaft two months more: fo that there will not be above three months in a year that they can be at fhort water, and probably much lefs than that at the extreme of fcarcity.

Let us therefore examine in the firft place what our fupply will be when the mills are fully fupplied with water, which takes in at leaft feven months in twelve :

	Cube feet.
1st. Achinclough Burn - - - - -	476,496
2d. Redburn in this state supposed equal - - -	476,496
3d. The coal-pit level at Wineford - - - -	25,200
4th. The Garron Burn which turns Kilsyth mill - -	370,388
5th. Culeam Burn, being one-third of the former - -	123,463
6th. Shawend Burn, in this state as good as Garron Burn -	370,000
The ordinary supply per day -	1,842,043

Hence it appears, that in the ordinary ftate of the fupply, the Achinclough Burn alone is fufficient, and that they altogether afford as much water, in one day, as will laft almoft four and a half days, exclufive of the times of great rains and floods, when it is certain as much water may be collected in one day, as will ferve the canal fifty.

Secondly.

Secondly. When the burn affords as much water only as to turn the mills twelve hours in twenty-four, then the account will ftand thus:

		Cube feet.
1st. Achinclough Burn	- - - - -	238,248
2d. Coal-pit level at Wineford	- - - -	25,200
3d. Redburn, supposed now one-third of Achinclough	- -	79,416
4th. The Garron Burn	- - - -	196,088
5th. Culeam Burn, one-third of the above	- - -	65,363
6th. Shawend Burn, equal to Culeam Burn	- - -	65,363
Common supply in dry weather	-	669,678

Hence it appears, that in common dry weather, the burns afford as much as is wanted for the fupply of the canal, and more than half as much to fpare, fo that three days water is then fupplied in two.

Thirdly. We come now to extraordinary droughts:

		Cube feet.
Achinclough Burn, as per former report	- - -	89,343
The Wineford coal-pit level	- - - -	25,200
Redburn, then supposed the same	- - - -	25,200
Garron Burn, which turns Kilsyth mill	- - -	65,365
Culeam Burn, one-third of the above	- - -	21,788
Shawend Burn	- - - - -	27,432
Supply in great droughts	- -	254,328
Computed expense of water per day	-	415,074
Deficiency per day during great droughts	-	160,746

It is to be remarked, that the above deficiency entirely arifes from the allowance for exhalations and foakage, neither of which was neceffary to be brought to account in the fcheme; for exclufive of thefe, there would be a redundancy of 34,614 cube feet per day.—It is however to be noted, that when the mills have but about feven hours water in twenty four, there will be a full fupply for the canal, and that it feldom happens that any drought continues in that degree above feven or eight weeks. It muft alfo be noted, that as the deficiency by exhalations and foakage gradually increafes from

the

the canal of partition toward the extremities, the supplies for this purpose are not necessary to be conducted to the canal of partition, but may be occasionally taken in by the way: and therefore, that these deficiences may be supplied by the several burns whose names have been mentioned, but not brought to account for supplying the canal of partition; yet I shall shew, that independently of these burns, and others that may be brought in to the same purpose, the deficiences above stated may be most certainly and amply supplied as follows:

It is mentioned, page 33, of the former report, that Dolater bog may very easily be put under water by way of reservoir: on this occasion it has been measured, and found to contain upwards of 220 acres English measure. Respecting the twenty acres for other works, we will suppose 200 acres of Dolater bog, by a dam at each end, laid six feet under water; out of this for supplying its own exhalations, we will allow one foot of perpendicular, we can therefore reckon, in time of need, to draw five feet of water from the Dolater reservoir; which will contain at that depth 43,560,000 cube feet, and consequently as much as will supply the aforesaid deficiencies for 270 days, or almost three quarters of a year, or would alone supply the whole quantity for 105 days!

Hence it appears that Dolater bog being turned into a reservoir, would, if necessary, supply the canal during the time of great droughts, without taking any water from the burns; or with their assistance as above, would navigate more than double the quantity of goods required even in the dryest seasons

Nor yet are those the only resources that could be procured if occasion required. In the reservoir way, it is remarkable that the whole face of the country, abounding in hollows, affords frequent opportunities for making them; in particular, a dam not above fifty yards long, where the dam-head of Townhead lint-mill is now fixed, would lay a valley under water four or five fathoms deep, and thereby form a reservoir equal in capacity to that of Dolater, already proposed.

This reservoir would not only be supplied by Shawend Burn, which takes its course through it (and which, though small in dry times, seems to be the best of them all, for supplying reservoirs by downfalls of rain) but will also receive the waters of Garron Burn, Culeam Burn, and Achinclough, all of which may be diverted thereto. The waters of Bishop's Loch and four others in the same neighbourhood, forming altogether a surface of between three and four hundred acres, and which might be pent up from four to six feet higher than at present, may also be brought to Dolater bog. But
without

without infifting further on refervoirs, many more yet of living waters may be applied to our purpofe by bringing them from a fomewhat greater, though no remarkable diftance; in particular, on the fouth fide of the canal, the Loggie water may be intercepted and turned to Dolater bog, from a point where it is as confiderable, if not more fo, than any of the burns mentioned.—On the north fide the water of Glazert, a confiderable rivulet, and along with it all the burns weft of Kilfyth, viz. Weft burn, Arnig, and Guine burn, before-mentioned, may be brought fo as to fall in with Garron burn and Culeam burn into Shawend burn, and then proceed altogether into the Dolater bog. Furthermore, into the head of the Glazert may not only be turned the water from the fpout of Ballagin, and a great number of fprings that feed the head of the water of Blain, but even a great part of the river Enrick, (I believe quite as much of it as can be carried to the bog of Bollat) may be brought round the hills upon a fufficient elevation to fall into the head of the Glazert, and be therewith carried to the point of partition: in fhort, were ten times as much water neceffary for the canal as what appears to be, there are the evident means of bringing it, and amaffing it without making any ftrain upon nature: I therefore beg leave once more to repeat a paffage of my former report, page 42: "it does not admit of a doubt but that nature has furnifhed a practicability of bringing as much water to the point of partition as can poffibly be wanted, and that by means fufficiently fimple and eafy," and again, "fo that there is all poffible certainty that the canal will be amply fupplied from end to end."

As the Glafgow branch will afford the means of navigating from Glafgow to Port Glafgow, and back again, in neap tides and dry feafons, it will undoubtedly be made ufe of for that purpofe when the river Clyde will not ferve; there will therefore be water expended by lockage in this paffage of veffels not navigating from fea to fea; but as the branch canal to Glafgow is propofed to be carried on upon a dead level, and without communication with the river Clyde at its upper end, only one lock full will be expended on the paffage of each veffel into or from the Clyde.

I am in no condition to judge what quantity of water for lockage may be wanted in this branch of trade; but if as much water were to be ufed in this branch as will be ufed in paffing from fea to fea, which will navigate four times as many goods upon the branch up and down as upon the tranfverfe paffage from fea to fea, there are fo many rivers which fall in below the level of the canal of partition, and yet fufficiently above the level of the branch canal, as are fufficient to fupply it ten fold, without letting down any water from the canal of partition at all on this account.

In like manner it is to be fuppofed, that at the Carron end of the canal many cargoes of goods will pafs backwards and forwards without croffing from fea to fea, all of which will confume a lock full of water each at their entry and departure from the canal, and yet more if a branch be carried off to Carron fhore, as it eafily may; but were this trade alfo equal to the trade acrofs, there are likewife the means of fupplying it with water, without letting any down from the canal of partition.

Having now fhewn that there is a practicability of producing a moft ample fupply of water, not only to anfwer the greateft poffible exigencies of trade, but alfo the greateft poffible errors that may be made in computation, it now remains again to fhew that we are in equal capacity of repaying the water that fhall be neceffary to be taken from the rivers that now fupply the valuable iron works, &c. upon the Carron, and the valuable mills and iron works that are towards the foot of the Kelvin.

Here I muft obferve, that though I have fhewn how it is poffible to bring a vaft abundance of water to fupply this navigation, yet I have equally fhewn how fmall a quantity really will fupply it.—It is fufficient therefore to repay to the full what we really are obliged to take.

Suppofe, therefore, the navigation is fupplied as firft propofed, viz. by the fix burns and a refervoir upon Dolater bog; it is manifeft that we take nothing from the mills but the value of the fix burns, in the dry feafons; for in all others, they are fufficiently fupplied without them.

From what has before been ftated, it appears, that 415,000 cube feet of water may be expected to fupply the canal per day, but that from this what is allowed for foakage, viz. 97,000, ought to be deducted; becaufe, as the foakage returns into the rivers, and will be maintained by means of the refervoir as great in the dryeft times as in ordinary, they will get an over proportion when moft wanted, and therefore be rather benefitted than hurt by this article: yet, not to be niggardly of a little water, nor to infift upon nice difquifitions where there is no neceffity; for the prefent, we will allow that we are to repay 415,000 cube feet of water whenever the burns will furnifh it, and in the dryeft feafons whatever they do furnifh, which at the leaft we fuppofe to be 254,000.—Now this is fcarcely three times as much as the Achinclough burn yields in the dryeft feafon fingly; a burn, therefore, capable of running three times as much as Achinclough in the dryeft feafons, and at other times in proportion, will afford as

much

much water as will repay all the mills both on Carron and Kelvin, and as somewhat about half that quantity is due to one set, and half to the other, it will follow that a burn half as big again as Achinclough will repay the Carron mills, and another of the same size will repay those of Kelvin.

I have mentioned in the former report, page 44, that the river Enrick, at Randeford, may be easily diverted into the Carron, and that it appeared to be there more than double of Achinclough: this I thought sufficient to make good my argument; but had I said four times as big, I believe I should have been nearer the mark: the burn of Ganakin I then but rarely mentioned, but it appeared to be near its fall into Enrick, full as big as the Enrick at Randeford, and having its rise in the Campsie Fells, it cannot be reckoned less than Achinclough at the level at which it may be diverted; hence we cannot reckon these burns together, at less than three times the estimated quantity of water necessary to be repaid to Carron, and therefore by half more than sufficient for the whole re-payment of vessels not passing from sea to sea; and will include the lockage at the Carron end of vessels not passing from sea to sea, were that supposed equal to the lockage of those that are; for it is to be remarked, that let the lockage be increased in whatever degree, the other articles of leakage, exhalations, and soakage, which make two-fifths of the whole stated quantity, will still remain the same, and therefore it is extra lockage only that is to be provided for on these extra accounts; yet to this purpose of extra lockage at the east end, (if we could possibly suppose it needed) not only the Grange burn, but a considerable part of the river Avon, could be brought round to supply the canal; and if this be not enough, as we have already said, the Loggie, the Glazert, the head of the Blain, and even the Enrick, can be brought round to the canal of partition, when their overplus waters may be sent down the rivers either way.

With respect to the re-payment of the Kelvin mills, I am of opinion that the waters that can be turned into the Glazert from the spout of Ballagin and head of the Blain, will of themselves be found sufficient: but not altogether depending on this, it turns out from computation, that as much water can be accumulated in Bishop's Loch, and the four others in that neighbourhood, by penning them to the depth of three feet nine inches only, as is equal to the proposed reservoir of Dolater, and consequently will of themselves repay the Kelvin's proportion of supply to the canal of partition for 210 days or seven months, which is two months more than it ever happens that any of the mills can be the worse on account of the supply taken for the navigation.

Besides

Befides the above, the following is a lift of fuch lochs as might, in like manner, be made ufe of for treafuring up water, which can be brought into the Kelvin in order to repay the mills, or fupply the lower part of the canal on the weft fide, viz.

Loch Umphrey*,	-	-	-	320 acres, at twelve feet deep,	$1\frac{6}{10}$
Cockney Loch,	-	-	-	80	
Ditto, as may be made by a dam,		-		640 at twenty-four feet,	15
Bardoure Loch,	-	-	-	82 at three feet,	$\frac{1}{4}$
Auchen Loch,	-	-	-	80	
Postel Loch,	-	-	-	80	

times as much as the reservoir of Dolater.

Alfo, a very large hollow in Kilmannan Moor, the land being of fmall value, that may be converted into a loch, and turned down the Allander.

Alfo, feveral fmall lochs, fuch as Kilmardonie Loch, Loch Green, St. German's Loch, and others.

Over and above all, we have the river Enrick for our fheet anchor, which being taken up, as already obferved, at a fufficient elevation to go over the low flat ground at the point of partition, between the Blain and the Glazert, is of itfelf fufficient to repay all that is taken for furnifhing the navigation; that is, in other words, that the whole navigation might be furnifhed with water from the Enrick, without taking any thing from any burn or rivulet; that at prefent feeds either Carron or Kelvin: indeed, this would have been the primary propofition, had not the thing been feafable by more direct and fhorter methods; I therefore only mention thofe things to fhew, that in cafe of any unforefeen occurrence or emergence, we are by no means limited to the fupply of the fix burns, or unable, in the nature of things, to repay any quantity that it may be needful to take.

It, perhaps, may feem that I have reckoned the Enrick twice, nay, three times over; firft, in making ufe of it for a fupply to the navigation; 2dly, in repaying the water borrowed from the Carron; and, 3dly, in repaying the water borrowed from the Kelvin.

But, in anfwer, I muft obferve, that if the Enrick be made ufe of as an immediate fupply, it muft be brought to the canal of partition; if that be done, it will be of itfelf a fufficient

fupply

* This is what it may be made to hold, more than what it now does as a reservoir to Dalmoore mill.

supply, and consequently there will be nothing to repay. If made use of to repay, it will do it to both without interfering; for the water of Enrick will be brought off so much lower down than Randeford, from whence the water is to be taken to Carron, when it has received several considerable rivulets and burns, that there will not only be water to repay what is due to the Kelvin, but water enough left for the mills, if properly applied, and, after all, water to supply another canal by the bog of Bollat, in case that also should be undertaken. But though it may be necessary to show from what abundance of sources water may be drawn for the supply of the canal, yet I do not apprehend it will be necessary to disturb the Enrick at all, except at Randeford, as first proposed. And I am so thoroughly satisfied that there is upon the whole so much more water to be procured, than is sufficient for all purposes, that I cannot hesitate to declare, that the promoters of this undertaking may, for the satisfaction of the country, very safely oblige themselves not to take any water that falls into the heads of Carron or Kelvin in times of scarcity, but what they do actually repay in equal quantity at the same time to the lower parts of the same river.

It may perhaps be doubted, whether there are any certain means of knowing when two streams are equal, so that the parties interested may be satisfied that they have their just equivalent: for the satisfaction of those, it is incumbent on me to say, that water-gages both may and ought to be fixed, so that any person by inspection shall be able to judge at any time of the equality thereof; furthermore, as all overflowings will fall into the rivers, and as the navigation will seldom be able to use all the waters allotted to it, the Carron and Kelvin rivers will of consequence be gainers by this overplus.

The next great point is the places of entry of the canal from the rivers Forth and Clyde, and in this affair it is not to be doubted, but that the greater the depth of water where the entries are, *cæteris paribus*, they are the better.

Conveniencies of trade, are matters that I never undertake to judge of. I am therefore happy that the places of entry have been pointed out by a committee of gentlemen appointed to make due enquiries thereupon; whose opinion, after having been reported before the convention of royal boroughs, has been by them approved, viz. " That the most proper place for the entry from the Firth of Forth will be " somewhere near the mouth of Carron, and on such part of the north-east side of " a farm called the Hewk, as Mr. Smeaton may think proper, after a due con- " sideration."

Now,

Now, having maturely confidered this matter, I cannot hefitate to fay, that the moft eligible place for the entry on the north-eaft fide of that farm, is that part which points down the fea reach whofe general bearing is N. E. by E.: this entry is on many accounts a very proper one ; for below this, there is no impediment in going out to fea, fave a flat or bar juft within the channel of the Carron's mouth at low water; upon which I found five feet water at low water, the 14th day of Auguft laft, being the 4th day after the full moon, within which there is a deep called Holemerrie, extending nearly a quarter of a mile, and reaching almoft to the place where the entry is propofed ; this deep has feven feet more water than the bar, and is therefore a very proper place for veffels to lie in when waiting for the tide.

It is to be obferved, that the tide preceding, and that fubfequent to the low water, when I founded, were the higheft of that fpring, which was a very moderate one, but which rofe above the faid low water full eighteen feet; fo that there would be at leaft twenty-three feet water upon the bar or flat. The fpring tides are faid fometimes to make twenty-feven feet upon the bar, fometimes to fall fhort to twenty-one or twenty-two ; the neaps make eighteen or nineteen feet upon the bar, and leave them about nine feet at low water ; hence it appears, that veffels drawing eight feet water may go in and out at all times of tide in neaps, and that at low water at fpring tides they may lie afloat in Holemerrie, till the tide ferves, and that veffels drawing four feet water, fuch as formerly propofed, may go in and out at all times, except at the low water of a very low fpring ebb. Thefe are certainly great advantages ; but, on the other hand, as there is not fufficient fhelter at high water for fuch veffels as come here to deliver their cargoes, and which are too large to pafs the canal, and as I do not appre-hend it would be practicable for fuch veffels as intend to enter the canal to do it with fafety when the wind is northerly, and blows hard ; it will, therefore, in this cafe, be neceffary either to dig an artificial harbour within land, or to make ufe of the turn of the river round the Hewk point, and the reach within the fame, as a natural harbour, as now it is; the former will be attended with very confiderable expenfes, and not without incon-veniencies, the latter would be attended with the following.

Firft. The bottom of the river in the Hewk reach, that is, above or up the channel from the propofed place of entry, particularly at a place called the Hewk ford, is $3\frac{1}{2}$ feet higher than the bar; in confequence, in all cafes, veffels will have $3\frac{1}{2}$ feet lefs water in going into the natural harbour, than in coming over the bar ; but as there is here no fcarcity of water at high water, this confideration, if alone, would be of lefs confequence. But what feems more material is, that when the wind blows at S.W. or W. which in thofe

climates

climates are the moft general of all winds, veffels may and frequently do lofe a tide in getting round the Hewk point; on the other hand, when in, an entry from about the Grange Burn foot would at all times be fafe, and with about four feet extra flow of tide, certain. In order, therefore, to take every advantage of the fituation, I would recommend it to confideration, whether it would not be advifeable to give the canal a double entry; one from the place already propofed, the other from the Grange foot; by this means the coft of a new artificial harbour would be avoided; and all veffels intended for the canal will, at all times, be fure of a fafe entry; for fuch winds as will render the outward mouth difficult, will carry them to the inner mouth; which latter will, at all times, be acceffible to goods and veffels from the material harbour.

The extra expenfe of the canal and interior lock, as well as fome addition of leakage, will be objections; but as to the firft, the coft will be very far fhort of making an artificial harbour; and, after what I have faid upon the article of fupplies, I flatter myfelf that the latter will totally vanifh.

With refpect to the weft end of the canal, it is certainly very rightly determined by the committee, " that the entry of the canal fhould not be higher up the Clyde than Dalmoore Burn foot," for higher than this there is no certainty of water for veffels of any greater draft than thofe fpecified in my former report; and even to Dalmoore Burn foot, there is not that plenty of water that could be wifhed. Having carefully founded the Clyde from Dumbarton to the Barns of Clyde, upon the 21ft day of Auguft laft, being the 5th day after the quadrature, there was at high water upon Dumbuck ford, $7\frac{1}{2}$ feet, and over Erfkine Sands above eight feet; all other places betwixt deep water and Dalmoore Burn foot founded above eight feet fix inches and nine feet. This tide fomewhat exceeded the dead of neap, being the firft tide that lifted; but as the ordinary neap tides are reckoned at feven feet fix inches upon Dumbuck, and but two feet more at ordinary fpring tides, and as the high-water mark at Dumbarton this tide fell fhort of the ordinary high water of a fpring tide by one foot $10\frac{1}{2}$ inches; it is very certain that it could have lifted not above two or three inches at moft above the dead of neap; and as I look upon it very practicable to deepen Dumbuck ford one foot, or even $1\frac{1}{2}$ foot, and by contracting the low-water channel of the river, oppofite Erfkin-houfe, by proper weirs to run down the loofe fands of which it is compofed, and conftantly maintain at leaft as good a channel as at Dumbuck, it follows that veffels drawing full eight feet water may, in moderate weather, and at high water, in all ordinary neap tides be able to enter the canal at the foot of Dalmoore burn. The day when the above foundings were taken was remarkably calm and ferene, fo that it was probable it was unaffected

by

by any winds; and, indeed, the near coincidence of our foundings, with the common eftimation of thofe that ufe the river, proves it. It muft, however, be obferved, that the tides of this river are very liable to be altered by the winds, which frequently make them exceed, or fall fhort, one foot of the ordinary quantity. In thefe cafes, it is poffible that veffels drawing eight feet water, coming at dead of neaps, and the wind in an un-favourable quarter, may have to wait a tide or two; but as S. W. winds make great tides, which are thofe which ofteneft blow, we may reckon thofe ftoppages to happen but feldom, and they may be placed among the unavoidable chances of navigation dependent on winds, which all veffels out of canals or narrow rivers muft be fubject to.

It were much to be wifhed, that the canal could as eafily be extended to deep water below Dumbuck ford, as it can at the eaft end; but having maturely confidered this matter, as defired by the faid committee, though I do not doubt of its practicability if expenfe be unlimited, yet, confidering how much more eafily the river may be helped in the degree before mentioned, which feems to me likely to anfwer the purpofe, I cannot recommend this part of the work to be undertaken for veffels of eight feet water, but for veffels drawing $9\frac{1}{2}$ feet water, it will be abfolutely neceffary, and of which I fhall, in the eftimate, give the beft account of I am able.

The proper method of effecting it feems to be, to avoid the fleechy bottom, and Whinftone rock, by making a ftrong fea bank within high water mark, and to build a wall where the fhore is rocky, fo as to confine a body of water for the navigation of veffels between that and the main land; but at Dunglafs the neck of Whin rock between the caftle and main land muft be cut through, fo as to afford a navigable paffage.

In regard to the general courfe of the canal, the ground has been carefully examined by boring, by Mr. Mackell, whofe abilities in engineery is well known, and who has delivered to me an account of the qualities of the foil for twelve feet deep, in no lefs than fifty-eight different divifions; together with many other meafures of the hills and hollows through which it has been defigned to pafs; alfo, remarks upon the great of lochs that can be made ufe of as a fupply, or for re-payment of others which are fo made ufe of; and to whofe care and induftry this bufinefs is much indebted. Having, along with him, carefully reviewed the courfe of the canal formerly traced out, it does not appear that any very material alterations or deviations from that courfe are neceffary, till we arrive at thofe places where we muft turn off on account of the extenfions. In re-gard to that at the eaft end, I find it advifeable to bring the canal as before, to or near Bainsford, (juft beneath the word Camelton, in fig. 2. plate 4.), where it interfects the

road

road from Falkirk to Carron: in its paſſage from thence to the Hewk, it may be carried on either ſide of Kerſe-houſe, the north ſide is the more direct, according to the dotted line P upon the plan; but as it meaſures but a quarter of a mile more, to go upon the ſouth ſide, by Mr. Longmuir's houſe, and without interfering with the policy, I have made the deſign and eſtimate upon that footing, the difference of expenſe for the 8½ feet canal being £558.

The moſt material deviation that I have made from the former deſign, is by taking the Allander paſſage in preference to the Canny's burn. Having originally given preference to the Allander paſſage; that part of the tract of the canal from the Allander to Garſcud bridge, was not ſurveyed, but was laid down by the eye; this alteration of my firſt intention, was what occaſioned the want of an actual ſurvey of that part; for finding, on computation from my notes, after I had left the country, that the expenſe of cutting the hill from Canny's burn into St. German's Loch, would be greatly ſhort of cutting that by the Allander; this occaſioned a correſpondent alteration in my deſign. But upon this laſt review I find it will be neceſſary, in paſſing from the Allander to Canny's burn, to paſs through two other hills as troubleſome as the former. This would not have been neceſſary for veſſels of the ſize I firſt computed upon; becauſe, at all events, I knew I could conduct them down the courſe of the Kelvin itſelf to Garſcud bridge; but this, for veſſels of the ſize propoſed, would be more troubleſome to do than even the impediment we mean to avoid. As, therefore, upon the footing of the preſent ſcheme, it will not, according to the beſt eſtimate I can make, coſt above £1000 more to go by the Allander, than by the Canny's burn paſſage, and at the ſame time ſhorten the diſtance one quarter of a mile, as well as avoid the policy of Killarmine-houſe, with a leſs chance of unforeſeen difficulties; on theſe conſiderations, I have now drawn and eſtimated the courſe of the canal by the Allander paſſage, at the ſame time ſtating the difference, in caſe the chance of the above ſaving is thought worthy of attempting.

The next conſiderable deviation is in order to avoid going through the policy of Graſcuddon, by cutting through a riſe of about eighteen feet extra height, which paſſage was pointed out by Mr. Mackell. This work I eſtimate to coſt £238, which, with ſome leſſer inequalities, may poſſibly amount to £300. This will, indeed, throw the general courſe of the canal about three furlongs further round; but, as it will ſhorten the Glaſgow communication by two furlongs, there is, upon the whole, only one furlong extra, the charge of which, for the 8½ feet canal, will be alſo about £300 more, but, as this will avoid at leaſt £300 extra charge in paſſing through the piece of water at

Grafcuddon, and furthermore avoid the difficulties that may attend a muddy bottom, in going down from the point of deviation to Grafcuddon; perhaps, upon the whole, there will be little difference in point of expenfe. I have, therefore, made the defign and eftimate according to this deviation.

Upon entering into the detail of this affair in the execution, it muft be expected that many deviations may appear eligible; it is, therefore, not to be expected the canal can, at prefent, be tixed to a point, and I can only fay in anfwer to thofe that might defire it, that I cannot take upon me to do it, and I think it would be very unadvifed in any man fo to do, even in works of much lefs bulk, extent, and expenfe than this. There muft be a degree of latitude in an affair of fuch great confequence, for fecond thoughts and improvements.

J. SMEATON.

Austhorpe, 8th October, 1767.

First FSTIMATE for a canal from Forth to Clyde, for veffels drawing eight feet water, and the locks, &c. to contain thofe of twenty feet wide and fixty feet long, with a branch of the fame fize to Glafgow.

£

The canal, for vessels of twenty feet wide, and drawing eight feet water; if made of full width, so as to draw easy, should have a twenty-seven feet bottom, and be $8\frac{1}{2}$ feet depth of water, which will require the canal to be $10\frac{1}{2}$ feet deep at an average; this made with slopes at a medium as three to five, will, at 5d. per yard cubic, come to £1903. 12s. $2\frac{3}{4}d.$ per mile; but, for even numbers, say £1904. per mile, and the canal, including the Glasgow branch, with its extensions, as per plan, being thirty-seven miles, in the whole, will come to - - - - - - - 70,448

This canal will admit of sea-built vessels of the above dimensions to pass in every part; but yet to allow for turning places and other extra widths, allow upon the whole one mile extra, - - - - - - - - 1,904

The width of the canal at top being sixty-two feet, and the same width being allowed on each side for depositing the earth, and for back-drains, fore-shores, &c. the whole width will be sixty-two yards, which takes $16\frac{3}{10}$ acres per mile, and for thirty-seven miles 603 acres, which, if purchased at a medium at £20, will come to - - 12,062

Besides what is immediately occupied by the canal, it will be necessary to allow for aqueducts, trenches, reservoirs, and other conveniencies, the value of 100 acres more, which, at the mean price of £20. per acre, is meant to include Dolater bog, - - 2,000

Carried forward 86,414

The

£

Brought over,   86,414

The perpendicular height before ascertained, from neap tide high-water mark to the surface of Dolater bog, is 147 feet; but, according to the present scheme, it will not be necessary to enter from the level of the bar, which is reckoned nineteen feet below high-water mark of neap tides; and as it may be adviseable to keep the canal of partition two feet above the level of the surface of the bog, the whole perpendicular will be 168 feet, which at eight feet each lock, makes twenty-one locks upwards, and the same downwards, and will in the whole be forty-two locks; which, for vessels of the dimensions specified, ought to be 20½ feet wide, and seventy-two feet long, which will cost £1000. each,   42,000

To an extra double lock, for a double entry at Grange burn foot, with extra expenses in diverting the burn, - - - - - - - 2,000

To making an aqueduct bridge over the Grange burn, - - - 500

To extra digging in passing a narrow gripe, between two rising grounds, west of Mungill's house, - - - - - - - - 100

To extra digging through the summit of the ground, between Glenfour and Newhall, above Camelton, - - - - - - - - 115

Extra banking across the hollows betwixt Newhall and Bonnie mill, and one small river,   600

To an aqueduct bridge in passing Bonnie mill burn, - - - 1,500

To extra work in passing Seabegs wood, Trannock burn, Acre burn, and some other hollows, and rises from thence to Castle Cary bridge, - - - - 400

To an aqueduct bridge for passing the river Bonnie, near Castle Cary bridge, and banking there, - - - - - - - - 1,200

To extra cutting in the canal of partition, so as to make it a fifty feet bottom, - 1,478

To making two dams, one at each end of Dolater bog, so as to form the whole into a reservoir, - - - - - - - - 300

To defending the dam-heads and banks of the reservoir with stones against the wash of the waves, - - - - - - - - 1,408

To making a sluice for drawing the water from the reservoir into the canal of partition, and over-falls for discharging the water each way, - - - 100

To building a dam and other works at Inch Belly bridge, for the passage of the canal there,   500

To extra expenses in passing the Loggie at Kirkintulloch, - - - 300

To a dam and works at Calder bridge, - - - - 500

To extra expenses in cutting the hill near New Kilpatrick, in the Allander passage, computed to be thirty-six feet above the bottom of the canal, the digging at 6d., and walling at 5s. per cube yard, - - - 9,166

To extra expenses in cutting the hill by the weavers house, above Grascuddon £258., this, with some other inequalities, may be called - - - - 300

To extra work in making a jetty to defend the mouth of the canal, from the lodgment of sand and mud, and clearing a passage into Clyde, - - 150

To extra cutting in passing the rising ground opposite Blart hill, and in altering the turnpike road, upon the Glasgow branch, - - - 700

Carried forward - 149,731

To

		£
Brought over -		149,731

To building an aqueduct bridge over the river Kelvin, in order to carry the canal to Glasgow, including other extra expenses in passing that river, - - -   **4,000**

To making aqueducts, trenches, with such tunnels and bridges as may be wanted, for bringing the several supplies proposed into the Dolater reservoir, - -   **500**

To six public roads, draw or turning bridges, viz. at the road from Falkirk to Carron and Stirling, Camelton, Inch Belly, Calder, and Brick-house, upon the road from Glasgow to Dumbarton, and one upon the branch canal, where the same road intersects it near Blart hill, at £600. each, - - - - - -   **3,600**

To ten draw or turning bridges of lesser dimensions, where the lesser roads intersect the canal, at £400. - - - - - - -   **4,000**

To twenty-one carriage bridges, of the draw or turning kind, for communications between the lands, which, with the great and small bridges, make at the rate of one in a mile, at £250. - - - - - - - -   **5,250**

To fifteen large tunnels, for communicating the lesser brooks under the canal, at £60. each,   **900**

To fifty-nine small tunnels, for preserving the present water-courses, making, with the large ones, at the rate of two in a mile, at £20 each, - - - -   **1,180**

To making towing-paths, back drains, gates, towing bridges, &c. at £25. per mile, -   **925**

To bringing the water of Enrick from a little above Randeford into Carron, -   **150**

To bringing the burn of Ballagin, with some springs rising out of the hills to the west of the spout of Ballagin, that now fall into the water of Blain into Kelvin, - -   **100**

To making trenches, sluices, &c. for drawing off the water of Bishop's loch, and three others in that neighbourhood, in case it shall be required, - - -   **100**

To expenses in making weirs to contract the channel at low water for deepening Erskin sand, - - - - - - - - -   **1,000**

To deepening Dumbuck ford, so as to make eight feet six inches water at common neap tides,   **1,000**

To temporary damages, unforeseen accidents, impediments, and works, engines, utensils, and surveyors' salaries, supposed at £15. per cent. - - - -   **25,865**

               198,301

N. B. The above is exclufive of all expenfes in making harbours, warehoufes, quays, and other conveniencies in landing, depofiting, and fhipping of goods, and of all expenfes attending getting an act of parliament, and of all furveys, &c. previous thereto; and for all allowance of intereft of monies, and of falaries to the principal engineers and accountants. Nor would I be underftood to afcertain any thing as to the value of the lands, the number of acres excepted.

N. B. Whatever waters may be wanted, further than the expenfe above computed will bring, muft be in confequence of fuch an addition of trade as will very well pay all additional charges.

Second

Second ESTIMATE of the probable expenſe of making a canal, of ten feet depth of water, from the Hewk farm, at the mouth of Carron upon the Forth, to below Dumbuck ford, upon the river Clyde, with a branch of the ſame ſize to Glaſgow.

£

The width of the vessels being twenty feet, as before, the same width of bottom will serve, viz. twenty-seven feet, and the depth of water being ten feet, we may call the whole depth twelve feet, the slopes being three to five, as before, will at 6d. per yard cubic, come to £2757. 6s. 6d. which, for even numbers, call £2758, this for thirty-seven miles will come to - - - - - - - 102,046

According to the above dimensions, the mean breadth of the canal at top will be sixty-seven feet, and allowing the same breadth on each side, for the banks, back-drains, and fore-shores, the whole breadth will be sixty-seven yards, this will contain 19. $\frac{4}{10}$ acres Scotch measure, per mile, which, if purchased at a medium at £20 an acre, comes to - 14,356

The length, breadth, and rises at the locks, being the same as before, they will be deeper by eighteen inches; which, together with a greater degree of strength necessary, not only on that account, but on account of the greater weight of vessels running against them, will produce a difference of at least £100 in each lock; there will, therefore, be forty-two locks, at £1100 each, - - - - - - 46,200

162,602

In the foregoing estimate, for 8½ feet water, the whole cost is - £198,301

Deduct, viz. cutting - - - - £70,448
Land - - - 12,062
Locks - - - 42,000
Fifteen per cent. - - - 25,865

Total deduct - - - - 150,375

Remains for extra works, lands, &c. - - 47,926
To this add one third for increase in these articles 15,975

The sum will be the supposed expense of the same articles upon the increased plan - 63,901

From the Forth to Dalmoore burn foot - - - - £226,503

The

The ESTIMATE for carrying the canal from Dalmoore burn foot, to below Dumbuck ford, being four miles, is as follows:

	£
To making an aqueduct bridge over Dalmoore burn	667
To making the canal within land three-quarters of a mile below the foot of Dalmoore burn; that is, cutting and land, at the price mentioned of the first and second article	2,369
To raising a strong bank within the tide's marks, and thereby inclosing a space to be used as a canal; this, at sixpence per yard, is £3,300 per mile, and for one mile and three-quarters	5,775
To defending the same towards the Clyde, with stones; these at two shillings per ton, come to £704 per mile, and for one mile and three-quarters	1,232
To building an aisler wall, where the Whinstone rock interferes, upon a sloping shore, being about half a mile in length; the solid being computed at five shillings per cube yard, will come to	1,173
To aisler facing for ditto, at eight pence per yard	1,320
To forming a slope of earth within the same, at sixpence per yard, cubic	352
To cutting the neck of Whinstone rock, behind the castle of Dunglass, so as to make a passage twenty-four feet wide, containing 64,000 cube yards, at five shillings	1,600
To one mile of cut below Dunglass, in order to clear the ford, and get into deep water; the cutting and land as per first and second article	3,146
This being done in the tides way, we may add for drainage of water, and other extra expenses	400
From Dalmoore burn foot, to below Dumbuck head	18,034
From the Forth to Dalmoore burn foot	226,503
The whole work being attended with greater hazard of rocks, mud, and many other difficulties, I think it necessary to allow twenty per cent. for contingent expenses upon the whole estimate	48,907
	£293,444

N. B. Exceptions as before.

Third

Third ESTIMATE of the expenfe of a canal of feven feet depth of water, from the Hewk farm, at the mouth of Carron to the foot of Dalmoore burn, upon the Clyde, with a branch of the fame dimenfions, to Glafgow:

£

As I suppose a canal of seven feet deep, vessels of about eighteen feet will be the ordinary size, a twenty-four feet bottom will be sufficient, and the mean depth being supposed nine feet, this with battens or slopes in the proportion of three to five at a medium, will, at 4*d*. per yard, cubic, come to £1,144 per mile, and this for thirty-seven miles, including the Glasgow branch and all extensions as per plan, comes to - - - 42,328

To allow for passing places, turning places, &c. I allow the value of one mile extra - 1,144

The width of the canal at top being fifty-four feet, and the same width being allowed on each side for depositing the earth, back drains, fore shores, and other occasions, this will make a breadth of fifty-four yards, which is $15\frac{6}{10}$ Scots acres per mile, and for thirty-seven miles, $577\frac{2}{10}$ acres, which, supposing the purchase made at a medium price of £20, will come to - - - - - - - - 11,544

Besides what is immediately occupied by the canal, it will be necessary to allow for aqueducts, trenches, reservoirs, and other conveniences, the value of £100 acres more, which at the mean price of £20 per acre, is meant to include Dolater bog - - - 2,000

The perpendicular height of 147 feet before ascertained from neap tide high-water mark, to the surface of Dolater bog, must be added according to the present scheme, the height of the said neap tide high water above the level of the bar, viz. nineteen feet. And as it may be adviseable to keep the water in the canal of partition two feet above the level of the surface of the bog, the whole perpendicular will now become 168 feet, which, at eight feet each lock, makes twenty-one locks upwards, and the same downwards, that is, in the whole forty-two locks, which, if made $20\frac{1}{2}$ feet wide, and seventy-two feet long, to hold vessels of twenty feet wide, and sixty feet long, will cost £900 each - - 37,800

To an extra double lock for a double entry, at Grange burn foot, with extra expenses in diverting the burn - - - - - - - 1,800

To making an aqueduct bridge over Grange burn - - - - 400

To extra digging in passing a narrow gripe between two rising grounds, west of Mungull's house - - - - - - - - 100

To extra digging through the summit of the ground between Glenfour and New Hall, above Camelton - - - - - - - - 100

To extra banking across seven hollows betwixt New Hall and Bonnie mill, and one small rise - - - - - - - - 600

To an aqueduct bridge in passing Bonnie mill burn - - - - 1,250

To extra work in passing Seabeg's Wood, Trannock burn, Acre burn, and some other hollows and rises from thence to Castle Cary bridge - - - 400

Carried forward - 98,466

To

	£
Brought over	98,466
To an aqueduct bridge for passing the river Bonnie, near Castle Cary bridge, and extra banking there, - - - - - - -	1,000
To extra cutting in the canal of partition, so as to make it a fifty foot bottom, -	1,478
To defending the dam-heads and banks of the reservoir with stones against the wash of the waves, - - - - - - - -	300
To making two dams, one at each end of Dolater bog, so as to form the whole into a reservoir, - - - - - - - -	1,408
To making a sluice for drawing the water from the reservoir into the canal of partition, and over-falls for discharging the water each way, - - - -	100
To building a dam and other works at Inch Belly bridge, for the passage of the canal there,	500
To extra expenses in passing the river Loggie at Kirkintulloch, - - -	300
To a dam and works at Calder bridge, - - - - -	500
To extra expenses in cutting the hill in the Allander passage, near New Kilpatrick, computed to be thirty-five feet above the bottom of the canal, the digging being estimated at 6d., and the walling at 5s. per cube yard, - - - - -	8,166
To extra expenses in cutting the hill at the weaver's house, above Grascuddon, with some other inequalities there, - - - - - -	250
To extra work in making a jetty, in order to defend the mouth of the canal from the lodgment of sand, and clearing a passage into Clyde, - - - -	150
To extra cutting in passing the rising ground opposite Blart hill, and in altering the turnpike road upon the Glasgow branch, - - - - -	600
To building an aqueduct bridge over the river Kelvin, in order to carry the canal to Glasgow, including other expenses in passing that place, - - -	3,500
To making aqueducts and trenches, with such tunnels and bridges as may be wanted, for bringing the several supplies proposed, into the Dolater reservoir, - -	500
To six public road draw bridges, or turning bridges, viz. that at the road from Falkirk to Carron and Stirling, at Camelton, Inch Belly bridge, Calder bridge, and Brick-house, upon the road from Glasgow to Dumbarton; also one upon the branch canal, where the same road intersects it near Blart hill, at £600. each, - - -	3,600
To ten draw, or turning bridges, of lesser dimensions, where the lesser roads intersect the canal, at £400. - - - - - - -	4,000
To twenty-one carriage bridges for communication between the lands, - -	5,250
To fifteen large tunnels for communicating the lesser brooks under the canal, at £60. each,	900
To fifty-nine small tunnels for preserving present water-courses, making, with the large ones, at the rate of two in a mile, at £15. each, - - - -	885
To making towing-paths, back-drains, gates, towing bridges, &c. at £20. per mile, -	740
To bringing the water of Enrick from a little above Randeford into Carron, -	150
To bringing the burn of Ballagin, with some springs rising out of the hills to the north-west of the spout of Ballagin, that now fall into the water of Blain into Kelvin, -	100
Carried forward -	133,843
	To

		£
Brought over -		133,843

To making trenches, sluices, &c. for drawing off the water of Bishop's loch, and three others in that neighbourhood, in case they should be wanted, - - - — 100

To temporary damages, unforeseen accidents, impediments, and works, engines, utensils, and surveyors' salaries, supposed at £10 per cent. - - - - 13,394

N. B. Exceptions as before, - - - 147,337

J. SMEATON.

Austhorpe, 8th October, 1767.

A REVIEW of several matters relative to the Forth and Clyde navigation as now settled by act of parliament, with some observations on the reports of Messrs. BRINDLEY, YEOMAN, and GOLBURNE, by JOHN SMEATON, civil engineer, and F. R. S.

I NOW fit down, at the recommendation of the committee of the Forth and Clyde navigation, to confider the reports of Meffrs. Brindley, Yeoman, and Golburne, and, (as defired), to make fuch obfervations thereon and anfwers thereto, as to me fhall feem proper. I fhall, therefore, endeavour to acquit myfelf of this bufinefs, not altogether by purfuing the queftions and anfwers in the order in which they have been produced, but by endeavouring to bring the main things as much as poffible into view.

The moft material point, as it occurs to me, is concerning the place of entry at the Carron end of the canal; and the queftion feems not whether the entry ought to be at the Hewk farm, or further up the river, at or near the Carron eftablifhments : but whether an entry at or near the Carron eftablifhments alone, will not be preferable to an entry at each of thofe places.

It has been given out, (and I find Mr. Yeoman touches upon it), that the places of entry were not of my choofing; and as they differ from the places pointed out in the plan contained in my firft report, that this deviation is contrary to my own opinion, and wrong in itfelf.

In order to clear my way through the whole of this queftion, I beg leave to premife, that when I made my firft report, the two great difficulties that were apprehended were money and water; of the latter I found a fufficiency, but of the former I had great reafon to be apprehenfive there would be no fuperabundance; and as the fort of veffels, fize of the canal, and fpecies of navigation, were entirely left to my choice by the Board of truftees, who then did me the honour to employ me, my endeavour was to produce a fcheme which was likely to anfwer the end propofed, viz. a general communication between the two feas, and which as fuch could be executed at the fmalleft expenfe.

In this view of a general communication between the two feas, it feemed to me a *sine quâ non*, that it fhould be of a fufficient fize to carry fuch veffels as could freely
and

and securely navigate the two Firths, whereon the two ends of the canal must terminate; for by this property every port on each of the Firths will in a manner become a harbour to the canal, and partake of the benefit thereof, in proportion to their more or less advantageous situation: whereas a canal of a less size, capable of admitting such vessels only as would navigate the canal, without being capable of navigating the two Firths, could never be considered as a general communication between the two seas, but only as a particular communication between two places, one at each end of the canal, and each communicating with the respective seas; and which places would thereby reap, in a manner, the whole benefit; and more especially so, if the points of termination were drawn into such shallow water, that no other port or place could avail itself of the passage without the use of mean or intermediate vessels to carry goods from the principal sea-ports, where large vessels could be received to the respective entries of the canal, which double transhipping would lay such a weight upon the out-ports, as would in effect become an entire monopoly of carriage between the two places of termination.

As I neither was at that time, nor profess now to be, so far master of the general art of constructing vessels for navigation, as to assign the measures most proper for each particular service, and finding the artists themselves in these branches to differ in their opinions, those who have been chiefly in the practice of building flat vessels for shallow water, attributing properties thereto which are by no means allowed by those who are in the habit of building sharper vessels for deeper water; and being willing to found myself upon secure principles, I took the ga boats or galberts of the river Clyde for my model: knowing that the daily practice of these vessels was to carry goods between port Glasgow and Greenock to Glasgow, and I was informed they occasionally went to much more distant places upon the Clyde, and even to Ireland; knowing also from a former acquaintance with the river Clyde, the difficulty of getting up these vessels to Glasgow in dry seasons and short tides: I took for granted that these galberts were constructed upon as flat a model as would answer the purposes of the navigation in the open Firth; and as I found they drew at a medium about four or $4\frac{1}{2}$ feet water, I concluded that a canal of five feet depth of water would give such latitude in point of construction, that my principal view could not miss of being completely answered, that of a free, open, and general navigation, between every port on the Firth of Clyde, to every port on the Firth of Forth.

It is true I know of no such vessel on the Firth of Forth; the flatness of the river Clyde, some miles below Glasgow, having produced the necessity of using

galberts

galberts on that Firth; yet, by parity of reafon, I fuppofed them as capable of navigating one Firth as the other.

Having thus found myfelf in poffeffion of a veffel, which, though the fmalleft that would effectually anfwer my view of a general communication, was yet confiderably more bulky than thofe commonly ufed in the artificial navigations in England; and as the bulk, weight, and expenfe of thofe works muft increafe in proportion to the weight of the veffels that are to navigate through them, this occafioned me to conclude that thefe were the largeft fize that would be convenient for an artificial navigation. See firft report, page 40, and fecond report, page 73.

The twofold view of a general communication, and at the leaft poffible expenfe, determined me likewife with refpect to the places of entry. I was not infenfible at that time that if the places were fo chofen, that the veffels could enter at all times of tide, it would be preferable to their entering at high-water only, at neap tides; but the extenfions neceffary to produce this, feemed to me likely fo much to fwell the expenfe, that I laid afide every thought of that kind, and contented myfelf with propofing fuch places as afforded a fecure entry at the high-water of neap tides. This produced the propofition of an entry a little above Abbotfhaugh, at the Carron end; for though the tide navigation might have been continued a little further up, yet the ground for a departure appeared to me far lefs proper. With refpect to the entry at Barns of Clyde, that could not properly be carried higher on account of a fhoal juft above, where the galberts are ftopped at fhort tides; and though the city of Glafgow had at that time an act of parliament which authorifed them to deepen the river, which might have enabled the entry to have been made nearer Glafgow, in cafe that had been performed; yet, as the magiftrates of that city had laid afide their profecution of that work for fome years, after it had been put in hand under my direction, I could not think it right to advife my employers to build any part of their fcheme upon one which another fet of gentlemen had it in their power to execute, or not, as they thought proper.

Thefe were the general views and ideas upon which my firft plan and report were built, which were delivered to the honorable Board of truftees for fifheries and manufactures in Scotland, in March, 1764; it now comes in courfe to fhew my reafons for the deviations from the above plan, as per fecond plan delivered in the month of October, 1767.

But

But first I must premife, that in whatever character my brethren may be ambitious of acting, I leave that to them; for my own part, I shall take this opportunity, as I find gentlemen apt to mistake the character in which I am defirous of appearing, to declare that I do not act in confequence of any public authority or commiffion, to enquire, determine, and declare what is best for the public, or what in its confequences will affuredly promote its good. I should indeed be very forry to have fo great a weight laid upon my shoulders, which neither my studies nor inclinations have qualified me to undertake. I confider myself in no other light than as a private artist who works for hire for those who are pleafed to employ me, and thofe whom I can conveniently and confiftently ferve. They who fend for me to take my advice upon any fcheme, I confider as my paymafters; from them I receive my propofitions of what they are defirous of effecting; work with rule and compafs, pen, ink, and paper, and figures, and give them my best advice thereupon. If the propofition be of a public nature, and fuch as involves the intereft of others, I endeavour to deliver myself with all the plainnefs and perfpicuity I am able, that those who may have an intereft of a contrary kind may have an opportunity of declaring and defending themfelves. I do not look upon the report of an engineer to be a law, at leaft not my own. As a propofition, every one has a right to object to it, and to endeavour to prevent its paffing into a law, or to oppofe its execution in fuch a way as may be detrimental to others: had, therefore, Meffrs. Brindley, Yeoman, and Golburne been fent for in confequence of my fecond report, publifhed in 1767, in order to have examined the matter thereof, and to have enabled those who might think themfelves concerned to oppofe the propofition, nobody could have objected to fuch a procedure; but after the propofition has been fuffered to pafs into a law, and thereby tacitly affented to, to fend for engineers to determine whether the law itfelf be rightly founded, and to prevent the execution till that be determined, is indeed to me a very extraordinary evolution, and a fort of proceeding which, as I think, every well-wifher to the fuccefs of the fcheme should fet his face againft. Had the matters objected to been different in the act from the fcheme, or had the execution of the act differed from the act itfelf, there had been fome colour for this interpofition; but, if not, and if it was as well known that the Hewk farm belonged to Sir Lawrence Dundas, in October, 1767, as that the eftate of Abbotfhaugh belongs to Meffrs. Garbett and Co. in 1768, thofe manœuvres cannot now be looked upon other than an unfeafonable interpofition to ftop the progrefs of the work.

Refpecting the two fchemes, I have no hefitation in declaring that my fcheme for a general communication between the two feas is contained in my firft report, that is, with fuch corrections as the difference of circumftances that have arifen fince my firft report

was

was drawn up, and better information might fuggeft; and that the enlargements upon my firft plan and report were propofitions given me by my employers, and I fhould be furprized, if this fhould be matter of news to any one concerned, fince I have fully declared the fame in my fecond report, to which I refer; nor could I poffibly fee the matter in any other light, fince the revifal of my firft plan was undertaken in confequence of a compromife between the contending parties in parliament at the clofe of the preceding feffion; one of which articles was, a branch to Glafgow of the fame dimenfions as the main canal, which, 'tis very probable, I might never have advifed as a thing to be undertaken by the proprietors of the main work. The places of entry were alfo pointed out to me by the report of the committee of royal boroughs, which I have ftated in my fecond report, pages 85 and 87, a copy of which I received from Mr. Chalmers, who alfo furnifhed me with the greateft part of the data, as ftated in my fecond report, upon which I proceeded; but I beg leave to fay it does not follow, that becaufe I did not form the propofitions for enlarging my firft fcheme, I did not approve of them; the material queftion, I think, is not who formed the propofitions, but whether the matter therein contained be right or wrong.

In regard to theincreafe of depth of water beyond my firft propofition, I take it as a felf-evident truth, that every increafe of depth muft be an increafe of advantage to trade; but as trade is not my profeffion, I neither do, nor ever did, take upon me to determine whether any given increafe of depth would be attended with an advantage equal to the increafe of expenfe, and alfo length of time attending the execution: this I muft leave to men of trade to determine, 'tis fufficient for me if I can compute what the difference of expenfe will be. It was, therefore, given me as a propofition to compute the expenfe of a canal of feven feet, eight feet and a half, and ten feet deep, in order that thofe whom it concerned might make their choice.

Refpecting the extenfions, it is impoffible for me now to fay what my ideas of the neceffity thereof might have been, had not this part of the work been undertaken and prepared for me before my arrival, by the convention of royal boroughs. That " the " greater the depth of water where the entries are made, *cæteris paribus*, they " are the better," I could eafily fee, without the help of the committee of royal boroughs; but in what degree trade would be benefited by a given difference of depth of water at the entries; or to fay what difference of depth is to trade worth what fum of money; how far the cap ought or ought not to be put off to certain very refpectable bodies of men, or the line drawn afide from the more general convenience, are matters that might have puzzled me, had they been left naked to my determination. I therefore

fore thought myfelf very happy, that this part of my tafk had been brought to a point by fo refpectable a body of men, who, if I underftand the affair, are a body incorporated by act of parliament, and have the fole power of regulating public matters of trade throughout Scotland, and whofe authority therefore in this kind of matters I had reafon to think fo much better than my own. I can therefore only now declare the reafons that induced me to acquiefce in the opinions of thefe gentlemen refpecting the places of entry; which, if I could not have done I fhould certainly have thought it my duty to my employers to have declared.

In the firft place, I found the number of veffels fo much encreafed at Carron, beyond what I had obferved at Carron fhore, when I was upon my furvey in Auguft, 1763, owing not only to the increafe of trade occafioned by an augmentation of iron works there, but more efpecially by the eftablifhment of the Carron Ship Company fince that time; that had my firft idea of a five feet canal, to be done at the leaft poffible expenfe, been adhered to, I fhould certainly, on this review, have advifed the entry to have been made below all the Carron eftablifhments, in order that fuch veffels as wanted to pafs the canal, independent thereof, might have room to go quietly by them. This, I fay, muft have been the cafe had the five feet canal been adhered to; but as the propofition now was for a canal of at leaft feven feet, and of a depth of $8\frac{1}{2}$ feet, at a medium, as a much greater quantity, and bulk of veffels muft be expected in confequence of its admitting thofe of a fea conftruction, I did not, nor do I now think, that the new cuts through which the river is turned, which begin juft below the elbow where the grounds of Abbotfhaugh terminate, of fufficient width to afford a convenient refting place for fuch kind and fort of veffels as were now intended to navigate the canal: add to this, that but a little way below the termination of thefe cuts, there is the moft remarkable fhoal or ford upon the river commonly called Jemmy Reay's Ford, upon which, at high water at a neap tide, according to all accounts there is not above eight feet water, and in low neaps fhort of that. As therefore the veffels propofed could not fometimes, even at high water, pafs this fhoal, the great and uncertain expenfe of lowering thereof, fo as to make it certain and convenient for veffels requiring any of the depths of water contained in the new propofition; and I may add, the not having a perfect certainty of its remaining fo, even after cleared; made it feem to me very advifable, in a work intended for a general communication, to avoid all uncertainty of expedients, and to carry the Canal below Reay's Ford. But, fuppofing that done, we cannot go above a quarter of a mile before we come to one of the moft remarkable loops in the whole river, which either muft be cut, or the navigation would remain inconvenient; and as my idea of the facility of cutting thofe loops, fo as to make them fuitable to the river,

did

did not, nor does now, happen perfectly to correspond with those of my brethren, who have since been consulted, I was of opinion that the canal could be continued superficially upon the banks of the river at far less expense than this neck could be cut; and with this further, and very material advantage, that there is at least four feet more water to Grange burn foot than if dropped into the straight reach below Reay's Ford; this in consequence carried me down to Grange burn foot, to the place where the interior entry is now proposed. And as I neither did, nor do approve of making a cut for the river Carron across the Hewk farm, in order to make the entry of the canal perfectly convenient; and upon a supposition that money was not wanting, it did and does to me seem advisable to extend the canal across the Hewk farm to Holmerrie, because here vessels of seven, or even eight feet water, instead of being confined to half an hour at the high water of a neap tide, may go in at a neap tide low water, which, in my opinion, is not only a great advantage, but I take the liberty of asserting once more the outward entry will be attended with this further convenience, that a vessel which can go into the outward reach of the Carron, may enter the canal at the exterior, or if not there, at the interior entry : for these reasons I concurred with the committee of royal boroughs, that if general convenience be the object, and not saving money, which is a language not spoken of to me till lately, the canal ought to be extended to the Hewk farm.

It is very certain that if the proprietors will abate of the conveniencies of the canal, the expense may be considerably shortened : but I am not convinced that the expense will, upon the whole, be shortened by executing the alterations of the river Carron proposed by the three gentlemen, whose reports I have now before me.

I should be much surprised if any of the proprietors who have attended the business of the canal, should be uninformed that it would save expenses by beginning the canal higher up the river than the Hewk farm; but it is certain the conveniencies are far from equal, as is apparent by the gentlemens' taking so much pains to form expedients to remove even in part, the natural difficulties which make the difference; and as to that part which they cannot remove, they tell you, that you will save money : they cannot say, they do not say, that it is as good to enter where you can go in at the high water of a neap tide only, as it is to enter where you can go in at the low water of a neap tide; but they tell you, that if you will give up this advantage, you will save some extent of canal, (made upon the superficies of the ground, where it is most easy;) and you will save the locks whereby you would be enabled to go down into deep water : why truly we knew all this, without Mr. Brindley, Mr. Yeoman, and Mr. Golburne. How much

much of convenience to trade, is worth how much money, I have already faid, is not in my power to eftimate ; 'tis enough if I can make an eftimate of the coft of what, I beg leave ftill to fay, is the moft complete ; whether this be, or be not worth the money I leave to men of trade to judge. For my own part I choofe to build my defigns upon fure grounds, whenever fure grounds are to be had ; 'tis time enough to apply to expedients when we can do no other ; that what is propofed are but expedients the gentlemen them-felves, who propofe them, apprehend ; becaufe they defire the works may be ftopped till it is feen whether they fucceed or not.

In regard to the Clyde, I think it ftands confeffed that in its prefent ftate an entry fuitable to a canal of even feven feet conftant water cannot be made materially higher than Dalmoore burn foot ; and even thither, there is not that ample fufficiency of water which is to be wifhed ; could it have been extended from Dalmoore burn foot, into as good water as is at Hewk farm, at an equal expenfe to the probable difference between beginning at Abbotfhaugh and Hewk farm, I fhould not hefitate to advife it, if the money could be procured to do it. This, I fay, is the cafe according to the ftate in which I found the river Clyde in the years 1763 and 1767, and to have built any part of a fcheme of this ftill greater extent, upon what the Magiftrates of Glafgow had it in their power to do, or not, at their pleafure, and which they might or might not be able to do, would, in my opinion, have been very indifcreet in me to have recommended. Mr. Golburne had not been upon the Clyde at that time ; and if he had, it would have been no demonftration to me that making a channel for veffels drawing fix feet water was making feven feet water, fince the galbert men could have told him that if they have an inch to fpare they never ftick faft upon a fhoal : be this as it may, if he can, as he fays in his report, bring up veffels drawing fix feet water to the Broomie law at all tides, (and as is faid for him, without lock or dam,) for the fum of £7000, he will do a thing remarkable in itfelf, and a good thing for the city of Glafgow, and in which I wifh him and the city of Glafgow much fuccefs : when that is done, the branch canal may be fhortened feveral miles, by dropping into the river Clyde juft below Blart hill : but as I like to work upon fafe grounds when I can, Mr. Golburne muft excufe me if I cannot recommend to the proprietors to give up their entry near Dalmoore burn foot, or ftop their proceedings in joining the two feas, until he has brought his fcheme to bear. Having now fhewn the reafons why my plans have been what they are, and in particular why I did not diffent from the committee of royal boroughs in regard to the places of entry ; and in the courfe thereof why I think a compound entry at Hewk farm, at Grange burn foot, and one nearer the Carron wharfs, preferable to a fingle entry near the Carron wharfs only; I shall now proceed to make some observations upon the most material points wherein I differ in opinion from my brethren.

1st, 2d, and 3d. With respect to the answers to the 1st, 2d, and 3d queries, I do not find that Messrs. Brindley and Yeoman have controverted any material point by me laid down; I have only to observe, that though they both in effect allow that they have not examined the sources of water with that exactness that I have done, yet they admit that 100,000 tons of merchandize may be carried annually from sea to sea each way, whereas the quantity I have estimated is no more than two-thirds of 100,000 each way. See 2d report, page 74. I do not mean however to contradict them.

4th. Mr. Brindley recommends a canal of four feet deep, and vessels drawing three feet water to be seven feet wide, and seventy feet long: this would be a very effectual way of securing the use of the canal to the two terminations; no port but that of Carron could have the least chance, for since this kind of vessels could scarcely venture out of the narrow cuts, not even below Jemmy Reay's ford, in a windy day; the large vessels that now come to the Green brae, (that is, the natural harbour, constituted in the reach within the Hewk point,) would be obliged to employ a mean or middle sort of vessel to carry the goods from the ship to the entry of the canal, and the same at the other end thereof: as to the ports of Leith, Borrowstonness, and many others upon the Firth of Forth, together with Greenock, Port Glasgow, and many others upon the Firth of Clyde, they would have no chance, even Glasgow would stand but a bad chance without a side cut of equal dimensions, or the main canal itself terminated there.

In this case Mr. Brindley would have done well to have recommended the sort of restraints that he would advise, to prevent monopolies and impositions at wharfs and warehouses, since no body but those at the terminations could have any business there. Mr. Yeoman agrees in thinking the seven feet canal to be most proper, and Mr. Brindley, that it should not be exceeded.

N. B. As no difficulty is too great for Mr. Brindley, I should be glad to see how he would stow a fire engine cylinder cast at Carron, of $6\frac{1}{2}$ feet diameter, in one of his seven feet boats, so as to prevent its breaking the back of the boat, or oversetting.

5th, 6th, 7th, 8th, 13th, and 14th. There is no doubt but that the cutting of the neck of land, referred to in the 5th question, will be a very great improvement to the navigation of the river Carron, up to Carron shore; because I look upon it that vessels find far more embarrassment and difficulty in getting round this loop, than in all the rest:

and

and I am alfo of opinion, that this may be done without prejudice to the canal, as now going on according to act of parliament, provided proper caution is ufed, and the undertakers laid under proper reftrictions; and it is a work that I am told Sir Lawrence Dundas did propofe to have done at his own charge; but with refpect to the cutting of the point of land from Grange burn foot to Holemerrie, or taking the two necks together in one fweep from above the upper Salmon Zair to Holemerrie, I cannot by any means agree with my brethren in opinion; but that inftead of making an excellent harbour of two miles, they will entirely fpoil a good harbour, which is now furnifhed by nature ready made.

I believe it may be laid down for a general rule, that all fmall harbours have fome inconveniencies, (fmall I call that at Green brae in comparifon of Milford Haven), for the fame means that render them more fecure for veffels when in, in general render them lefs acceffable, efpecially by fome particular winds.

With refpect to veffels intending to enter the canal, there is no difficulty; they can always enter by either the exterior or interior mouth of the canal, or one of them, whenever they can get up the fea reach, or mouth of the river Carron, as before taken notice of I muft obferve that the fea reach is wide enough for veffels to turn to windward, but that in the reach next above, leading to the Hewk point, the channel is too narrow for large veffels to do it, and troublefome for fmall ones, unlefs we except at or near high water fpring tides. It will therefore happen, that when the wind blows right down this reach, which bears S. S. E. and N. N. W. or indeed within fix points of the compafs, that there will be a difficulty in getting veffels, efpecially large ones, up this reach; but then it is to be obferved that this reach, lying nearly parallel to the Firth of Forth, nearly the fame winds that hinder them failing up this reach will alfo be againft them failing up the Forth to Carron mouth; however, as there is ample room in the Firth of Forth for veffels to turn to windward, I do admit that it may happen that veffels may get into the Carron mouth, when they cannot get up this reach; but in all thefe cafes, as the wind will blow over land, this reach will be a place of fafety for fuch veffels to lie in till the wind changes favourably; and indeed there is no occafion for their lying here, unlefs it blows hard againft them, becaufe the length of this reach is not fo great, but that they may one tide of flood warp themfelves in round the Hewk point, where they meet a fafe harbour, and in every refpect, except the circumftances above mentioned, an excellent harbour, as is already found by the large veffels that frequent it, and deliver their cargoes at the Green brae, a harbour capable of taking veffels at fpring tides of five or fix hundred tons, and of three or four hundred tons, (according as they are built), at neap tides. The author of a tract now before me, feems to have been fully

fenfible

senfible of the importance and excellence of this harbour; when fpeaking of the river
Carron he fays, " within the mouth of the river there is a moft capacious ftation for
" fhips of any burthen, of about half a mile in length, having from twenty-two to
" twenty-four feet water in fpring tides, and from fixteen to eighteen in neap tides.

" Further up the river, (near the place where the canal is intended to come in, and
" where the Carron company have built wharfs or quays for the export of their iron and
" other manufactures), the harbour has been greatly improved by cutting one large
" loop, and there is another now cutting in the river, (both at their own expenfe),
" which will make twelve feet water in fpring tides, and eight feet in neap tides.

" The harbour of the river Carron is capable of containing 1000 fail of fhips, and
" the boats ufing the canal may fail to any one of them, whether they be at the entrance
" of the canal, (where fhips ufing the coafting trade do now come up), or at the
" mouth of the river at Green brae, (where fhips, drawing from twenty to twenty-four
" feet water, of 500 or 600 tons or upwards may lie), and this without any expenfe of
" lock dues, &c. as the river is a free navigation."

Who was the author of this tract I know not, but when I was upon the view of the
river Carron, in Auguft, 1767, it was put into my hands by Mr. Gafcoigne, as a thing
whofe truth was not to be queftioned: it is entitled, " A fhort defcription of the harbours of Borrowftonnefs and Carron, with a comparative view of their importance with
regard to the communication between the eaft and weft feas." Moved with the truths
above referred to, I was very unwilling to propofe any thing to the detriment of this
moft ufeful and excellent harbour, and being at the fame time as great an enemy to monopolies and impofitions at wharfs and warehoufes, as any of my brethren, I was willing
to take the advantage of this harbour, in fuch manner, that a great part if not the
greateft part of all the goods that will require tranfhipping in the Carron, may be done
without the ufe of wharfs and warehoufes at all. This part of the river, which alone
can be properly called the harbour of Carron, is of a confiderable width, the bottom
every where a foft mud, eafy for veffels of all kinds to lie upon, and is divided into two
channels by a bank alfo of foft mud, upon which there is fourteen feet water at fpring
and nine at neap tides, according to foundings marked thereon in a plan annexed to the
tract above mentioned; the two channels are on each fide near the fhore, and are rather
capacious than confined: it is not eafy to fay how wide they are, for as the whole bottom
is foft mud, which in courfe forms itfelf upon very gentle flopes, a very little difference
in the flow or ebb of the tide makes a confiderable difference in the width; however,

as

as the whole bank is covered at, or soon after, the first quarter's flood, the river I suppose then to be at least 600 feet wide; what the precise width is, I cannot at present say, having not yet got any plan sufficiently exact to determine it; this, however, I know from inspection, that it is a great width, and apparently the widest part of the river.

The depth of the canal, as it is settled by act of parliament, and the width thereof, and of the locks as intended in consequence, are capable of carrying vessels built galbert fashion, (and capable, in like manner, of navigating the two Firths) of 100 tons burthen; such sea vessels, therefore, as cannot pass the canal, but want to tranship in Carron water, will lie in one of the channels, where they now do, and where one of these large galberts may lie alongside of them; or if need be, one on each side at a time, from whence the cargoes can be transhipped from vessel to vessel, without the use of wharfs and warehouses, while the other channel and bank are left clear for the passage of vessels to the canal, or to Carron shore: nor can vessels for the canal be in the way of those going up the river; for, besides room for several that will be made in the tail of the canal, exclusive of the river, whenever they are opposed by wind at the inner entry, they have a fair wind at the exterior: nor can there be any objection to this method of transhipping, since, in a manner, the whole of the coals of Newcastle and Sunderland are shipped in this way, and unshipped at London; and not only coals, but the India, and, in general, almost all the ships of large burthens, that use that port. I beg leave further to observe, whenever vessels happen to be stopped from going round the Hewk point into the harbour, that they are at safe mooring in the Hewk reach, where they need not lose time, if this way of transhipping is used, as the galberts can go to them there, and make their entry good at the Hewk lock.

Let us now examine what my brethren would give us in lieu of these advantages. That the making a straight or curved cut, as they have proposed, will enable vessels to go up to Carron with the same wind that enables them to enter the Carron's mouth, or otherwise to be towed up by horses, I can in part grant; and that this will be an advantage to the Carron wharfs and warehouses, I can grant also; but it happens here, as in many other cases, that while one thing is mended, another is made worse; and it seems to me, that by the gaining this one advantage, of a more general entry, we shall lose all the rest. I wish my brethren had been kind enough to tell us, how wide at bottom and top they proposed to make this cut; because I must tell them, unless it is of very different dimensions from what is already executed as a specimen in the last new cut, I shall not like their project any better than they like mine. Mr. Golburne, who seems

to

to be the oracle in this part of the work, fays, " and by confining the new river on each " fide, the current will act upon the bottom, and grind it down to a proper depth." This proper depth he no where tells us; however, it is plain enough he does not intend it to be very wide, and, indeed, that is fhewn by his eftimate for doing it; for to make the bottom fuitable to the bottom of the river at head and tail, it muft be dug or wafhed away to the main depth of twenty feet below the furface of the prefent foil of the lands through which it is to pafs, the broken ground excepted; but fuppofe there fhould happen to intervene fome hard matter in the bottom, the current may then not act thereon, fo as to grind it down to a proper depth, but, on the contrary, turn it afide, and make a frefh crook inftead of the old one: fuch things may happen.

Now, I would be glad to appeal to any fkilful and unprejudiced mafter of a fhip, a perfon who has had the care of one of 300, 400, 500, or 600 tons, whether he would choofe to lie in a gullet, where he cannot turn himfelf about, but if he goes in head forward, he will be obliged to go out fternforward, with large lighters or galberts by his fide (or be obliged to lofe this way), to tranfhip his cargo, and where other veffels of equal fize will be obliged to rub by as well as they can, where he is fubject to fpeats and floods coming down, which run like a fluice, as may be feen in the laft made cut, into which floods in winter frequently bring down fhoals of ice, and where, if the wind blows in or near the mouth of the gullet, the fea will have a fetch of three or four miles: I fay, I would be glad to afk any experienced and unprejudiced fhip-mafter, whether he would look upon this as any harbour at all, efpecially for large veffels? Certainly, he would anfwer in the negative: then pray, gentlemen, where is your harbour? Is it in the cuts now made above? no; thefe have the fame objections, except that arifing from the fwell, with the addition of one ftill more in their disfavor, which is, that there is not water to get a fhip of any burthen into them: the harbour muft be then in the open reach of the river, between the tail of the laft new made cut, and the head of the new propofed cut; but it muft be below Jemmy Reay's ford; in fhort, it muft be confined to a fpace in the natural river not above 200 yards in length, and where there is not above eleven or twelve feet of water at a neap tide, this very difficult of accefs in time of floods and fpeats, occafioned by a long narrow entry, and this is your excellent harbour of two miles in length! excellent, indeed, for the Carron fmacks, as it will be the moft likely means of banifhing all larger veffels. Oh! but Mr. Golburne propofes to clear away the " ftones and hard gravel near Reay's ford," by " dredging or ploughing, as fhall be moft convenient;" but I can tell Mr. Golburne, that he muft dredge or plough, which ever he finds moft convenient, and that pretty deep too in fome places, a great part of the way between Reay's ford and Abbotfhaugh, in order

to

to get two feet more of water than there now is. Mr. Golburne fays, that the expenſe of cutting and completing the new channel will not exceed £3000 *; but he does not fay how much the dredging or ploughing Reay's ford, &c. will coſt, nor whether the dam to be erected acrofs the river Carron, in order to force the water through the new cut is included, nor how often that dam may chance to break down, before the Carron will fubmit to be controuled by it; nor how long the navigation may chance to be ſtopped after the dam is erected, while the channel is grinding down to a proper depth.

Mr. Golburne computes that there muſt be ſeven feet water in the channel of the Carron at neap tides, ſo that veſſels drawing ſix feet run up the river at low water. Does Mr. Golburne mean that they now do ſo, or that they will do ſo when the cut is made, and Reay's ford dredged? does he mean at the Carron mouth or at Reay's ford, and quite up to the Carron wharfs? for I own myſelf perfectly at a lofs to underſtand both the logic and the arithmetic of this paragraph. This however I am certain of, excluſive of theory, that inſtead of three feet water at low water, upon Reay's ford, I have been carried over the river upon a man's back, when he was not wet above half-way up the leg; and as there is not above eight feet water at high water over Reay's ford, I do know by the help of a little theory, that Reay's ford muſt be lowered about ten feet to make ſeven feet water at low water neap tides in dry ſeaſons, which I believe is far lower than any part of the bottom of the river above Reay's ford, and will not cut itſel out till a great way below it. Reay's ford is, I believe, (unleſs it is ſome accidental matter that may have been thrown out of the new cut which I reckon not upon), the ſhalloweſt part of the river, but I did not find, when I founded it in 1767, above ſix inches of clear navigable water more above than at Reay's ford.

I can readily agree with my brethren, that this new reformed Carron will afford an excellent harbour of two miles for a particular ſort of veſſels, and had I with them viewed thoſe veſſels through the ſame optic glaſſes, I might have thought all others of little conſequence, when ſeen through the oppoſite end; ſo far ſo good; but what it was that could provoke my brother Brindley to call the preſent intended canal, a canal without a harbour! ſurprizes me. Had Mr. Gaſcoigne put the ſame little tract into Mr. Brindley's hands that he did into mine, I hope he would have had a better opinion of the harbour of Green brae, but perhaps the wind may be changed, and now blow
right

* Mr. Golburne allows £100 a year for maintaining the new cut, which in effect is equivalent to a capital of £2000, so that the capital laid out upon completing this new cut will be £5000, a sum very far exceeding the extra cost of conveying the canal upon the surface.

right down the Hewk reach, and very hard, so that no veſſels can enter: however, if he has no opinion of that, I muſt beg leave to inform him that I look upon the harbours of Leith and Borrowſtonneſs, Port Glaſgow, and Greenock, as well as many other harbours upon the Firths of Forth and Clyde, to be harbours to the canal, where they can tranſhip their goods on board of veſſels; that whenever they can ſail up the Forth and enter Carron mouth, they can enter the canal by one of the Hewk entries, or, if they like it ſtill better, by the Carron entry, without troubling any wharf or warehouſe, and by this means many trading places will partake of the benefit of this public ſpirited canal.

Mr. Brindley and Mr. Yeoman both agree in ſaying that the alteration of the river will not produce a bar: I cannot ſay it will, but this I ſay, that I do not know any thing more difficult to pronounce upon than the effects that attend alterations of natural rivers; and I appeal to them, whether they do not, as I do, find, that even alterations of the ſimpleſt kind are often attended with effects that were not foreſeen; this I know from experience, that whatever increaſes the rapidity of a river, tends to form it into hills and holes; that the great quantity of matter that will be driven out of the new cut will be lodged ſome where, that the rapid water iſſuing from the tail of a contracted cut, will have a tendency to deepen Holmerrie, and depoſit the matter at ſome diſtance beyond it, as happened on opening the laſt new cut, and therefore will naturally lodge it where the bar now is, conſequently will have a tendency to encreaſe it; this is nothing but plain reaſoning from plain matters of obſervation; let my brethren ſay otherwiſe.

I cannot quit this ſubject without taking notice of an aſſertion of Mr. Yeoman, on the 14th article; he ſays, " there is but one wind that can carry a veſſel round the " Hewk point, or even abreaſt thereof, after allowing every poſſible advantage of tide " and fair weather." The harbour of Green brae is indeed very difficult of acceſs, if there is but one wind by which it can be entered, and even not that " in a gale of wind:" very hard indeed brother Yeoman; I marvel much how the Carron Ship Company carry on their very extenſive buſineſs under theſe difficulties.

As to the ſea reach or Carron mouth, this is common to each project, and it is agreed that veſſels can turn to windward therein; I have already ſtated that to veſſels going upward, the Hewk reach will bear N. N. W. neither brother Yeoman nor I, are obliged by profeſſion to be complete ſeamen; but I would be glad to aſk any one that is ſo, whether he cannot ſail up ſuch a reach with the wind at W. and alſo with the wind at N. E.; and with all winds from N. E. round to W. taking in all the

eaſtern

eastern and southern points, in short, taking in twenty parts in thirty-two of the whole compass; and as this reach has the Hewk farm on the larboard side, and a high steep bank of mud on the starboard, which is hardly covered at high water neap tides, whether even he cannot sail in with those winds, even in a gale of wind? also whether when at the head of this reach, a ship sailing in with a fair wind and tide of flood, may not shoot up sixty or seventy yards, so as to come abreast of the Hewk point with the wind right a-head in the turn of the point, where she will be at least in perfect security, and from whence she may if she pleases, warp herself to her proper birth, if she wants to stop there? nay further, whether small vessels drawing seven or eight feet water may not, where there is a breadth of four or five hundred feet, and a depth of at least $8\frac{1}{2}$ or nine feet at a neap tide high water, turn to windward, and weather this point; or work up the Green brae reach, if the wind being favourable in the former, and the turn of the Hewk should be right down that; and this even though a tier of ships and lighters should occupy the channel on one side? Those that deny the practicability of doing this in water perfectly land-locked, on all sides a clear shore, and nothing but soft mud to run upon; let them attend to the narrow passage often left between the tiers of vessels in the port of London, through which vessels of large sizes are obliged to work when the wind is foul, or to the narrow passage into the port of Shields, from the Black Middins to the Hird Sand; nay, further, whether a vessel before or after it has gained the Hewk point may not as easily be towed by horses through a large open river to the Grange burn foot, as it can through a narrow cut from Holmerrie to the same point; especially with a flood in the river?—As to any improvement above that place I oppose it not.

9th and 10th. I do not materially differ from my brethren, in answer to these questions; I have only to observe, that the land cannot be gained at a small charge, nor in my opinion in a small space of time.

11th. If cutting through the very tract proposed for the canal, or interfecting of it twice with the new cut, will be no obstruction to it, I do not know what will; I am therefore perfectly at a loss to know what my brethren mean, by asserting that it will be no obstruction.

12th. No more than will be according to the present design, as every one will lie as commodious for the Carron entry if they do not choose the other.

15th. The greatest part of what can be said on this head, has been already said; I

Vol. II.   Q   must muft however here take notice, that though both Mr. Brindley and Mr. Yeoman join in afferting that there will be no more obftruction from floods than at prefent, yet I would afk them whether the fame quantity of water, paffing through a fmall channel, muft not move with a proportionably greater velocity, than in paffing through a large one, and defire them or any one elfe to look at the laft new cut in time of floods, and then tell me whether there is not more fhelter for veffels from floods and from ice in the Eddys, occafioned by the turns of the river in its prefent ftate, (and a greater length too) than can be in a confined cut, in a right line, or fo nearly fo, where they can efcape no part of it; I do not contend however that the river fhould be crooked, otherwife than to preferve the natural advantage of the harbour of Green brae.

16th. Some attention and fome expenfe may be neceffary, and I dare fay any of my brethren can tell me how to do it whenever it becomes neceffary, but upon what foundation Mr. Brindley fays, that in cafe the river-fhould break through this point, " the lock intended to be built at Grange burn would then be left inacceffible to " veffels drawing fix feet water," I cannot conceive ; had he founded the Grange burn for a quarter of a mile upward, he would have found fo much water, as, with his management, would have been able to have procured a paffage of twice fix feet of water.

17th. The connecting any fide cuts, or cuts of communication, with the main undertaking, which I always underftood to be a direct communication between fea and fea, was never according to my fentiments or advice. The Glafgow branch was a matter of neceffity, arifing from prior ftipulation between parties, the end of which I never hefitated to fay would, in my opinion, be more properly anfwered by the magiftrates of Glafgow putting their act in execution for improving of the river; but who can compel them? With refpect to the Carron entry it was no part of my plan; nor the connecting of it, as to proprietorfhip, with the main canal, any part of my advice; if therefore the gentlemen have, without being fully advifed of the nature of the ground, inadvertently agreed to a thing that will lay any degree of embarraffment upon the main defign, it is no fault of mine. In regard to the agreement referred to, I never had any copy of it, except one delivered me fome months ago by Mr. Garbett, and which I profefs myfelf not to underftand. I have, at a public meeting of the proprietors, defired an authentic copy, with the precife terms of its meaning, but have not yet received it ; and till then, I do not think myfelf juftified in making any alterations in the track of the canal, or in taking notice of any paper that fhall be delivered me by any perfon, whether proprietor, or not, unlefs it comes to me authenticated, as by

the

the act is directed, and therefore till I do receive an authentic copy, and be made to underftand what is meant to be executed by thofe alone whom I acknowledge to have the power to direct me, I cannot advife what is moft proper to be done in the cafe as it is now circumftanced : but thus far I can advife, that if the proprietors do not think proper to confine themfelves to a fole entry at Carron, I do not think the track pointed out by Mr. Brindley to be the advifeable one.

18th. Refpecting this article Mr. Brindley fays, in his preface, that his time would not permit to make " minute examination of the ftrata of the earth," yet, under this head, he fays, " as I find the materials at the fummit (or point of partition) very " favourable, I recommend deep cutting at that place, not only for the faving of " lockage, but with a view of acquiring very confiderable quantities of new water." Mr. Yeoman recommends to cut as deep as we can, not fo much with a view to fave lockage, as to acquire more water. Mr. Mackell reports, from his examination in the fummer, 1767, that the cruft of peat earth is generally not above five feet thick, and under that, in every place he tried, there was quick mud; and as Mr. Brindley acknowledged verbally, that in fuch a foil deep cutting was not to be recommended, or any material quantity of water to be expected, there is an end to this bufinefs of deep cutting ; I wifh it may prove more favourable ; but unlefs we can cut fo deep as to fave a lock's height, the reft will be no faving at all; and as to water I fhould be fur- prifed if we did not raife fome, to which I hope we may be juftly entitled in confe- quence of what is remarked in pages 75 and 76 of the 2d report, but I think much water is not to be expected ; becaufe, if any communication with the hills produced any confiderable pen, it would foon find its way through fo thin a cruft, and fhew itfelf, efpecially as this cruft has frequently been perforated, in digging peats. It is probable a quantity may be lodged under the cruft, which may lie there inclofed, like water contained in a bladder, but which, when let off, there is an end of.

19th. Upon this query the two gentlemen feem divided; Mr. Brindley recom- mends to begin at the point of partition, becaufe, he fays, it is his " conftant " practice to do fo, and, in the prefent undertaking, it feems particularly advifeable " on many accounts :" but pray, Mr. Brindley, is there no way to do a thing right but the way you do ? I wifh you had been a little more explicit on the many accounts ; I think you only mention one, and that is to give more time to examine the two ends ; but pray, Mr. Brindley, if you were in a hurry, and the weather happened to be bad, fo that you could not fatisfy yourfelf concerning them, are the works to be im- mediately ftopped when you blow the whiftle, till you can come again, and make a more

mature

mature examination? and further, do you ufually begin at the moft difficult part of a work firft, with raw hands, before they are trained to bufinefs? I have fometimes done fo, and repented it. I think it no difficulty to make a hole in a hard or foft rock, becaufe it is every day's practice; but I do apprehend fome difficulty in getting even the neceffary depth into Dolater bog, notwithftanding you find the materials there fo very favourable. Mr. Yeoman very judicioufly and very candidly obferves, (and in that fpirit all his remarks and anfwers are drawn up), that as he apprehends " in a country " like this, where the canal is to pafs through, it will be utterly impoffible, at leaft it " will be very inconvenient, to begin any where, and carry on the works in a progreffive " manner, without fome intervals, becaufe of the want of accommodations for work- " men, and efpecially at the firft fetting out." I have often told the proprietors that nothing of confequence could be done this year, but merely to make a beginning, and get the workmen into training, fo that they may afterwards be branched off to different parts; that for the fake of getting accommodations for workmen, without laying a diffi-culty upon them and the country in fupplying and lodging them, it would be right to put on workmen at both ends, the middle, and alfo to the neck of the hill in the Allander paf-fage, by which means the work may be carried on in five different places: befides, men may be fet to different ftations to cut the trenches, aqueducts, &c. but all this cannot be done at once, nor is there any neceffity of pufhing, 'till the plans can be made out, feveral of which are already done, and in Mr. Mackell's hands, before I went laft into Scotland; and one very material and much wanted, viz. for the locks, would have been done, had I not been employed in anfwering thefe reports. As to what Mr. Brindley throws out about making dry cuts, I will be bound to fay that he will not finifh the canal in four years, as he afferted, nor as I think in twice four, if he carries his work progreffively each way from the fummit, in order to take the water with him; and as to fuch opportunities of filling the cuts with water from the fummit, or as offer themfelves from the rains in the winter, we fhould have done that, though he had not told us. Something of this kind is to be found at page 76, of my fecond report.

As to traffic from the fummit, I have heard of none except Mr. Caddell's iron ftone, which, as it only pays a penny per ton, I muft be fo juft to the intereft of the proprie-tors who employ me, as not to tell them that I think this an object which fhould induce them to put their works out of the regular courfe of proceeding. I am obliged to Mr. Yeoman in giving his fentiments of what he has feen done, and that whether in favour or disfavour.

20th and 21ft. I have reported, and the engineers have confirmed it, in their replies to the firft and fecond queries, that there is water enough for the navigation, and to fupply what fhall be taken from the mills; the act fays exprefsly, that all fuch water as fhall be taken from the fources of the rivers Carron and Kelvin, to the detriment of the mills fpecified, fhall be replaced in kind: this being the cafe, I do not fee why the proprietors fhould be prefcribed to, as to their mode of doing it. If they are the fhowers the Carron company fo much covet, we can give them fufficient of that from the head of the Enrick, to compenfate the drainage of Dolater bog, and the hill fides along which the canal is carried; but as we propofe to make tunnels and aqueducts to give the burns and runners their natural courfe, and to take nothing into the canal that we can help, beyond our proper fupplies; and fince all fuperfluous water taken in muft be difcharged at the over-falls, as it cannot be ufed, I would afk my brethren, whether, after this, a much more material drainage is not likely to arife from the foakage of the canal, and the Dolater refervoir, notwithftanding all the care and pains that either they or I can take; and fince I have demonftrated, fee report fecond, page 80, that the Dolater refervoir alone will be capable of holding water that will fupply the canal one quarter of a year, and all deficiencies for three quarters of a year; I hope we may take the opportunity of filling it in winter, or great downfalls of rain, when twenty times as much water as they can ufe at Carron is running wafte over the dam-heads, after all their refervoirs are full: it is our bufinefs only to give them as good means of filling their refervoirs, as they now have, not to enlarge their refervoir at Larbert, nor procure them Loch Coulter from Stirling mills; in fhort, I can fee no other ufe of thefe two queries, but to raife difficulties where there are none.

I come now to take notice of fuch fupplemental articles as my brethren have thought proper to add, which have not fallen in with fome part of the foregoing ob-fervations. The principal is, they have found out a new road, more near and convenient for the city of Glafgow; but I muft beg the gentlemen to reflect, that whatever they have been upon, my object was a communication between fea and fea, and a branch to Glafgow as per treaty. I have as high an idea of the trade and importance of Glafgow as they can have, and wifh as well thereto; and for this reafon, fhould have been moft heartily glad that Glafgow had ftood at Dalmoore burn foot: yet, notwithftanding all they have faid, I can by no means give up the entry there; but as Glafgow ftands not there, I further wifh, that at any moderate expenfe, the canal could have been extended below Dumbuck.

The

The paffage they hint at, and which is in fome meafure defcribed by Mr. Yeoman, is due to the affiduity and fagacity of Mr. Mackell, who found it out after he left London, in the fpring, 1767, and during the time he was employed in reconnoitring, previous to my review in Auguft, 1768, and who then fhewed it to me ; and it was fhewn to Mr. Brindley by Mr. Laurie, the furveyor, and which paffage, on confideration, I rejected for the following reafons.

1ft. The ground from Calder bridge to the paffage over Kelvin, at the Printfield, is in general rough, and fome part of it very rough and crooked.

2dly. A rife of more than twenty feet for near a quarter of a mile above the level of the canal on one fide, and almoft as high on the other, but for a fhorter length.

3dly. A very deep and wide valley to crofs the Kelvin, by a very large aqueduct bridge.

Thefe appearances made me judge that the expenfe of this paffage would confiderably exceed the expenfe of cutting the Allander fummit, efpecially as I had then in view the carrying the canal upon a higher level, and faving a confiderable quantity of cutting in that fummit, and which I have now ftill more reafon to think is practicable. Add to this, as we had no other view than to go to Dalmoore burn foot, it appeared to me not only confiderably more winding, but further about from fea to fea, though Mr. Mackell thought otherwife refpecting the diftance.

The communication to Glafgow would, indeed, be fomewhat fhorter, but then it parted from the main canal at fuch an elevation, as to require feveral locks to go down to Glafgow, which would not only be expenfive; but require water to work them ; whereas, in the way the Glafgow branch was planned, it not only required no extra lock or water, but laid fo nearly parallel to the river, that it might be ufed for a communication from Glafgow to Port Glafgow and Greenock, and back again ; which, though no part of the ftipulation, as I confidered the branch canal as a dead weight upon the main undertaking, this circumftance might make it anfwer better, and be a proportionable advantage to the trade of Glafgow, there being no appearance then of the Clyde's being improved. It was in vain to fay—gentlemen, you may make a branch canal of your own river, by putting your act in execution.—No, we muft have a branch canal of equal dimenfions ; but if Meffrs. Brindley, Yeoman, and Golburne, have

again

again caufed the magiftrates of Glafgow to refume the improvement of the river, as I am convinced in my own opinion, that the track marked out in the act of parliament for the main canal, is upon the whole the preferable track; the truly wife, and fenfible thing will be to give the Glafgow gentlemen a branch cut, from near the three part mill dam into the Clyde, below Blart hill, as hinted at before; and though this will coft them at leaft three locks, with water to work them, yet the proprietors had better be at the expenfe of thefe locks, and procuring this water, than carry the canal over the Kelvin through the valuable grounds near Glafgow. Where Mr. Brindley's canal was to terminate does not appear; this he has referved for the next time he comes; but Mr. Yeoman's termination at the Holme Sands I can by no means approve, not only becaufe that is the moft precarious part of the whole river, and becaufe it lies in the power of other perfons to remove the impediment or not, but alfo becaufe there is another circumftance, that Mr. Yeoman poffibly was not acquainted with, which is, that if the magiftrates of Glafgow only order the ftones to be thrown into the river, that John Adam laid down at Marlin ford in the year 1760, for building a dam there, they may levy one fhilling per ton upon all veffels paffing that place, and thereby lay the canal under a heavy contribution.

That my brother Brindley fhould prefer the Printfield paffage I can readily comprehend: a late author has very folidly demonftrated, that every man, how great foever his genius, has a certain hobby horfe that he likes to ride; a large aqueduct bridge over a large river does not happen to be mine, who am of opinion, that a given fum of money is as folidly laid out for pofterity in cutting through the neck of a hill, as in building a bridge to carry water over water, though the admirers of the wonderful may not be fo loud in their applaufes.

As to Mr. Brindley's 4th and 7th fupplemental obfervations, which have not been touched upon before, I could have told him the fame things if I had met with him in Staffordfhire.

I afk Mr. Golburne's pardon for not attending before to one of his obfervations: he finds Holmerrie not above 100 yards long, and not above five feet deeper than the bar, whereas I have ftated it at feven feet deeper than the bar, and a much greater length. I did not meafure the length; but I founded the depth, and found the deepeft part feven feet deeper than the bar: this was during the very firft fpeat that happened after the laft new cut was opened, out of which firft and laft there has been driven a great quantity of matter: I therefore would afk Mr. Golburne's opinion, whether fome

of

of the matter out of the new cut may not fince that time have been lodged in Hol-merrie, or that my pilots were as folicitous to find out the greateft depth, as his were to find the leaft ? be this as it may, the channel of the river in this place is for a great width deeper than the bar, and the bottom foft mud, fo that veffels cannot be hurt in lying thereon, in three or four feet water at low water, though Holmerrie fhould be quite filled up; or if two or three at a time, weaker than the reft, cannot bear it, they may ftill find a birth in Holmerrie, in cafe Mr. Golburne will not quite fill it up with the matter out of the new propofed cut.

Having now gone through every article that feems of confequence to the right un-derftanding of the bufinefs in hand, I have only further to add, that for my own part I fee nothing to oppofe this great work's being brought to a timely iffue, but the dif-ferent opinions of men, and I fhould not defpair of feeing it executed, even with the remnant of time that I have in my power to apply to it, in cafe I am fuffered to go on with the work according to my own experience and ideas; but if engineers are to be conftantly brought down to infpect and fee how the pot boils, I think neither I nor any other man can go on with it, to the advantage of the proprietors, under fuch circum-ftances, any more than I could fit down at the crofs of Edinburgh, and write this anfwer to my brethren, while every one at pleafure had an opportunity of overlooking and afking me why I begun this paragraph in this manner, or treated that fubject thus.—If, inftead of making plans, I am to be employed in anfwering papers and queries, it will be impoffible for me to go on with the bufinefs, and therefore perhaps thofe that wifh no good to the undertaking may be defirous of furnifhing me with this kind of materials. All the favour I defire of the proprietors is, that if I am thought capable of the undertaking, I may go on with it coolly and quietly, and whenever that to them fhall appear doubtful, that I may have my difmiffion.

28th October, 1768.   J. SMEATON.

COMPARATIVE VIEW of the expenfes upon the works of the Forth and Clyde canal, as per general account, down to Chriftmas, 1771, with the original eftimate :

	£	s.	d.
By account of materials, freight and carriage thereof, workmanship, digging, and temporary damages - - - - - - -	71,747	12	2
Estimate of the whole, per fecond report - - - £147,337 0 0			
Carried forward -	71,747	12	2

Estimate

	£	s.	d.	£	s.	d.
Brought over	147,337	0	0	71,747	12	2
Estimate of additional expense of the new line, neat valuation -	1,734	0	0			
Ten per cent. thereon for contingencies - - -	173	8	0			
Total sum estimated	149,244	8	0			
Total land purchase - - - £13,544 0 0						
Ten per cent. on ditto - - - 1,354 8 0						
Deduct ————————	14,898	8	0			
Total of materials and workmanship, valued per estimate	134,346	0	0			
Ditto of the Glasgow branch, deduct -	13,007	7	4			
Remains for the main canal as now ———————				121,338	12	8
Remains of the estimate to be expended - -				49,591	0	6
Difference of past expense and of that to come				21,155	11	8

N. B. If it is supposed that the works were half done at Christmas last, then the expense will exceed the estimate by about $\frac{1}{8}$. But if the expense of the Glasgow branch be given to this work, then the exceeding will be but about $\frac{1}{16}$th.

(Sic. subtr.)　　　　J. SMEATON.

Edinburgh, 3d August, 1772.

A PLAN or model for carrying on the mechanical part of the works of the canal from Forth to Clyde, by J. SMEATON.

Engineer in chief.
Engineer refident.

| Surveyor of the eaftern department. | Surveyor of the middle department. | Surveyor of the weftern department. |

To each of the furveyors, a foreman of the $\begin{cases} \text{Digging,} \\ \text{Carpentry,} \\ \text{Mafonry.} \end{cases}$

Their duty as follows :

Engineer in chief.

1ft. To make plans for execution of fuch works as fhall be directed by the committee.

2dly. To pitch upon the ground whereon the faid works are to be conftructed.

3dly. To correfpond with the committee upon fuch points as they fhall think neceffary.

4thly. To correfpond with the engineer refident, and fend fuch directions from time to time as himfelf or the refident fhall find neceffary.

Engineer refident.

1ft. To attend fuch meetings of the committee as he fhall be ordered to attend, or fuch meetings as he fhall think neceffary to procure directions.

2dly. To fee the plans and directions of the engineer in chief put into execution.

3dly. To mark out the grounds to be purchafed, and to enter into fuch treaty with the proprietors as he fhall be directed by the committee.

4thly. To

4thly. To fupply fuch plans and directions for the leffer part of the work as he fhall be defired by the engineer in chief, or which his abfence, or the neceffity of the cafe fhall make expedient.

5thly. To correfpond regularly with the engineer in chief.

6thly. To give a monthly account of the ftate of the works to the committee, or oftener, if required, and to fend a duplicate to the engineer in chief.

7thly. To attend each part of the work upon any emergency or difficulty, and alfo to fee all new matters or methods put rightly in hand.

8thly. To furvey the materials, and make fuch purchafes thereof as directed by the committee.

9thly. To vifit all the works from end to end as often as poffible.

N. B. It is fuppofed that the refident engineer ought to have power of employing a land furveyor, if occafion, to meafure and furvey the lands, and to have power of difcharging any officer under him on neglect of duty.

Surveyors of particular diftricts.

1ft. To attend to the orders of the engineer refident.

2dly. To fee the proper quantities, qualities, and converfion of the materials committed to his care.

3dly. To order the foremen concerning the works whereon they are to employ themfelves.

4thly. To tranfmit an account weekly of the progrefs of the works to the engineer refident, and of what materials, &c. are wanted.

5thly. To receive the neceffary wages from the pay-clerk, and account with him for the money.

6thly. To

6thly. To refide as near as may be to the place where the principal works are under his care, to vifit them daily, and the moft diftant once a week, or as often as poffible.

Foreman of particular branches.

1ft. To obferve the directions of his furveyor.

2dly. To fee all his men as often as poffible, and to fee that they keep to their labour.

3dly. To receive money from the furveyor, and pay the men, of whom he is the overfeer, keeping a daily account for each particular man, and delivering a duplicate to the furveyor, in order to be tranfmitted to the committee.

N. B. As gangs or fetts of men muft neceffarily be employed at different places under the fame foreman, fome one of thefe muft be appointed overman of a gang: it is prefumed that the overman may act as clerk of the check to his gang, and give an account of their time to his foreman.

To prevent abufes, the engineer refident to have infpection of all papers and accounts of the officers under him, whenever he thinks it necefiary, or is ordered fo to do by the committee.

(Signed)   J. SMEATON.

London, 14th March, 1768.

CALDER

CALDER NAVIGATION.

The REPORT of JOHN SMEATON, engineer, concerning the state of the Calder navigation, from a view thereof, taken the 25th and 26th days of November, 1767.

See a map of the Calder, vol. 1. p. 21.

GREAT damages having happened, particularly to the banks and spade work of the Calder navigation, from a flood in the night between the 7th and 8th of October last ; on a view thereof, I made the following observations : That the river was in this flood, by marks at Horbury mill, two feet ten inches perpendicular higher than any flood which has happened since the river was first surveyed by me in the year 1757 ; that at Minfield Low mill it was above four feet higher, and at Brighouse three feet eleven inches ; that, in fact, it was higher than any flood in man's memory, or of which there is any tradition, being universally allowed to be higher than an extraordinary flood which happened about forty years ago ; which, in the neighbourhood of the Calder, is distinguished by the name of Bowman's flood, and which was always reported to me to be the highest ever known : but the flood of October last, by the marks at Horbury mill, (and the testimony of John Horn, a millwright, who has worked there and at Wakefield mills ever since Bowman's flood, and remembers it), appears to have been four inches higher at Horbury mill, than Bowman's flood, and as in the late flood the rise was more considerable in the highest part of the river than below ; these facts being considered, it seems more marvelous that the works have not been wholly swept away, rather than that they have been broken in particular places.

In order to guard the navigation as much as possible from such an extraordinary event as Bowman's flood was reported to be ; it was always my intention, and which I have verbally mentioned to the commissioners, while I was concerned, that as soon as the navigation was opened to Brook's mouth, to build flood-gates like those at Dewsbury upper mills, at the head of Thorn's, Batty's, and Kirklee's cuts, and also to make up and defend the banks in particular places to an extra height and strength ; and, had this been put into execution, I cannot help saying, that the damages on the present occasion had laid in much less compass, though probably not entirely prevented.

It

It appears from this view that the banks for by very far the greateft part have with-ftood the flood, and that in thofe places where they have failed, it has not been by their burfting, but by the waters overflowing their tops, and have been either cut down in confequence of a current overflowing them, or beaten down by the waves raifed by the wind, which, in the decline of thefe floods, was very violent, and which laft caufe of difafter might be prevented by covering them over with rubble ftones, where fo par-ticularly expofed : it appears alfo, that nothing of confequence has happened to any of the capital works, viz. the locks and dams; for though two or three of the locks have fuffered fome damage, yet this has not been occafioned by any failure in point of conftruction, but by an attack from without, arifing folely from the failure of the banks : the worft of thefe being Batty's lock, one of whofe chamber walls was wafhed down piecemeal into the lock, by a current of water fetting acrofs the top of the wall, from a breach in the bank at the lock-head; yet even in this fituation of diftrefs, the pillars to which the gates were hung were left ftanding.

From hence it appears practicable, by guarding againft the difafters that now have happened, partly by increafing the height and ftrength of the banks in particular places, and partly by the erection of flood-gates, not only in the three cuts above mentioned, but in fome others that now appear to be much expofed, that this navigation might be made as fafe againft fuch floods as the great flood of October laft, as it appears to have been before againft all fuch as happen in the common courfe of nature.

The great difficulty and rifk attending the navigation is during the courfe of the prefent winter, till the feafons comes in which the neceffary works of fafety may be per-formed; for till the breaches are made up, the navigation is liable to be interrupted, and what is already done in the way of repair, deftroyed by common floods overtop-ping thofe places not fully made up, and this can only be prevented by a timely and fpeedy application of a fum of money to effect thofe purpofes; and which, if fpeedily applied, and the feafon continues favorable, I hope will not be confiderable in proportion to what has been already laid out : but as it depends upon circumftances of weather not to be forefeen; and when things are obliged to be done at a pufh, are neceffarily done at a great difadvantage in point of expenfe, there is no poffibility of making any regular eftimate. But, fuppofing the navigation to be faved by a feafonable application till the winter is over, I am of opinion that the fum of £3,000 directly laid out during the courfe of next fummer in erecting the neceffary works and defences, will give a reafonable expectation of feeing fuch another flood pafs over without interruption to the navigation, at leaft, by the failure of any of its own works; for it is to be obferved, that by the failure of

Dewfbury

Dewſbury dam, belonging to Mr. Greenwood, the navigation would have been interrupted the greateſt part, if not the whole time, that it may probably now be, had nothing particularly happened to the Calder works.

It appears to me, that eight, or at moſt nine pair of gates, with proper bankings in particular places, as before mentioned, will be ſufficient, and all the materials being prepared and laid down upon the reſpective places, and the works divided into two, or perhaps three claſſes, that the works of each claſs may be got above water, without interrupting the navigation above ten days at each time, which, as it will be at intervals known beforehand, will, I apprehend, be no material detriment.

Gates upon Thorn's Ledger, Batty, and Kirklee's cuts, appear the moſt neceſſary ; but I ſhall not prolong this diſcourſe at preſent, by entering into the detail of what ought to be done in future ; for in caſe this propoſition is reſolved upon by the commiſſioners, which ſeems to me the moſt likely, and perhaps the only way of giving a laſting ſecurity to the good effects of this navigation to the public, and to the property of thoſe who have ventured thereupon : though I both am, and expect to be, much hurried in buſineſs, yet I ſhall not be wanting in giving my beſt advice concerning the proper placing and conſtruction of the propoſed works, if thereto deſired.

J. SMEATON

Austhorpe, 30th November, 1767.

CALDER

Mr. SMEATON, having viewed the river Calder, the 26th and 27th of December, 1770, reports as follows:

HE has the pleasure to inform the company of proprietors, that he finds the river now put into as good a state of security as could possibly be expected in the time, and is, indeed, in general, in a very defensible condition; but to give the works every possible degree of security against the extremes of floods, he begs leave to subjoin the following remarks.

WAKEFIELD.

The heels of the flood-gates to be breasted up about one foot, and the banks made good to the bridge.

THORN'S CUT.

The flood-gates to be breasted and continued one foot higher to the high bank.

LUPSIT DAM.

The old setting in the body of the dam to be taken up and used as footing stones to a rubble body; the footing stones to be disposed in the arch of a circle, like the original dam, and from thence lagged up with an even slope to a proper height; the materials there rather want putting in order, than any great addition.

HORBURY BRIDGE.

The bottom, under the navigation bridge, to be filled with rubble, to such a height as to allow navigation water only; this will strengthen the bottom when the water goes over the bank below next the land, and prevent so much from flowing through the bridge. The bank next the river from the bridge to the lock to be heightened by a new cover, and also a little above bridge to be raised, and to be rubbled next the river where worn.

LONG CUT.

The banks to be raised and strengthened a little above and below Mr. Walker's bridge.

The

The bank on the weft fide of the cut, for forty or fifty yards above where the old bridge ftood, at Noel pond, to be a little raifed with lagged rubble, fo as to prevent the water going over that part before it does over the lagging at Noel pond.

BATTY MILL CUT.

About fixty yards of the bank, next the land below the oridge, to be lowered fo as to be eighteen inches lower than the reft of the banking below bridge, and to be lagged over, that the back water which, unavoidably, will get into the cut above, may be difcharged there without breach of banks; the place in the bank next the river, juft below the flood-gates, to be well fecured by an addition of rubble, where it has been attacked by the late floods.

KIRKLEE'S CUT.

The bank on the fouth fide the cut, above the flood-gates, to be rubbled over on the back fide with fmall rubble, to prevent the wind wafh of the water from taking it down when the banks belonging to the farm break above.

ANCHOR PITT.

Some rubble wanting to fecure the tail of the dam's end wall, on the Bradley fide.

LILLANDS.

The bank above the lock on the fide next Lillands, to be made good with rubble behind the hedge.

GOOL CUT.

The bank next the river below the flood-gates to be made good to prevent the rivers breaking into the cut.

SOWERBY BRIDGE CUT.

When the water is off, to examine the places whereabouts the runs are, and to carefully tread and beat the earth that fhould line the fame; if this is not effectual, a further depofition of afhes will be of ufe, and probably in time ftop of themfelves.

The method now taken of increaſing the weight and ſtrength of the banks, which ſlide on the ſide next the river, and diminiſhing thoſe on the oppoſite ſide, is the moſt likely way to make them effectual.

It ſeems to me that water will be beſt brought into the cut from the river Calder, by carrying a ſough through the high ground behind the houſing of the town of Sowerby bridge; its dimenſions and depth will depend upon the level, whether it can come from Sowerby bridge dam, or thither by an open caſt from Holling's mill dam, and if the latter muſt be brought by a ſough through the high ground contiguous to that mill, which level being taken, Mr. Smeaton will further adviſe.

J. SMEATON.

Austhorpe, Jan. 16th, 1771.

To Mr. Thomas Simpſon, clerk to the company
    of proprietors of the Calder and Kebble na-
    vigation.

The

The REPORT of JOHN SMEATON, engineer, upon the means of improving the navigation of the rivers Aire and Calder, from the free and open tidesway to the towns of Leeds and Wakefield respectively.

FROM a careful infpection of thefe rivers, it appears to me, that the original pro-jectors of the navigation, not having had any notion of the extenfive trade that was likely to be carried on by means thereof, have formed their plan upon too diminutive a fcale, and particularly with refpect to depth or draft of water. This appears plain from this circumftance, that feveral of the threfholds of the locks have been laid fo high in fome of the moft critical places, that with full ponds in the ordinary ftate of the river, there is not above two feet fix inches of water over the threfholds; and alfo not being aware that it would become the practice of millers to draw and keep down their ponds, in order to levy contributions on the navigation; though as to what concerns the working of the mills, it is manifeftly their intereft to endeavour to keep them full; the navigation has always laboured under difficulties on thefe accounts.

The level of the lock threfholds evinces what the views of the firft projectors were refpecting the conftruction of the navigation, and had this been the only fault, it would have been poffible, by the re-building thefe locks, to have amended it; but the fame idea, by endeavouring to fave locks in point of number, and to fave length of cutting, has tempted them to carry the locks too far upwards, fo as to leave fhoal fpaces in the river for a confiderable diftance below.

Nor do thefe impediments laft defcribed occur only within the compafs of the locks, or the artificial part of the navigation, for they equally fubfift below the loweft lock, which is at Haddlefey; and though the fpring tides there in ordinary make fix feet additional water, yet the neaps do not commonly reach thither, fo that in thofe cafes in ordinary dry feafons, there will not be two feet of water up to Haddlefey lock at high-water neap tides: this is the cafe, not only within the limits of the prefent navigation, as fubject to act of parliament, that is, down to Weeland, but there is alfo a confiderable fhoal equally as bad as any of the reft, that lies about a mile below Weeland, called Stock Reach, over which, though the neap tides fenfibly flow, yet they do not make, in the whole, above two feet depth of water, and it is plain that no very material change for the worfe has happened in the river fince the navigation was begun, becaufe the

depth

depth of water upon the lower gate threſhold of Haddleſey lock correſponds with the preſent depth over the ſhoals aforementioned, nor would there be the water below the ſaid lock, within the limits of the act of parliament, that there really is, were not the bottom of the river worn down at a place called the Hurſt, by weirs, which at beſt do their buſineſs imperfectly, and are ſubject to be torn away by every land flood.

To carry on a navigation under theſe diſadvantages in dry ſeaſons, they have early had recourſe to flaſhes, that is, by drawing down the water out of ſuch of the dams as were built by the navigation, and by doing this as it were altogether upon the two rivers there is created in the ſpaces below an artificial freſh, which enables veſſels drawing $3\frac{1}{2}$ feet water to move for a time; but as the creating this freſh below drains the ponds above, they require in dry ſeaſons ſuch a length of time to fill, that it frequently happens they cannot give above two flaſhes per week, and thoſe flaſhes laſt but for a ſhort ſpace of time, ſo that veſſels going down therewith, being crowded together, are unavoidably ſtopped in paſſing every lock; ſo that while that is doing, the flaſhes are ſpending, thoſe that are foremoſt ſtand the beſt chance, while thoſe behind are frequently catched by the keel: ſo that taking the chances they ſeldom get upon an average above two ponds per flaſh, going down, and in coming up, or meeting the flaſh, they ſeldom get, at an average, above one; wherefore, veſſels will be frequently from Stock Reach to Leeds or Wakefield, a week or more in making good their paſſage, that otherwiſe would be performed in fifteen hours. Hence, there is a neceſſity of a greater number of veſſels being employed upon the river than would otherwiſe be ſufficient to do the buſineſs, and, in conſequence, a great deal of time neceſſarily rendered vacant to the boatmen, &c.; hence ariſes a manifeſt detriment to the public from ſo great a number of veſſels and men lying unemployed, whoſe hire and wages upon the whole muſt be made good by the conſumers, and hence complaints of the boatmen, who rather than be perfectly idle, often employ themſelves in doing miſchief.

It is manifeſt, that theſe ſtoppages, which laſt in a greater or leſs degree for ſeveral months in the year, muſt in their own nature produce the neceſſity of employing a much greater number of boats and men than would otherwiſe do the buſineſs; yet, if a number were employed ſufficient to do the buſineſs in the flaſh ſeaſons, there would not be half employ for them the reſt of the year; it follows, therefore, that in thoſe ſeaſons, the buſineſs done, notwithſtanding every effort, will be far ſhort of the demands of the public: hence great complaints, for want of diſpatch, againſt the rivers, and againſt the undertakers, and from thoſe who, if they were apprized of the difficulties and diſadvantages under which the buſineſs muſt be done, would be ſurprized how ſo much is performed in ſuch a way.

The

The mills alfo complain of the water in dry feafons being brought down in flafhes, that a great part of it runs wafte over their dams, and that they are thereby deprived of a regular fupply, and are obliged to wait a confiderable time to get their ponds filled again, which complaint, though juft, I apprehend has originally been brought upon themfelves by their drawing down their ponds, which in return fuggefted the practice to the navigators drawing their ponds in order to fill the others : thus the mifchief by one party's endeavour to draw down, and the other to fill, becomes accumulated, and which, with the prefent trade upon the river, it feems to me that nothing can remedy but a different mode of navigating.

To propofe an adequate remedy for thefe obftructions and impediments with as little detriment as poffible to private property, the proprietors of the navigation applied to me, and of this I fhall now endeavour to acquit myfelf in the beft manner I am able. In doing this I fhall not take upon me to confider this matter in fo extenfive a light, as if the navigation had never been made ; but keeping to the prefent tract and works wherever it can be properly done, I fhall content myfelf with fuch alterations as appear neceffary to procure the effential of a navigation, viz. the means of keeping veffels always afloat, fo as to move freely in either direction when in the proper channel of navigation, and when not loaded beyond a proper draft of water.

Having carefully viewed and founded both rivers from the towns of Leeds and Wakefield, to Carlton Ferry, near Snaith, I find on comparing thofe foundings that the rivers are capable (after making fuch alterations as in any view of improvement feem indifpenfible) of being made to carry veffels drawing three feet fix inches water in dry feafons, and at other times four feet in its ordinary ftate, which according to the prefent fizes of veffels will carry thirty tons, and forty five tons refpectively.

Experience teaches us, that one of the greateft impediments to navigation is that of navigating in mill dams, yet without forming this navigation wholly anew it is hardly poffible to avoid it totally. I therefore lay it down that in order to fecure three feet fix inches water in mill dams in dry feafons, that there ought to be made five feet at a full head ; this will allow the miller to draw off eighteen inches, which will give him a very fufficient fcope, and yet referve three feet fix inches to the navigation: this will be confiftent with the millers' intereft, as mills expend more water as they draw down to do the fame bufinefs, and few on this river will do any good execution when drawn down this quantity, the more they are drawn down, the more time they will take to fill, fo that what is more than the fcope abovementioned cannot in my opinion be confidered

fidered in any other light than as means of hurting the navigation; and if it can poffibly be fuppofed (which I cannot fuppofe) that any advantage can arife from drawing
lower than eighteen inches within head, yet this will be compenfated ten-fold, by
faving all the water now expended in flafhing, which I look upon to be far more than
what is confumed in the neceffary and proper lockage of the veffels. Drawing therefore
the line here, as perfectly reafonable, in order to prevent difputes with the millers,
that the navigators be at liberty to diminifh the depth of the river in fome proper place
between the head of the navigation cut and the mill dam refpectively, fo as to form
an artificial fhoal, the top of which, in the loweft part, not to be lefs than two feet
beneath the water's furface at a full head.

ALTERATIONS PROPOSED.

Beginning therefore at Leeds, I propofe the following principal alterations. Juft below
the warehoufe a fhoal will require to be dredged, and another formed below the head of
the cut: the cut above the lock to be dug deeper, fo as to make five feet water at a
full dam, as far as the elbow, and all the threfholds being confiderably too high the
lock muft be rebuilt, and as the tail cut and river below are too fhallow, and the
deepening both is expenfive and uncertain, I propofe to extend the cut, and place the
new lock lower down, which will drop the veffels into deep water; the river then holds
good, or may be made fo, with a little help, to Hunflett mill.

Hunflett mill lock, New mill lock, and Twaite mill lock lie all too high, and at
all events conformably to what is above ftated muft be rebuilt; the cuts are all too
fhallow, both above and below, as well as intermediate fhoals in the open river, and as
the ponds are fhort, and thereby the mills can the more readily draw them down, this
fpace is fubject to every fpecies of embarraffment, to avoid which I propofe to take a
cut out of the river a little above Hunflett mill, and forming a fhoal (as before mentioned)
to crofs a corner of the Breaks, to crofs a clofe of Sir William Milner's, another of
Mr. Cookfon's, Sir William Milner's again, and laftly through a pafture of the Right
Honourable Lord Irwin's, commonly called Clark's pafture, fkirting the high ground,
and then falling into the river; this cut will fcarcely be a mile in the whole, and will
pafs all the three mills at once. Flood-gates will be neceffary at the head, to keep the
floods out of the cut; two locks will be fufficient, with bridges and tunnels, to communicate the roads and water-courfes, the hawling tract being upon one of the banks
may be fenced off, if thought neceffary.

There

There are three fhoals between the elbow and the tail of the prefent cut; as they are rocky fcalps they may be taken up without interrupting the prefent navigation, or to avoid them, the cut may be extended about 150 yards, fo as to fall into the river oppofite the prefent lock.

Ryder fhoal will need a little clearing, as alfo one below Mr. Fenton's ftaith : the dam boards will be neceffary, as at prefent, in dry feafons, but may be made to ftrike at any time of flood, with far more eafe than at prefent.

Cryer cut will want a little deepening towards the head, and the north bank making good.

Woodlesford lock, its bottom lies too high by about $2\frac{1}{2}$ feet, but the river being embarraffed with fhoals below it, the laying it deeper would be to no purpofe; it will be neceffary therefore either to build an intermediate dam and lock above Swillington bridge, or to continue Cryer's cut till it can be dropped into deep water above Swillington bridge, the latter, as it will leave the river free from additional dams, is in many refpects to be preferred, but, as it would occafion the cut to crofs the broad flat part of the valley, and to join the river where there is little flat ground on the oppofite fide, the water would in time of great floods become too much confined. What I think the moft eligible is to continue Cryer's cut along the face of the high ground below the houfing of Woodlesford, and croffing the road (from Wakefield to Swillington bridge) to carry it along the fkirt of the high ground of Woodlesford field to the Oak Tree clofe, then to crofs to the point of high ground belonging to Mr. Blades, and from thence to the tail of the cut of Fleet Mill lock, by which the fhoals at Woodlesford, and the draft of Fleet Mill will be avoided together. This new cut will be nearly a mile, and to be furnifhed with proper bridges and tunnels.

Mithley lock bottom is alfo too high, and there being fhoals all the way down, almoft to Sir William Lowther's ftaith, that cut to be lengthened, and a new lock and flood-gates at the head.

From hence to Caftleford, good water, with prefent dam boards there; but, if any alteration fhould take place, the navigation fhould have power to build a new dam below the joining of the two rivers, in order to keep the water to its prefent height up them both.

The

The bottom of Caftleford lock lies too high by two feet nine inches, and the river below is greatly embarraffed with fhoals both above bridge and below, infomuch that in the prefent ftate of things, veffels could not pafs Caftleford without flafhes, even if Knottingley dam were always kept full; thofe embarraffments will be all avoided together by a cut from the mouth of the two rivers acrofs Caftleford Ings to the elbow below Caftleford, being a little more than a quarter of a mile in length, with a new lock, road bridge, and flood-gates at the head.

From hence the river continues very good till a little below the tail of Brotherton weirs, where there is a fhoal having but four feet two inches water, and one below Ferry bridge, which has only three feet eleven inches, with a full pond at Knottingley, fo that with the great number of mill cloughs and wafte cloughs upon this head of water, it can, in all common feafons, be reduced fo as to lay the loaded veffels a-ground upon the fhoals laft mentioned. But this is not all, for even though Knottingley mill were always to keep a full head, the river, for about a mile below Knottingley lock, is fo greatly embarraffed with fhoals, that with a full pond at Beal dam (the next below) there will not be above a foot of water in fome parts of this fpace, and for the greateft part of that fpace not two feet, and at a medium about eighteen inches, and fo great is the tendency of the river to filt and form obftructions in this part, that though weirs for the whole length have been maintained for a long courfe of years at a very confiderable expenfe, yet this part of the river feems to be in a ftate of growing rather worfe than better, and even if it were, at a great expenfe, to be dug deeper, fo great a quantity of loofe fand is lodged by the floods, that there is no anfwering that it would not choak up almoft as faft as it can be deepened. The expedient for paffing thefe fhoals has been by putting boards upon Beal's dam, in two heights, fo as to raife the water two feet above the dam's crown; and as thofe boards cannot be fo eafily difengaged in two heights as in one, they have frequently been obliged to remain during the courfe of flood, which has produced complaints from the land-owners in regard to the overflowing of their lands, and from the proprietor of Knottingley mill, in regard to its being fooner put into back water, the only certain and adequate remedy therefore that I fee, is to form a new cut by all thofe obftructions from deep water to deep water, which will be done by departing from the river a little below Brotherton weirs, and from hence purfuing the general boundary of the north fide of Brotherton marfh, and falling into the river into the bend below Anfleet drain, from whence the river is very good to Beal, efpecially with one height of boards, which may be made to ftrike as before mentioned, in any time of flood, or it will do without the boards, by taking a cut out of the river, juft below the bridge-houfe, and joining it to the elbow of the

prefent

prefent cut above the lock; but I prefer the former expedient, as in this cafe the more eafy, and becaufe it will alfo retain more water at Knottingley weir.

The cut through Brotherton marfh to have flood-gates at the head, or joined to the great road-bridge for conveying the north road over the cut, a road-bridge anfwerable to Sutton-lane, and tunnels for the drainage water of the Anfleet. A fhoal two feet under dam's height, below the head of the cut, will be wanted to retain navigable water from Caftleford, and by way of communicating the navigation to Ferry-bridge, Knottingley mills, &c. a fide cut will be required by the faid fhoal, with a pair of provifional ftop-gates, to be fhut only in cafe of the water above being reduced till it is penned by the fhoal.

In order to prevent too much confinement to the water in high floods, I propofe that for fixty yards, near the bottom of the marfh, the banks of the cut to be laid low and fecured, fo that the water may find a paffage there as at prefent.

The lock of Beal is almoft the only one upon the river Aire that lies deep enough, and from thence to Haddlefey there is good water, with one height of boards at Haddlefey dam, which may be made to ftrike as the reft do; at prefent two fets of dam boards (as on Beal dam) are frequently found ufeful for penning a greater pond of water to produce flafhes below.

The open river below Haddelfey lock, for nearly three quarters of a mile is, embarraffed with continual fhoals, which again occur at the Hurft, and where, notwithftanding the weirs conftructed for that purpofe, the water is not better, for being the place whereabouts the neap tides generally terminate, the warp conftantly fluctuating in the tides way, has a tendency to fettle in this place, and confequently in dry feafons the bottom is the higheft, when it is moft wanted to be low. Thefe obftructions are within the limits of the prefent act of parliament, but the fhoal at ftock reach about a mile below Weeland, being without the limits of the prefent act, admits of no palliation, but as it has nearly the fame water upon it at high water as the fhoals above, it has no other ill effect than increafing the number of places where the veffels are fubject to be ftopped: but, however, as by this means a tide is frequently loft, it certainly, even in the prefent ftate of the navigation, is a confiderable object.

The moft obvious way of curing thefe fhoals is by building a lock and dam at, or below, the Stock Reach, fo as to pen a fufficiency of water over them altogether, but befides that dams fhould always be avoided where they reafonably can be, there is in this place this objection, that as the fpring tides would flow feveral feet over the dam, and bring with them a great quantity of warp, it muft be expected to fubfide in the ftill water made by this dam, fo as probably to increafe the prefent or produce new fhallows, as bad as thofe intended to be cured; befides, as in the prefent ftate of the river, there are fhallows ftill below, particularly at the top of Snaith's marfh, and Carleton ferry, where there is but juft a fufficiency of water at neap tides, it is no ways certain, that the ftopping of the neap tides more early than ufual by the new dam above, may contribute to the raifing thofe fhoals ftill higher, and thereby to create new obftructions, which in the prefent fettled ftate of the river, is not to be expected.

What I therefore propofe is to take a cut or canal out of the river, juft above the prefent lock of Haddlefey, and croffing a fmall part of Haddlefey marfh, to fall in with the hedge of Longcroft, and croffing Egbar Lees, the old Eye, and boggy ground below Sherwood, to make towards the high ground, oppofite to Tranmore houfe, and there fkirting the border of the high ground near the arable fields and marfhes to go on to Henfall common, and keeping nearly the direction of the fea bank (on the marfh fide of it) till we get to Gowdale marfhes and the Intake, and purfuing that boundary nearly, to fall into the river a little above Gowdale lodge clough, where the river being deep, the veffels can be made to lock out and in at low water, and lie ready to fave their tide over the fhoald at Snaith's marfh and Carleton ferry, over which fhoals, according to my obfervations and information, veffels loaded as before mentioned, are always fecure of a paffage at high water.

In the cut before mentioned, a large tunnel will be required for the paffage of the water of the old Eye under the canal at Edgar Lees, and to give the ufual paffage to the water in high floods and breach of banks from above, I propofe the banks of the canal near the old Eye to be made low for 100 yards, (and to be fecured) that the back flood water may go over them, with ftop gates at the high grounds to prevent the floods from making their way down the cut, and alfo at the head to prevent the river from flowing in till it goes over the marfh banks, &c., ftop gates alfo to be placed at the crofs bank between Henfall marfh and Heck Ings, and likewife between Heck Ings and Gowdale Intake; the following tunnels will be wanted for communicating drainages, viz. Haddlefey marfh, Henfall drain, Heck Ing clough, and Gowdale marfh; proper bridges will alfo be placed to join the roads and communications to grounds; the principal will be to Haddlefey marfh, Sherwood road, to the

Ings,

Ings, Egbar ditto, Henfall ditto, Henfall road to Weeland, Heck lane to the Ings, and from Gowdale marfh to the Intake.

I have alfo examined the grounds on the north fide of the river, and find a cut equally practicable there, and in fome refpects preferable, but being wholly through inclofed farms, I have caufed the firft furvey to be made as above defcribed, which will pafs chiefly through the open Ings and marfhes, and for the greateft part through grounds of very little value, being the feparation between high grounds and the low ; the whole length will be about three miles and three quarters.

Though I do not fee an abfolute neeeffity of extending the cut below Gowdale lodge clough, in order to procure the effentials of a navigation ; yet as obftructions at prefent unforefeen may arife, it feems to me neceffary that the undertakers fhould have a power of removing obftructions, and making horfe towing-paths from the outfall of the river Aire into Oufe, to Gowdale lodge clough, as well as above.

Having now difpatched what feems moft neceffary refpecting the river Aire, it remains to make the requifite obfervations upon the river Calder.

The moft embarraffed part of this river, and the only one dependent on mills, is the cut from the warehoufe to the lock. Were the mill at the tail of this cut to keep always a full head there would need little alteration, but as this is fcarcely to be expected, the moft practicable expedient is to dig the cut eighteen inches deeper, and to form a barrier as before mentioned, fo that the mill cannot draw above eighteen inches within head. This however will not be perfect, as the withholding the water at the Soak mills at Wakefield, from whence alone this cut is fupplied, will occafionally and temporarily ftop the paffage of this cut. The navigation therefore in order to make it complete to Wakefield fhould have a power either of filling the cut from Wakefield mill dam, or to purchafe ground to make warehoufes near the tail thereof, with a road to lead to them.

The dam boards at Caftleford being preferved as at prefent, the whole pond from Penbank lock to the river Aire will have a fufficiency of water, a place called the Meer Round excepted, with fome little places that may eafily be helped near the tail of Penbank : the bottom of Penbank lock however lies too high by nearly a foot, and as it muft therefore be rebuilt, the proprietors fhould have it in their power to rebuild it in a new place and lengthen the cut, fo as to drop it into deep water about 100 yards lower, or to rebuild it in its prefent fituation, as may ultimately appear moft advifeable.

The

The dam boards of eleven inches being preferved and made to ftrike with eafe, at Kirkthorpe Lake and Penbank dams, will make the neceffary water from Wakefield lock to Penbank, except one fhoal, called Holford, about a quarter of a mile below Kirkthorpe lock, which being a rocky fcalp, I advife to be lowered by hand labour, rather than make any lengthenings in the cut of Kirkthorpe.

The Meer Round is a fandy fhoal of no great extent: it is probable it may be helped by proper weirs, without making any new cuts, and hence we pafs into the river Aire without further obftructions.

It is neceffary to mention, that for the fake of preferving the courfe of the prefent navigation from Knottingley townfide upwards, it will be proper to keep Knottingley lock in repair, and alfo for the fake of preferving the communication of Mr. Brandling's coal ftaith, it will be neceffary to keep the Thwaite lock in repair; and in general where there is no particular objection, I would advife, that all the prefent locks be kept up, as by this means when any thing happens amifs with one lock, where there is a duplicate, the navigation can be carried on by the other while it is thoroughly and fubftantially repaired, nor will the mills lofe any thing by a double leakage, as it will be very eafy to keep the locks, where there are duplicates, in far better order than is practicable in the prefent ftate of bufinefs.

J. SMEATON.

Austhorpe, 28th December, 1771.

The

The REPORT of JOHN SMEATON, engineer, upon a view of the Calder navigation, the 25th and 26th Feb. 1779, in company with Messrs. Royd, Charlesworth, and Waterhouse, of the committee, and Mr. Walpole, agent.

RESPECTING matters of common repair, those having been pointed out on our paſſage up the river, I ſhall at preſent confine myſelf to ſuch things as tend to improve the navigation, by removing thoſe obſtacles that length of time and experience have pointed out after ſucceſſive repeated floods, of which the following ſeem to occaſion the moſt frequent ſtoppages, and the moſt frequently need temporary helps.

1ſt. Horbury mill paſture cut head, now called the Broad cut head.

2d. The Long cut head, at Dewſbury.

3d. Batty's mill cut, tail and head.

4th. Kirklee's cut, head and dam.

5th. Tail of Anchorpit lock.

6th. Brighouſe cut head and dam.

1ſt. BROAD CUT HEAD.

The head of this cut, from the gradual wearing away of the land by floods, has always been very ſubject to ſand, therefore often required to be raked, and become very troubleſome ; there was once a rubble weir thrown in from the up-ſtream point, which ſeemed to anſwer very well, till the ſucceeding floods took away the land, leaving the weir diſjoined, which then coming in the way of veſſels, has been totally removed. The only complete remedy will be to make the entry from the cut from the deep water higher up, and as it appears from obſervations upon the whole river, that thoſe cut heads preſerve themſelves beſt from ſanding, whoſe upper point of land projects further into the ſtream of the river, than the lower point, this cut would beſt commence from the ground above broken by the floods making the projecting point, the upperſide of the

cut's

cut's entry, and this being preferved as it is, by being footed with rubble, the annexed fketch, plate 5. fig. 1. will give an idea of the alteration; it will be proper to leave a fpace between the fouth bank, and the foot of the high land, otherwife all the flood-water will be ftopped from going that way : and to prevent the flood-water from going in at the new entry, and out again at the old one, and thereby hauling in fand and filt, it will be proper to put ftop-gates upon the new cut head, at A; and laftly, as this new cut and works will be dug and founded below the level of the water of Horbury mill dam, with the water of which and Broad cut it will be in a great meafure furrounded; therefore, unlefs a trench can be cut eaftwards, under the high ground on the fouth, thefe works muft be bottomed by artificial drainage, which, as it will be in an open gravel, will be very expenfive. If, therefore, the execution of this fcheme fhould be thought too expenfive, I would propofe, for trial, that contained in fketch, fig. 2, where, to prevent the lands going away any further, and thereby widening and filling up the river, I would begin to foot it up from the broken ground above, with a flope of rubble, and laftly carry out its lower end into the ftream as is fhewn at B, in the fketch, fig. 2, which being raifed but a little above the common furface of the water will, as far as it reaches, in time of frefhes and floods, act as a dam, and inftead of a fand in its eddy, will make a deep, and thereby keep a channel open, and being carried out to a fufficient diftance what matter goes round its end will be carried down the ftream without being depofited, as it is now upon the eddy fhore. Doublefs this will make the entrance more awkward, but any thing is better to a veffel than being catched by the bottom, and particularly in the dam of a mill, or to need being raked off after every flood and frefh : the original weir or jetty was conftructed on this principle, but the ground not being footed up and fecured for a length above it, it was wafhed away fince my attendance, and the place left as before mentioned.

2d. THE LONG CUT HEAD.

This cut head needs frequent raking, and is more troublefome than it fhould be, though it opens into the navigation's own pen of water. I do not find that internally it is attended with much trouble, and as the down-ftream point of the entry projects further out than the up-ftream, I would recommend to try the effect of cutting and reunding off as much of the down-ftream point, as can be done with fafety, facing up the lower part of the new dug furface with rubble, to prevent the action of the floods upon a new furface, from taking away what may be prejudicial; and if the fimply taking away from the down-ftream point fhould not be fufficient, it may be further helped by being rendered more prominent by a rubble jetty from the up-ftream point,

somewhat

SKETCHES of the Improvements to be made on the RIVER CALDER.

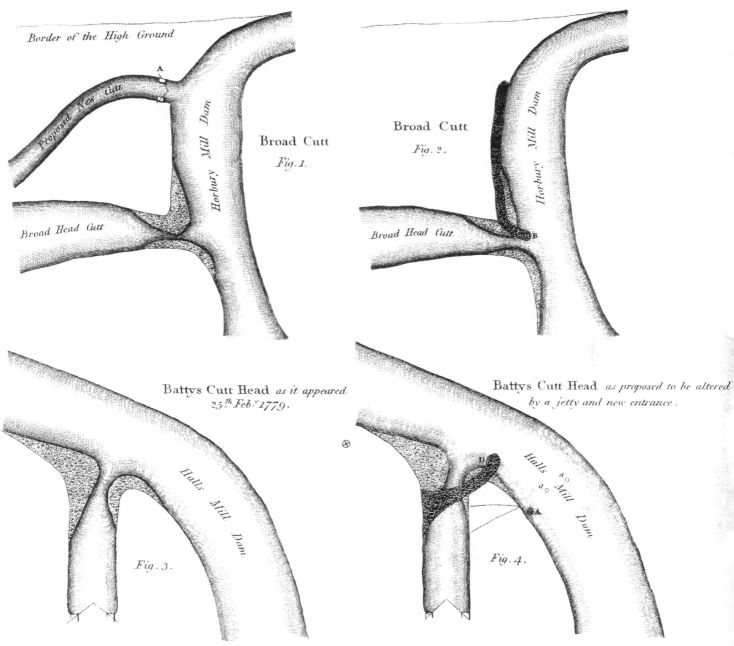

Border of the High Ground

Proposed New Cut

A

Horbury Mill Dam

Broad Head Cutt

Broad Cutt
Fig. 1.

Broad Cutt
Fig. 2.

Horbury Mill Dam

Broad Head Cutt

B

Battys Cutt Head *as it appeared* 25.th Feb.r 1779.

Halls Mill Dam

Fig. 3.

Battys Cutt Head *as proposed to be altered by a jetty and new entrance.*

Halls Mill Dam

D

A

Fig. 4.

A PLAN of HEWICK BRIDGE and part of the RIVER URE.

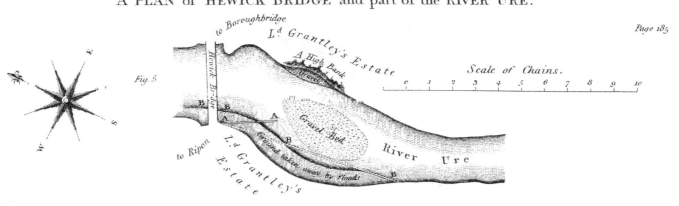

to Boroughbridge

L.d Grantley's Estate

A High Bank

Fig. 5

Hewick Bridge

B

A

to Ripon

L.d Grantley's Estate

Gravel Bed

Ground taken away by Floods

River Ure

Scale of Chains.

0 1 2 3 4 5 6 7 8 9 10

Page 185

Published as the Act directs 1812, by Longman, Hurst, Rees, Orme and Brown, Paternoster Row, London.

W. Lowry sculp.

del.

somewhat in the nature of that proposed for Broad cut head. If nothing of this kind should fully answer, an effectual cure can be made by a new cut following the skirt of the high land under Mr. Wilcock's house, and coming into the river a little higher than the opposite point of watergate lock; but as the navigation cannot now be stopped, the digging of the cut under level of the natural drainage, and the necessity of a very stout pair of flood-gates at the head, or rather a flood-lock, are considerations that in my opinion render it adviseable to try what effect will follow from the more simple methods above pointed out.

3d. BATTY's MILL CUT TAIL AND HEAD.

In regard to the tail, which was sanded on this view, and is now liable to sand; it wants nothing but a low rubble weir or jetty from the up-stream point of the tail of the cut, exactly like that at Wakefield, which was constructed by my directions, and the cut's tail has never sanded since. It is very possible that the sharpness of the up-stream point of Batty's Tail might for years be preserved by a bush or root, a few rubble stones, or other accidental matter, which being now worn and taken away, may render it more liable to sand. A good deal of the upper part of this cut is rather shallow, but as what is below the upper gates may always be cleared at pleasure by drawing off the water; I shall not dwell upon it. The principal grievance is a great tendency of this cut's head to sand, so that it needs often raking, and seldom continues long quite free. This cut going both by Batty's and Hall's mill, its head opens into Hall mill dam, and as this dam continues good when vessels are fairly into it quite away to Cooper's bridge lock, with a good entry there, and as a new cut cannot be taken upwards to much more advantage, without being carried through ground to a considerable height, and where a new pair of stout flood-gates will be wanted, I would advise to attempt to remedy this entry, in a way somewhat similar to that proposed for the Broad cut head; because, exclusive of this single circumstance of its great tendency to sand, the rest can scarcely be better.

The sketch, fig. 3. which is merely done from memory, is supposed to represent the cut head, as it appeared sanded upon this view, and in the narrow of which there was but barely three feet water at dam's height, though every other part was unexceptional. In order to remedy this, I propose to cut a new entry, pointed square to the stream, as shewn in sketch, fig. 4, its up-stream side and point, defended by a rubble jetty, raised but a little above dam's height, that the water may fall over it, and which upon the same principle as laid down respecting the jetty or weir for the Broad cut head;

head, inftead of a fand on its down-ftream fide, will produce a deep, and by projecting it a little into the river, will alfo produce a deep round its termination, and thereby
continue a deep paffage from the cut to the river. This again will doubtlefs render the
entry more awkward; but the very beft in every refpect, and the cheapeft are not
always compatible in an artificial inland navigation. However, to render the entry
more fafe, in time of frefhes, when alone in reality there can be any difficulty, I would
advife a couple or three ftrong piles to be driven in the courfe of the river, as fhewn
in the fketch at *a a*, and that a very ftrong mooring-poft be fixed upon the land at
A, by which veffels may be lowered gradually down the ftream, and by a crofs rope
from the veffel's head, fwung into the cut before they quit their hold at A, and if in
going up in a frefh they get faft to A before they go out, their getting out will be perfectly eafy and fafe, though there were to be no piles in the river at all.

4th. KIRKLEES CUT HEAD AND DAM, and
5th. THE TAIL OF ANCHOR PIT LOCK.

If Kirklees cut head were the only defective part, it would be worth while to try a
remedy upon the fame principles as the laft, but as the dam itfelf is too fhallow, in
feveral places, which have grown worfe rather that better of late years, in confequence
of which frequent ftoppages happen that require the reach above to be drawn down,
which of courfe produces a lofs of time and hindrance there, to wait for its filling; and
as the tail of Anchorpit lock, though very good formerly, has become much obftructed
by ftones, &c. being brought down from the tail of the dam, and therein depofited;
to remedy all thefe evils at once, I would propofe to take a new cut out of the Anchorpit dam on the fouth fide, and following the verge of the high ground to carry
the level of Anchorpits dam water along, till near the flood-gates of Kirklee's cut,
and there, by a new lock, to drop the veffels into Kirklee's cut, juft below the prefent
flood-gates; by this means the navigation by thofe mills will be rendered independent
of them; a circumftance, I believe, very defireable to all parties.

A pair of flood gates will doubtlefs be required at the entry from the river into the
new cut, but the land at this entry being bold, the placing of flood gates will be very
convenient, and as the whole will be dug and founded, (the lock pit excepted) above
the level of Kirklee's cut water, the expenfe will not be enhanced by artificial drainage,
for the work being bottomed from the tail towards the head, the water will be drained
by the cut, and nothing of the prefent paffage difturbed till the new one is opened.

6th.

6th. BRIGHOUSE CUT HEAD AND DAM.

Since the flood-gates near the head of this cut were erected, that part of the cut acrofs the low meadows from the flood-gates to the river, has become very fubject to fand, which being of fome length has become too troublefome. The river has alfo become more fhallow between this cut head, and Lilland's lock tail fince the breaches at Lilland's. Thofe evils will admit of a complete remedy only one way, that I can fee, and that is, to carry the cut from the prefent flood-gates up the meadows, to a point nearly oppofite to Lilland's lock tail, and to fecure the water there by a catch dam, which will alfo render the navigation by Brighoufe independent of the mills, which is indeed very defirable.

Flood-gates at the head will certainly be needed, and the digging of the cut will require artificial drainage, but as the cut will be fhallow, and I have hopes that the foil will not produce much water, and that this may be done at a moderate expenfe, as no new lock will be required; and by thefe means the navigation will be rendered independent of all the mills from Cooper's bridge lock to the head.

Above Lilland's nothing of moment occurs till we leave the river at Brook's mouth, and there the double lock is a very great evil, on account of its great confumption of water, at a place where it is moft fcarce, but as no alteration in it can take place for the prefent year, I fhall referve my opinion upon the particular mode of remedy, till a further opportunity, only intimating that the double lock fhould be converted into three fingle and detached locks, of which I expect that the uppermoft may be converted fo as to make one.

The runs in the Sowerby bridge cut fhould certainly be ftopped the enfuing feafon. The mode of which being fufficiently pointed out and underftood upon this view, it need not enter into this report.

J. SMEATON.

London, 22d March, 1779.

RIVER AIRE DRAINAGE.

The REPORT of John Smeaton, engineer, upon the drainage of certain low grounds lying upon the river Aire, in the parish of Kildwick, from a view thereof, taken 31st May and 1st of June, 1764.

THE river Aire feems in general to run very dead, and in a remarkably meandering courfe in the neighbourhood of Kildwick, fo that the lands lying thereupon, are much liable to floods in the winter feafon; but the lands which were the immediate object of this view were many of them under water, at the time of taking it, and the reft were fo fpoiled by the long continuance of the water upon them, that their value is greatly diminifhed, and all further improvements rendered difficult and precarious, without immediate help from drainage. I have alfo been informed, and my own obfervation agrees therewith, that thefe lands are now in a much worfe ftate with refpect to drainage, than they formerly were, infomuch that feveral parts are now below the level of the furface of the adjacent river, into which they drain, (when in its ordinary ftate) that formerly were above it, and then were firm and dry land. The evident caufe of which alteration is as follows: On the fouth fide of the river Aire, comes in a beck, which rifing more immediately from among the hills in rainy feafons comes down with great rapidity, and in its courfe, wafhing the foot of certain cliffs, or beds of pebbles, gravel, and fhingle, brings with it a great quantity thereof at fuch times, which matter being difturbed or moved by art, for the fake of wafhing or collecting fuch pebbles as are of the lime-ftone kind, for the purpofe of making lime, about two miles above the mouth of this beck, as I am informed, muft of courfe greatly contribute to increafe the quantity that otherwife would be brought down its courfe: the quantity of this matter already brought down is fo great as to have in a manner filled up its own channel, fo that for nearly a mile in croffing the plain, from the hills to the river, its bed is in general higher than the adjacent lands, the waters being confined between artificial banks; but thefe banks having been raifed only occafionally, and by piecemeal, as the bed of the beck has grown higher, are in general too weak and flight to confine the waters in time of high-floods, fo that the furface of the adjacent lands is covered with pebbles and gravel in many places, where it has been driven through the branches.

The

The channel of the river Aire is in this neighbourhood naturally deep; but the great quantity of gravelly matter brought down by the beck above mentioned in the channel of the river, being carried downwards by the violence of the floods of the Air, in rainy seasons, has in a great manner filled up, or obstructed the channel of the river for half a mile below the mouth of the beck; and it has been remarked, and, in my opinion, such an effect must naturally follow from such a cause, that the shingle above mentioned, travels every winter lower and lower in the river, as the quantity brought in increases, and at the same time near the mouth of the beck grows higher and higher, so that it raises the water in the river above these obstructions, in the manner of a dam, and at present pens it over the surface of several of the lands, to the great detriment of others with which it is almost on a level; and as this evil has come on gradually increasing, it is probable that the surface of the river has gradually risen several feet within the compass of a moderate number of years, as is confirmed by persons now living, and, therefore, if some method be not taken for remedy thereof, in all probability it will grow to such a pitch as to render useless all the low grounds upon the river Aire, between the said beck and Coningley mill, unless the river should force itself a new course through the meadows below, as it probably will do (such instances being frequent) unless timely relieved by art.

As the effect of the aforesaid obstructions is to pen up the water of the river above the same in the manner of a mill-dam, which reaches nearly upon a dead level as far as Coningley mill, which I compute to be about three miles, by the course of the river; the meadows on both sides lying low, are immediately put under water by a small rise of the river by rain. The meadow in going up the river towards Coningley mill gradually rises higher and higher above the surface of the river, as it was in the dry season, when I viewed it; yet, as these lands lie further from the outfall of their water, the water advances upon them in proportion to their distance, when it is raised by a flood, so as to make them equally liable to floods, though not of so long a continuance. The land lying on the south side of the river, between the aforesaid beck and Kildwick bridge, and for about a mile above the said bridge in the same side, seems most oppressed with water. On the north side, between the bridge and Coningley mill, something has been attempted by way of banking; but as a great part of the embanked land was not at the time of this view above a foot higher than the surface of the river, and a good deal only a few inches, it follows, that though these banks may be of use in keeping out sudden and casual floods from spoiling their crop in summer, yet the continued floods in winter, as there will be no passage for the

waters

waters from the furface of the lands into the river through the tunnels, the river being then generally for a long time together higher than the furface of the lands, there can be no drainage, and confequently the banks will be of little ufe, as the land will be ftarved and impoverifhed by the continuance of the downfall and foakage water upon them; and as the quantity of fhingle keeps increafing, and the furface of the river will be raifed higher and higher, it is very probable, that in the courfe of a few years, the furface of the river will be higher than the furface of a great part of the faid embanked land, in which cafe the partial drainage which they now enjoy will ceafe even in fummer.

For a general remedy of the drainage of all the low lands that adjoin the river Aire on both fides, between the faid beck and Coningley mill, I would recommend the making a new cut or river from that part of the river ftagnated by the faid obftruction, to a convenient part below them, by which means feveral feet of water may be drawn off from the faid ftagnated pool, and thereby the river, as well as the lands dependent hereupon, be put into their ancient courfe of drainage and improvement.

Having carefully viewed the courfe of the river from the commencement of the obftructions to the termination thereof, being in length about half a mile, there appears to me by eftimation to be at leaft fix if not eight or nine feet of fall. I did not take a level, becaufe the time previous to the meeting of the gentlemen concerned would nt allow it; but as there is an ample fufficiency of fall for the purpofe, it is of lefs confequence what the level exactly is. Now it is evident, if there be fix or even five feet of fall, that by a fufficient cut the water will be drawn off five or at leaft four feet below its prefent furface, and as the channel is in general fix or feven feet deep above the obftructions, the water will reduce itfelf, in dry feafons, a proportionable quantity all the way as far as it pens upwards, that is near to Coningley mill, and, confequently, take off the furface waters from all the adjoining lands, as before afferted.

This cut I would not advife to be at a medium of lefs dimenfions than as follows, viz. twenty feet bottom, forty-five feet top, and eight feet deep.

The proper place feems to me to be on the north fide of the river, from or near the head of Mr. Richardfon's corn farm, thence almoft ftraight through the low end of his grafs, turn thence through Mr. Baldwin's Garforth Ing, and thence through Mr. Baldwin's Holme, into the prefent courfe of the river, where grows a pretty remarkable clump of afh trees. This courfe being meafured, was, in length, three furlongs and

twenty-three

twenty-three yards. A cut of the above dimenfions, at 3*d*. per yard cubic, will come to £79. 9*s*. 6*d*. per furlong, and for three furlongs twenty-three yards £246. 14ˢ. 8*d*. The land cut will be two roods twenty-nine perches per furlong, and therefore in the whole, two acres and eight perches. The land covered on each fide will be nearly equal to the cut, which being fuppofed of half value, the damages to be paid for the cover will be equal to the value of the land for the cut; one bridge over the fame will be wanted, which, if built of ftone, may probably coft about £80.

A cut of the above dimenfions will vent as much water as the old river, when the flood rifes even with the grafs, but were it of ever fo great dimenfions, it cannot vent the water fafter than it comes.

The old river I would advife to be kept open, as a receptacle for the fhingle coming down from the beck: and as the old river will be lefs loaded with water in time of floods, it will afford a readier paffage for the water of the beck; and the force of the current being lefs violent in the old river, will not drive down the fhingle fo faft into the river below the tail of the new cut. Hence I difapprove of cutting for the beck a new courfe into the river to a lower point; becaufe, by giving it too much fall, it will bring down fhingle the fafter, and, by carrying it in at a lower point, the fhingle fo brought will fooner get down the river into the part now unobftructed; to do this would therefore be to deftroy the whole for the fake of a part.

In cafe the fcheme above propofed fhould meet with difficulty, a drainage from the lands in queftion may be obtained by bringing up a back drain from, or near the point above fpecified, between the river and the high land, and banking againft the river. This may be done on both fides the river; in which cafe the drainage on the fouth fide muft be carried underneath the beck by a funken funnel or fox: this method may be practifed with great convenience as far as the bridge, but yet is not obliged to terminate there, but may be carried forward through the low grounds on both fides, as far as ne-ceffary, towards Coningley mill. As it would greatly add to the expenfe to raife the banks fufficiently high to keep out the great winter's floods, and at the fame time, as I apprehend, impoverifh the lands for want thereof, I would advife not to raife the banks more than four feet above the low water's furface, laying them, where made new, very flat and floping, and well fodded, that the water, in going over them, might not make a breach.

Something

Something might be done by way of clearing the old river, but to be in any degree effectual, muſt be at a much greater expenſe. The parts moſt requiring clearing, are thoſe near the head of the ſhoal, that is near the mouth of the beck, which, of conſequence, would be again the firſt filled up; upon this ſcheme it would therefore be adviſeable to turn the beck to the tail of the ſhoal, becauſe there is moſt room for the depoſition of its contents, without immediately obſtructing the parts ſuppoſed to be cleared; but as this would in conſequence produce an obſtruction ſtill more extenſive, a radical cure will be rendered ſtill more difficult and expenſive.

J. SMEATON.

Austhorpe, 26th July, 1764.

P. S. For the evils ariſing to the lands from the flooding of the beck, I ſee no laſting remedy; they ariſe not for want of fall, but rather by hewing too much, ſo that let whatever courſe be made for it, it will be liable to be choaked with gravel, unleſs the coming down of the gravel could be prevented; if this cannot be done than by enlarging its channel, and affording it room for the depoſition of its contents, it may be effected by raiſing and ſtrengthening its banks, partly with the matter lodged therein, and faced up next the channel with new earth from behind to give it room, by raiſing a freſh bank parallel with its eaſtern one, at forty or fifty feet diſtance, taking the matter from between the banks to make the new one, which will, in ſome meaſure, make a new channel, letting the beck have both the old and new courſe to range in time of floods.

RIVER

RIVER AIRE CANAL.

The **REPORT** of JOHN SMEATON, engineer, upon the plan and projection of a canal on the north side of the river Aire, from Haddlesey to a place called Brier-lane End, in the township of Newlands, as laid down by Mr. JESSOP, engineer.

IN my report of the 29th Dec. 1771, upon the improvements of the navigation of the rivers Aire and Calder, from the towns of Leeds and Wakefield, to the free and open tides way, after setting forth the practicability and utility of a canal to be made on the south side of the river Aire, from Haddlesey to Gowdale Lodge Clough, and that it would be the means of making a communication free and open at high water in the shortest tides and driest seasons, between the upper parts of the river and the lower, without the use of flashes. I further mentioned, that having also examined the grounds on the north side of the river, I had found that a cut on that side would be equally practicable, and in some respects preferable, to that on the south, but which was not further pursued at that time, for the reasons therein set forth. The application to parliament being postponed to another session, having given opportunity to the proprietors of the rivers Aire and Calder, further to consider the utility that a canal on the north side of the river would be to them, and how the ground on that side would be affected thereby; they again applied to me for that purpose, but being myself altogether otherwise engaged the past summer, I recommended Mr. Jessop to make the necessary surveys, plans, and projection, which being produced, at the further request of the said proprietors I have taken a view of the line of the canal, as proposed by Mr. Jessop, and having also examined his plan, report, and estimate, I am of opinion, with Mr. Jessop, as follows:

1st. That by extending a canal from Haddlesey dam head to the river Aire, at or near a place called Brier-lane End, in the township of Newlands, being in length about $7\frac{1}{4}$ miles, the communication between the upper part of the river and Armin, will not only be shortened considerably more than by a passage from the said dam to Gowdale Lodge Clough, before proposed on the south side, but will be attended with this further great convenience to trade; that whereas by the Gowdale canal, there would, in dry seasons and short tides, be a navigable communication only at high water; by the present plan, there will be a navigable communication between the upper parts of the river

and

and Armin, at all times, except at low water in dry feafons, which would very often be a faving of twelve hours in the paffage.

2dly. That by raifing the fouth bank of the canal fo as to be flood proof, all the great tract of low grounds, lying to the north of the canal, will be entirely embanked and defended from the floods of the river Aire, arifing from breach of banks in that long tract of river between Haddlefey and Brier-lane End, being by the courfe of the river at leaft 12½ miles; and that, though the water being confined upon the lands on the fouth-fide of the canal may, in fome fmaller over-flowings, increafe to a greater depth; yet, in the great over-flowings of the river, as the water can never rife higher upon the lands than the water in that part of the river from whence it proceeds, there then can be no great difference, and in all cafes, as there will be a lefs extenfive body of water to be difcharged, it will be much fooner difcharged by its out-falls, which, in drainage, is of the greateft importance, as the waters ftanding long upon the grounds do more damage than can arife merely from the depth of it.

3dly. That the drainage of the whole country neighbouring upon the canal, will be greatly mended, for as the furface of the canal will not be liable to rife and fall like the river, being defended by gates at each end, all thefe grounds that lie higher than the furface of the canal will drain immediately into the canal, and all thofe grounds whofe furface lies even with or below the level of the canal, will drain into the back drains that muft, of neceffity, be cut behind the banks, which, by conducting the water from the furface of the adjacent country, by the moft ready paffage, to the main drains of the refpective out-falls, will be a great means of improving the prefent drainage, and as there will be capacious tunnels under the canal, where it interfects fuch drains with ftop doors upon the tunnels, to prevent the waters reverting where neceffary; this advantage will operate equally on either fide.

4thly. That the above means will put the drainage upon a better footing than it is at prefent, which is fufficient for a fcheme of navigation; yet, as the canal is propofed to be embanked flood proof on both fides, between Newland's bank and the out-fall at Brier-lane End, the drainage of Carlton might be improved by running off a part of their top waters by a fluice into this part of the canal, and yet Newland's fully defended by the faid ftop-gates propofed at Newland's bank, and which, at any rate, will be a proper fecurity.

Upon

Upon the whole, I fee no objection to this defign of Mr. Jeffop's, but what would be made to a canal paffing in any line through this country: on the contrary, I think it well adapted to anfwer the end, and to avoid the fevering of properties as much as poffible.

Refpecting Mr. Jeffop's eftimate, I beg leave to obferve, that the longer I live, I every year fee more into the reafons why eftimates are generally exceeded in the execution, and how impoffible it is, without repeated proofs from experience, to conceive how this can happen in fo great a degree; I therefore do not wonder, that he has come confiderably below the prices in the articles of conftruction, that I fhould efteem it neceffary, he not having had an opportunity of having recourfe to my eftimate for the performing of the Gowdale canal, which is fimilar to this, but which now being put into his hands, he will be enabled to re-model his eftimate according to the prices thereof. Indeed, refpecting the article of land, I have done nothing more than give the quantity of acres, cut and covered, as I apprehend that no engineer fhould venture to name a price for it.

J. SMEATON.

Austhorpe, 5th December, 1772.

RIVER LEA NAVIGATION.

AT a meeting of the trustees for the river Lea, at the Bull Inn at Ware, on Wednesday, the sixteenth day of July, 1766.

PRESENT,

Thomas Dimſdale, M. D.
Thomas Pleemer Byde, Eſq.
John Calvert, Eſq. M. P. for Hertford.
John Jenkinſon, Eſq.
John Rowley, Eſq.
Thomas Walton, Eſq.
Timothy Caſwell, Eſq.

John Ivee Gopp, Eſq.
Mr. Thomas Srolt.
Abraham Hume, Eſq.
William Cowper, Eſq.
Samuel Atkinſon, Eſq. Mayor of Hertford.
Paul Field, Eſq.
Richard Dickinſon, Eſq.

ORDERED, That Mr. John Smeaton ſhall forthwith take a ſurvey of the river Lea, and make a plan and eſtimation of the beſt method of making a navigation from Hertford to the river Thames, and that Mr. Thomas Yeoman ſhall aſſiſt him therein, and that Mr. Smeaton make a report thereof to the truſtees on or before the twenty-ninth day of September next, and Mr. Smeaton is deſired to be as particular as he can in the deſcription of ſuch parts of the courſe of the ſaid intended navigation, where he ſhall be of opinion that new cuts ought to be made, and what effect ſuch new cuts will have upon the lands they ſhall reſpectively paſs through, and the mill and other eſtates adjacent.

Examined by me,
HENRY THOROWGOOD,
Clerk to the truſtees,

The

The REPORT of JOHN SMEATON, engineer, upon the new making and completing the navigation of the river Lea, from the river Thames, through Stanstead and Ware to the town of Hertford.

THE prefent navigation of the river Lea, above the tides way, being in all probability formed upon fuch expedients as occurred before the invention of locks, or penfluices, with double pairs of gates, is now, when compared with the modern improvements in navigation, become very defective, notwithftanding many attempts to improve upon its prefent principle, which is that of pens and flafhes from eighteen weirs and turnpikes; befides a ciftern or lock properly fo called at Ware mill; and the lock at Bromley improperly fo called, having only fingle gates for penning in the tide, through which the veffels navigate in tide of flood, or about the time of high water, thefe, with a ciftern or lock occafionally ufed at Hackney water-works (lately erected) are the prefent means of navigation through a courfe of river of more than thirty-one miles from Bromley lock to Hertford, and having 111 feet fall or thereabouts, as appears from a furvey and level of the river, which feems to have been very well taken near thirty years ago.

In the month of July, 1766, having, with the affiftance of Mr. Thomas Yeoman, engineer, and by order of the truftees for maintaining and improving the faid navigation, carefully taken a view with many obfervations on the prefent ftate of the river and navigation thereof, and having compared the fame with the old plan before mentioned, I herewith prefent the truftees with a new plan of the faid river, with three fchemes for completing or improving the navigation thereupon. The firft containing every improvement that in point of expenfe I would recommend; the fecond containing the moft frugal fcheme that I can recommend; the third is a medium between the two extremes, and which, upon the whole, feems the moft eligible; the whole is founded upon thefe general principles:

1ft. To do no damage to the adjacent lands; after the lands required for the works, and temporary damages in conftructing them are paid for.

2d. To do no damage to the mills by diverting their water; and to preferve to them the feveral navigations in their own mill-ftreams that they now have.

3d. As

3d. As much as poffible to prevent the mills from injuring the navigation, by draw-ing down their ponds.

4th. To make a fafe and certain navigation in the dryeft feafons with three feet water in the general, and at leaft two feet fix inches in the fhoals of the river and ford-ing places made acrofs the cuts.

To execute this fcheme I propofe

1ft. To keep the river where it is tolerably ftraight, and fufficiently deep, for a length of courfe together.

2dly. To join thofe deep cuts and canals with locks thereupon, whereby the veffels will be enabled to pafs from one deep to another without any flafhes.

3dly. To make ufe of fuch of the old weirs or turnpikes as will be neceffary for keeping up a conftant body of water in the cuts and canals, in no cafe to pen higher than thofe weirs or turnpikes do at prefent; but in general not within four or fix inches of the prefent height, and if it fhall appear that any damage is done to the grounds by keeping up the neceffary body of water in dry feafons, to embank the river in fuch places, and make back drains to carry off the foakage water into the pool next below, at all other times, as there will be a fufficiency of natural depth of water, the gates of the weirs and turnpikes to be drawn, in order that the river may have its free paffage as at prefent. N. B. Three or four men that will be neceffary to fee that no damage is done to the locks and cuts, will alfo draw the gates as aforefaid.

4th. To navigate by the flow of the tides hitherto, till the barges are paft the falling in of Hackney brooks.

Advantages of thefe maxims.

From hence the mills will receive great advantage from no water being wafted in flafhes, which not only greatly diminifh their heads but obftruct them by tail water.

The lands will alfo be benefited by not being liable to an uncertain pen of water.

The navigation will be not only improved by being rendered certain at all times, except in frofts and extraordinary floods, and by the fhortening the diftance, but by na-vigating

vigating in more still water, and in the straighter courses much labour in towing will be saved, and will be rendered so expeditious, that a vessel at moderate work will make her way upwards at the rate of two miles per hour, passing the locks included; that is, from Bow bridge to Hertford in about thirteen hours.—Hence the ruin of horses straining of barges, and great wear and tear of the tackle will be saved, as well as greater expedition be secured.

The different schemes referred to, with the separate expenses of the divisions thereof, will be sufficiently explained and compared by inspection of the three abstracts, and the particulars of each division will be had by inspecting the two estimates of the different parts.

It is to be noted, that nothing is to be supposed as proper matter of my estimate, save what is contained in each abstract previous to the double line; the value of lands and weirs is thrown in, in order to make a comparison of the different methods; which will again turn out somewhat different, according to the different values that are put upon them in the opinion of different persons, and that no conclusion may be drawn from the values that I have placed to weirs, turnpikes, and cistern lock at Ware, I have mixed the uncertain value of repairs that will be wanted to each therewith.

As I have endeavoured to settle the land-marks or principal stamina of this under-taking, I am well assured that what remains (which could not be done before) will be properly enquired into and represented by Mr. Yeoman, that is, if any of the three or any part of the three schemes are thought proper to be carried into execution. It will now be proper that Mr. Yeoman be employed to look over the grounds where the works relative to such scheme are to be, and as he is a judge thereof, which I am not, to take notice of the probable value; as also how, by any deviations from the lines I have drawn for the cuts and canals, the divisions of properties may, as much as possible, be avoided; and in case any other way than what is pitched upon on the first face of the thing shall seem more proper on re-inspection, to make report thereof, with such other matters as shall be directed, or shall occur to him, by which means the trustees will have the whole matter fully and clearly before them.

I furthermore desire, that what I have set down in the abstracts, under the title of expense of the works, is to be considered as so much hard money, without defalcation on account of the charge of this scheme, the act of parliament, the interest of money,

or

or any falary or gratuities that may be thought neceffary to the principal director or directors, or the clerk or clerks of the works.

It perhaps will be wondered why, in my third fcheme, I have not preferred the keeping of that part of Enfield mill-ftream, contained in the fecond? my reafon is, that as that mill-ftream is at prefent fed by a limited quantity of water, and millers are perpetually over drawing their fupplies, in dry feafons the navigation would frequently be interrupted thereby.

I alfo entirely reject the new ciftern lock at Hackney as part of my fcheme, as neither its floor nor the river below is deep enough for navigation, without flafhes there, as at prefent.

On infpection of the plan, it feems very defireable that a cut fhould be carried from the four mill pond at Bromley to the Thames at Limehoufe hole, and which, from a general knowledge of the ground, I believe very practicable; yet, as this thought did not occur to me till I was compiling the plan, I muft refer the gentlemen to Mr. Yeoman's report thereupon, to whom the fame thought has occurred, and who has viewed and levelled the fame, and which I fhall infpect when I go up to town if thought eligible.

If the navigation below Bromley is thought proper to be continued through Bow Creek as at prefent, it will be proper to have a power in the act to lay **Bromley** lock lower, whenever it fhall need a thorough repair, and make it with two pair of gates.

The following eftimates are founded upon the fuppofitions, that the ordinary fize of the cuts and canals are a fixteen-feet bottom, with batter, as one to one on each fide, and to be three feet deep of water, but to allow for paffing places, four at leaft in a mile, making fords and extra depth of ground where it lies confiderably higher than the water; they are calculated upon a mean width of eighteen-feet bottom, and $4\frac{1}{2}$ feet deep, and to be done at $2\frac{1}{4}d$. per cubic yard, drainage of water, where neceffary, included; on thefe principles the cuts will be made for £27 10s. per furlong, and each furlong will, at a chain broad, including banks and back drains, contain one acre per furlong.

The

The locks to be made fit for the Ware barges at £100 per foot perpendicular, the extra charge of lock pits and drainage of water further included.

I have fuppofed all the weirs, with the ciftern lock, to be private property; all the turnpikes belonging to the navigation; and that Archer's weir muft be purchafed, in order to be removed; if I have miftaken in that point, my miftake therein can eafily be corrected.

J. SMEATON.

Austhorpe, 24th September, 1766.

An ESTIMATE for making navigable the river Lea, by fuch cuts and canals as were approved of by the truftees of the faid river upon their infpection of the fame, from Bow bridge to the prefent place of delivery at Hertford.

By JOHN SMEATON, engineer.

From Bow bridge to Lea bridge, inclufive.

	£	s.	d.
To clearing a shoal just above Bow bridge, called Bow Bridge Hill	150	0	0
To clearing two fords below Old Ford dock	50	0	0
To two miles of canal, which, at 3d. per yard, and with the dimensions specified in Mr. Smeaton's report, comes to £33 per furlong, and for two miles	528	0	0
To two locks penning 4 1 each	820	0	0
To two tail bridges to ditto	30	0	0
To bridge and flood-gates at the head of the cut	120	0	0
To making five fords for communication across the canal, which, including bridges, is at the rate of four passages per mile, at £10 each	40	0	0
	* £1688	0	0

To be purchased for this work, land sixteen acres

From Lea bridge to Newman's weir, inclufive.

	£	s.	d.
To cutting two loops of the river Lea, between Lea bridge and Hildyard's turnpike	30	0	0
To five miles and an half of canal, exclusive of Enfield mill-stream, at £33 per furlong	1452	0	0
To scowering Enfield-mill stream, with contingents thereon	330	0	0
To lockage upon this canal 33 4 perpendicular	3333	0	0
Carry forward	£5145	0	0

* The sum of 1688 *l.* is wrong, but it was so delivered in.

	£	s.	d.
Brought forward	5145	0	0
To seven tail bridges, at £15 each - - - - - - -	105	0	0
To a road bridge, answerable to Hildyard's turnpike - - - - -	80	0	0
To three large tunnels for the principal drains across, at £200 each - -	600	0	0
To nineteen fording places, at £10 each - - - - -	190	0	0
To altering bridges, and erecting a stanch for regulating the water in the mill stream -	600	0	0
To extra expenses in continuing the navigation to Stewardson mill, by erecting a lock, and making a cut from the canal to the river - - - - -	500	0	0
To ditto, on account of ditto, in making a stop in the river below the going out of Stewardson's mill-stream, and in the erecting gates for the continuance of the navigation -	500	0	0
	£7720	0	0

Land to be purchased for this work thirty-six acres.

From Newman's weir to, and including King's weir.

	£	s.	d.
To digging one mile and three furlongs of canal, at £33 - - -	825	0	0
To four locks, penning twenty-one feet - - - - -	2100	0	0
To three tail bridges to ditto, at £15 - - - - - -	45	0	0
To a road bridge to Waltham Abbey - - - - - -	90	0	0
To one ditto, answerable to Hollifield bridge - - - -	70	0	0
To one aqueduct bridge over the Cheshunt mill-stream - - -	200	0	0
To three large tunnels for draining, under the canal, at £150 - -	450	0	0
To a bridge and flood-gates at the head of the cut - - -	120	0	0
To eight fording places, at £10 each - - - - -	80	0	0
	£3980	0	0

Wanted for this work twenty-five acres of land.

From King's weir to, and including Carthagena turnpike.

	£	s.	d.
To a cut, one furlong seven-tenths, past Carthagena turnpike - -	56	2	0
To a lock, penning three feet - - - - - -	300	0	0
A tail bridge - - - - - - - -	15	0	0
	£371	2	0

Wanted to purchase one acre seven-tenths of land.

From

From Carthagena to, and including, new turnpike.

	£	s.	d.
To a cut by new turnpike, three furlongs in length	99	0	0
To a lock, penning five feet	500	0	0
To a road bridge, answerable to new turnpike bridge	80	0	0
To clearing a shoal between the two turnpikes	85	0	0
To a tail bridge over the lock	15	0	0
	£779	0	0

Land to be purchased for this work, three acres.

From new turnpike to, and including, Field's weir.

	£	s.	d.
To a cut at Field's weir eight-tenths of a furlong	26	8	0
To a lock, penning three feet nine inches	375	0	0
To a tail bridge to ditto	15	0	0
	£416	8	0

Land to be purchased, eight-tenths of an acre.

From field's weir to, and including, Ware weir.

	£	s.	d.
To clearing a shoal below Standstead bridge	100	0	0
To a cut from above Stanstead bridge to Ware weir, being one mile six furlongs and two-tenths	468	12	0
To three locks, penning eleven feet six inches	1150	0	0
To three tail bridges	45	0	0
To four fords	40	0	0
	£1803	12	0

To be purchased for this work, fourteen acres and two-tenths of land.

From Ware weir to, and including, Ware mill.

	£	s.	d.
To a cut, three furlongs, by a new passage, into Ware mill head	99	0	0
To a lock, penning eight feet eight inches	866	13	0
To a tail bridge	15	0	0
To a ford	10	0	0
	£990	13	0

To be purchased for this work, three acres of land.

VOL. II.   Y

From Ware mill to the prefent delivering place at Hertford.

	£	s.	d.
To clearing a shoal near the balance engine	150	0	0
To a cut from the elbow above Fordham's mill into Dicker mill head, five furlongs	165	0	0
To a lock, penning seven feet six inches	750	0	0
To a tail bridge to ditto	15	0	0
To a road bridge to the mill	80	0	0
To three fords	30	0	0
To deepening Dick's mill-head, proper for navigation	99	0	0
To altering bridges and other conveniences	180	0	0
	£1469	0	0

To this work, wanted five acres of land.

GENERAL

GENERAL ESTIMATE for new making and completing the navigation of the river Lea from Bow bridge to the delivering place at Hertford, by such canals as have been approved by the trustees on their inspection of the same; by JOHN SMEATON, engineer.

Divisions of the River	Canal (M. F. I.)	River (M. F. I.)	Total length (M. F. I.)	Locks No.	Rise (f. I.)	Expense of the works (£. s.)	Acres of land (A. P.)
From Bow bridge to Lea bridge inclusive	2 0 0	0 5 7	2 5 7	2	8 2	1688 0	16 0
From Lea bridge to and including Newman's weir	5 6 0	0 2 0	6 0 7	7	33 4	7720 0	36 0
From Newman's weir to and including King's weir	3 1 0	0 1 1	3 2 1	4	21 0	3980 0	25 0
From King's weir to and including Carthagena turnpike	0 1 7	1 4 0	1 5 7	1	3 0	371 2	1 7
From Carthagena turnpike to and including New turnpike	0 3 0	0 3 0	0 6 0	1	5 0	779 0	3 0
From New turnpike to and including Field's weir	0 0 8	0 6 5	0 7 3	1	9 6	416 8	8 0
From Field's weir to and including Ware weir	1 6 2	2 1 2	4 0 4	3	6 14	1803 12	2 2
From Ware weir to and including Ware mill	0 3 0	0 0 4	0 4 0	1	8 8	990 13	3 0
From Ware mill to the delivering place at Hertford	1 0 0	0 2 0	1 2 0	1	7 6	1469 0	5 0
	14 5 7	10 3 4	25 1 1	21	101 11	19217 15	104 7
Distance from Bromley lock to Bow bridge	—	0 4 8					
To making towing-paths, bridges, gates, and stiles, at £20 per mile	—	—	25 5 9	—	—	512 0	—
To repairing six weirs and turnpikes, to be kept up for the use of the navigation, at £200 each	—	—	—	—	—	1200 0	—
	—	—	—	—	—	20929 15	2093 0
Add £10 per cent. upon the works for contingent expenses, and upon the lands for temporary damages	—	—	—	—	—	2093 0	10 4
						23022 15	115 1

N. B. In this estimate there is nothing included in the above for surveyors', or clerks' salaries.

London, 26th February, 1767.

J. SMEATON.

NEW RIVER WORKS.

The REPORT of JOHN SMEATON, engineer, upon the improvements making by the New River Company, near Ware, in the county of Hertford.

AT the defire of Mr. Walton, on Saturday, the 21ft April, 1781, I viewed the works of the New River Company made and carrying on near Ware, and made fuch remarks thereon, as appeared to me neceffary to give an anfwer to the queftion propofed to me, viz. In what manner, and how far the alterations and enlargements that the New River Company are now making in the New River, near Ware, have a tendency to af-fect the mills and navigation upon the river Lea below Ware?

To anfwer this queftion diftinctly, I muft in the firft place obferve, that fubfequent to all outlets for wafte water to return from the company's works and ftreams back again into the river Lea, there is a fquare open tunnel of ftone, through which the water paffes, and is called the gage. I had not accefs to take the meafures with exactnefs, but was informed it was fourteen feet in length, fix feet in width, and two feet in depth. Now, if the furface of the water below the gage were of equal height, or perfectly level with the furface of the water above the gage, the water would then remain ftagnant, and none would run through the gage; but if the furface of the water above the gage be but a fmall, and even imperceptible difference, higher than the furface of the water be-low the gage, this difference forming a declivity, will caufe the water to run towards the lower part, and the velocity or fpeed wherewith it will run will have a certain relation to the meafure of this declivity; the quantity of water therefore taken or paffing through the gage in a given time, will be proportioned to the velocity wherewith it paffes. In order, therefore, that this conftruction called the gage may be really and truly fuch, or meafure of a certain given quantity of water flowing through per minute, per hour, or per day, thefe conditions are neceffary, firft, that the furface of the water fhould be kept at a certain conftant given height before it enters the gage, and at another conftant certain given height after it has paffed it, fo that there may be a conftant and equal declivity of fall from the water's furface above the gage to the water's furface below, and thereby alfo occupying the tunnel with the fame degree of fulnefs: for if either the furface of the water above the tunnel be raifed, while the furface of the water below remains at the fame height, or the furface of the water below the tunnel be lowered,

that

that of the water above ftill remaining at the fame height; in either cafe, the declivity or fall from furface to furface being greater, its velocity will increafe in a certain proportion, and confequently the quantity of water drawn through the gage in a given time will increafe in the fame proportion; and the fame thing will happen if the furfaces of the water above and below the gage are both raifed, but that above more than that below, fo as to form a greater declivity or fall; for by whatever means a greater fall is produced, a greater velocity and difcharge will enfue.

Something appears to have been attempted towards a regulation of the water's furface above the gage, but nothing appeared to me to have been done with that view below the gage; however, while the capacity of the river below the gage remains the fame, its capacity to convey water will remain nearly the fame alfo, and this will in fome meafure prove a regulation to the height of the furface of the water below the gage; but if below the gage in feveral reaches, impediments, and obftructions to the motion of the current are removed, which will alfo be the confequence of widening or deepening, or both, fo as to increafe the fection of the river, the natural confequence will be, that a given quantity of water will be run off as difcharged at a lower furface, and if the difcharge of water at the gage were entirely limited by a fluice gate, or otherwife, to a certain given quantity, the furface of the water below the gage would by this means be actually lowered confiderably; but the moment the furface of the water below the gage is in any degree lowered, the tunnel of the gage being open, a greater quantity will iffue in confequence of a greater declivity or fall, which, in confequence, will in part keep up the furface below to a higher pitch than it would have been, had the quantity been actually limited; that is to fay, the effects of removing obftructions are in part that of creating a greater declivity at the gage, and in part, that of conveying away a greater quantity thus drawn through the gage in confequence of that greater declivity: it therefore appears to me very evident, that however laudable it may be, and is, in the New River Company to improve the water paffage, confidered as an aqueduct, yet the effects of widening, deepening, cleanfing, and removing obftructions to the water's paffage, have a direct tendency to create a greater declivity at the gage, and thereby to draw down a greater quantity of water from the river Lea, even fuppofing the water's furface above the gage to be kept to a regular height, which does not appear to me to be always the cafe, becaufe there do not feem to me fufficient means of doing it.

I have intimated before, that the regulation of the water's height above the gage has been attempted, but this does not appear to me done in fuch a way as fufficiently to

anfwer.

anfwer the end. The ftream of water in paffing from the river Lea to the gage paffes by an over-fall, or tumbling bay, which feems intended to let the wafte water go back again into the river Lea, when the furface is above a certain height upon the gage : but this over-fall can by no means fully and accurately anfwer the end, as being very much too fhort and confined : when I was there, about one inch of water was cafcading over it, fo that if the crown of this tumbling bay be the proper height for the water above, or before it enters the gage, the furface was then one inch above its proper height, and what effect this may have will be feen hereafter. But this is not all, for I obferved a frefh water mark $2\frac{1}{2}$ inches higher than the furface then was, at which height the water had been, in my opinion, fometime in the morning of the fame day; and when the water's furface was at that mark, there would have been at or about $3\frac{1}{2}$ inches water over the tumbling bay, and confequently near upon the fame quantity, that is, above three inches, too much water upon the upper fide, or above the gage.

The water brought down to the works proceeds originally from the river Lea, where there is conftructed a piece of machinery called the balance-engine, which, if kept in complete order, accurately adjufted, and left untouched, might be a great means of keeping the water at the gage more nearly to a juft height above it; but from the very great difference in the admiffion of the water there in the morning and noon of the fame day, it muft follow, either that the balance-engine is much out of order and adjuftment, or that the company's fervants, in whofe power it is left, had drawn there an occafional quantity of water; and if the latter, of whatever ufe it may be to the company, it cannot be confidered, while in the power of their fervants, as of any ufe towards a regulation of the water between the proprietors of water upon the river Lea and the New River Company.

The reafoning upon the matters above ftated will equally hold, whether the quantity be ten cube feet, or a thoufand; it therefore now comes in courfe to give you fome idea of the quantity by which the ftream of the river Lea may be leffened by the operations above mentioned to the mill-owners and navigation. To do this, however, with accuracy and precifion, would require more obfervations and meafures than can be taken in one day; but yet, what may be deduced from what I have feen and obferved, may be fufficient to enable you to judge of the ground you ftand upon, and whether an examination more in the detail be neceffary.

When I was there, the difference, fall, or declivity at the gage, between the furface of the water above and the furface below the gage, I judged to be about an inch, or

rather

rather better; fuppofing the difference to be one inch, according to my computation this difference would produce a difcharge (from the meafures of the gage before ftated) of 1100 cube feet of water per minute, that is, above thirty tons, and nearly equal to the water expended by one of Mr. Walton's beft mills, carrying two pair of ftones, and would fill a ciftern lock of a middling fall in about five minutes.

Now, if the difference were four inches, the difcharge would be nearly double, and if the difference or fall were fourteen inches (which I apprehend it is very practicable to procure by a fuitable alteration of the New River), the quantity difcharged would then be near upon the greateft poffible, through that gage, viz. 3800 cube feet per minute. I would not, however, be underftood, that when the water was above three inches higher than it fhould have been, that there was a quantity difcharged due to a fall of three inches ; becaufe, for reafons already given, an increafed difcharge at the gage would raife the water below, and diminifh the fall, fo that it is probable, the true difference of fall was never more than two inches, in which cafe, the difcharge would be above 1500 feet per minute ; but which, by fuitable alterations in the river below, may be carried (as already mentioned) as far as 3800 cube feet.

J. SMEATON.

London, 3d May, 1781.

To

To the trustees of the river Lea.

Gentlemen,

WHEN I attended you at Ware, at your meeting of the 21ft May, by order of Mr. Walton, for the purpofe of explaining to you fuch matters as might arife from a confideration of my report made to him of the 3d May, which contained my opinion of the tendency of the improvements making by the New River Company, near Ware, a copy of which report he had caufed to be communicated to you; it was, among other things, obferved by you, gentlemen, on your view, that fince it appeared that the furface of the water, in paffing the ftone gage, was lower with refpect to the roof of the tunnel of the faid gage, than it was with refpect to the top of the two flat ftones placed as marks, and referred to in the act of parliament, that therefore the tunnel of the gage was placed too high by the difference. To this I obferved, that it did appear to me as if the gage was too high for the ancient marks, but that, on account of the current ftate of the water, this could not be fully afcertained, unlefs an accurate level were taken; and this you defired me to afcertain by taking a level accordingly.

I then obferved to you, on confidering the fituation of the two flat ftones, which feemed ancient marks, that they, probably, would be found not to be upon a level, as they have been generally confidered, and as they are fuppofed to be by the wording of the act of parliament, which refers them to the level of the fame head of water flowing down the Manifold ditch, and if not thofe two ftones or marks, (the water being directed to be kept level therewith), the one communicating with the head of water before it has paffed the gage, and the other with the tail after it has paffed it, will become *boná fide*, two feparate ftandards, one to regulate the height of the water above the gage, and the other below it, that have been originally placed with this intent, viz. to regulate the quantity of water that will pafs the gage with a given defcent or declivity, as is pointed out to be neceffary in my faid report of the 3d of May, and to which I beg leave to refer; which matters, with what elfe might occur as material to the fubject, in order to afcertain a juft meafure of water to be taken by the company, you defired me alfo to examine at the fame time.

Accordingly, upon the 29th of May, I took a fet of levels at this place, from which it appears——

1ft. That

1ſt. That the roof of the ſtone tunnel compoſing the gage is full five-eighths of an inch higher than the top of the old part of the flat ſtone mark, at the Chalk Iſland, which marks the height of the water before it paſſes the gage.

2dly. That the top of the Tumbling bay, acroſs the Manifold ditch, which ſerves as an over-fall to let the overplus water taken into the balance engine, and paſſing by the Manifold ditch, return into the river Lea, when it is too much in quantity to be taken in at the gage, is full half an inch higher than the top of the mark at the Chalk Iſland.

3dly. That the ancient flat ſtone or ſtandard, near Chadwell ſpring, which communicates with the water of the New River after it has paſſed the gage, is near upon three-quarters of an inch lower than the flat ſtone at Chalk Iſland, that communicates with the water before it has paſſed the tunnel.

From the above levels it appears, that if the ſurface of the water be kept continually to agree with the two marks, one for the head water, the other for the tail, (the difference of level being three-quarters of an inch), that a conſtant equable quantity of water will continually flow from the Manifold ditch through the gage into the New River, according to the apparent intent of the act of parliament, becauſe

2dly. That if the two ancient marks had been found upon a level, and the water's ſurface had been kept even with thoſe marks, the water would remain ſtagnant in the trough or gage, and none would have run from the Manifold ditch to the New River.

3dly. That if the water be kept even with the two marks above referred to, then the gage will not be filled ſo as to vend a full bore by about five-eighths of an inch in height or depth.

4thly. That if the gage be made to vend a full bore of water, then the water muſt be kept full five-eighths of an inch deep upon the Chalk Iſland mark, and if the water be at the ſame time only even with the ſtandard at Chadwell, there will then be a fall at the gage of $1\frac{1}{4}$ inch, which will be almoſt double to what it ought to be.

In expreſſing myſelf as above, I have referred my meaſures to the two ancient flat ſtones or marks; the one at Chalk Iſland, after the water has paſſed the brick arch from the Manifold ditch, the other near Chadwell ſpring; becauſe theſe, it appear to me, are the only marks remaining that have the appearance of having subsisted at the time of passing the act above mentioned; the gage itself, which at that time is recorded to be of wood, is now rebuilt with stone, and the constituent parts, viz. the tunnel, with white marble, which, in point of length, breadth, and height, agrees with the dimensions specified in the act.

The old flat stone near Chadwell spring appears to be entire, but one-half of the old flat stone at the Chalk Island seems to be gone, and to be replaced with one of a more modern date, which is at or near upon one-eighth of an inch higher than the old one.

Upon both the upper surfaces of the old flat stones there are grooves or channels, cut of about one-half an inch deep, round about near the border, with openings to the out-side, the intent of which, as it would seem to me, has been, that when the water comes near the level of the surface of the stones, it will, by flowing into these channels the more readily shew when the surface of the water is in reality even with that of the stones; which is to me a convincing proof that the surfaces of these two ancient flat stones are in reality the capped stones placed as standards, and referred to in the act. I must further remark, that the new part of the flat stone of Chalk Island mark is not grooved like the old one, and that upon both this and Chadwell spring, flat stones are placed round, or octagonal stones, of about two inches in height, which form a step or rise of two inches, and the middle of the step is covered with a rounding cap or head, reaching some inches higher, both which steps and their caps, appear to be of work-manship much more modern than the flat grooved stones; but for what purpose there placed does not appear to me.

It is said above that the top of the Tumbling bay, across the Manifold ditch, is full half an inch higher than the standard mark of Chalk Island; that is, than the surface of the old part of the flat stone there, from which all my levels were taken, but it is to be understood that the top of the Tumbling bay, as here taken, is the top of an iron bar which forms its crown, over which the water falls; but were this iron bar removed, so that the water might run over the stone upon which it lies, then the top of the Tumbling bay there would not be higher, but rather lower than the Chalk Island standard mark. Whether this bar was placed upon the top of the stone-work in the original con-struction of this Tumbling bay, or has been placed there since, or upon what occa-sion, does not appear to me, but it would seem that the whole of it has been done new since the date of the act, as there is a date upon it, I think, of 1746.

Upon

Upon the whole matter it appears clearly to me, that if the New River Company take such a body of water as will run through the tunnel of their gage, without raiſing the ſurface above the level of the roof thereof; that is, without any pen, and provided an equal depth of water flow over the two capped ſtones or ſtandards, at the two places above mentioned, ſo as to produce a declivity equal to the difference of elevation of the ſurfaces of the two ſaid ſtones, they would not take more water than they would by the act have been entitled to, if their gage had correſponded with the two ſaid ſtandards : but if the New River Company take no more water than will flow through their tunnel or gage, when the ſurface of the water is kept even with the two capped ſtones or ſtandards, then the difference will be in their disfavour, as they will then take leſs water than they would have been entitled to by the act; but yet, though the ſurface of the water above the gage were kept even with the Chalk Iſland mark or ſtandard, if the ſurface of the water below the gage were kept below the Chadwell ſtandard, the effect would be by increaſing the declivity to draw more water through the gage than the due quantity : and the widening and deepening of the river below has a manifeſt tendency to produce this effect, as is fully ſtated in my ſaid report of the 3d of May, therefore, though it now happily appears that there are diſtinct marks by which the quantity may be regulated, yet, by means of the improvements made, the whole being in the power of the company's ſervants, they are able to take a quantity far exceeding that ſpecified in the act.

J. SMEATON,

Austhorpe, 18th July, 1781.

RIVER

RIVER URE WORKS.

The REPORT of JOHN SMEATON, engineer, upon a view of the works upon the river Ure, taken the 27th April, 1770.

AT the requeſt of the committee of the Ure navigation, I proceeded this morning, attended by Mr. Myers, Mr. Paſhley, and Mr. Moon, from Bondgate green to Milby lock, and obſerved as follows :

1ſt. The baſon at Bondgate green ſeems to be ſet out properly enough, if any thing, too near the road ; it ſeems beſt to place the warehouſe at the head, or upper end thereof, and to bring in the water of the Skeld, in a ſubterraneous tunnel, through the little encloſure on the ſouth ſide of the houſe, near the ſouth end of the chain bridge, placing a draw-gate at the head to ſhut out the floods, and regulate the water, with a proper grate before it : the ſurface of the Skeld, as it may not be high enough to an- ſwer the level of the head canal, as now ſet out, and in part completed, may be raiſed by forming an artificial ſhoal or dam, a little below the flood-gate, with rubble, whoſe interſtices being wrecked full of ſand and gravel, may eaſily be made to pen the water to the height required, and be very durable. In ſome convenient part of the head canal, if beyond the high ground, the better, it will be proper to place an over-fall, of at leaſt twenty feet in length in the crown, and high enough to pen a mill's water over the artificial ſhoal above mentioned, before it runs over the over-fall, in order that in dry ſeaſons the water of the river may not run waſte down the canal, after the works are filled to a ſufficient height.

2dly. All the banks about Bondgate green to be made at leaſt one and a half, but better if two feet above the ordinary navigation's ſurface, ſuppoſed to make four feet depth of water, and to be at leaſt nine feet broad at top ; and in caſe this is not higher than the riſe of the floods of the Skeld, to be increaſed in height till the floods will be prevented from making their way into the canal.

3dly. I ſtaked out the remaining part of the canal from Bondgate green to Rhode's field lock, and ſet out the place for Rhode's field lock, and adviſe that the batter of the ſides of the canal that remain to be done, be made five feet back on each ſide, for every three feet perpendicular.

4thly.

4thly. The rife, according to Mr. Jeffop's levels, from the upper gate threfholds of the Oxclofe lock to the furface of the water in the head level, as now fet out, is twenty-five feet; but as it was propofed to raife the water in the reach above Oxclofe lock, one foot extra, that is, to five feet above the upper threfhold of the faid lock, there will then remain twenty feet for the difference between the furface of the water in the reach over Oxclofe lock, and the furface of the water in the head level, to be divided into two locks' rifes; but being informed by Meffrs. Myers, Pafhley, and Moon, that the bottom of the cut, as it now ftands, finifhed below Bell Furrows locks, is yet two feet higher than the upper gate threfhold of Oxclofe, it will follow, that either this cut muft be deepened from Bell Furrows lock tail, for a confiderable length downwards, that is, in all the deep part of it, or the water muft be raifed two feet extra, that is, to fix feet upon the threfhold of Oxclofe lock; and obferving upon view, that the fides ftand very fteep in the deep cut, and the bottom narrow; obferving alfo, that Oxclofe lock, is built fufficiently high and ftrong to bear an additional extra rife of two feet, I look upon it as far more eligible to do this, than to deepen the cut. In this cafe, the banks of the cuts in the low grounds next Oxclofe lock will require to be heightened and ftrengthened, and the wafte holes in the upper gates to be ftopped, and in lieu thereof, to place a fide wafh or over-fall, near above the lock, for difcharging the wafte water into the river, the crown to lie about two inches above the gage height of the canal's water, that is, fix feet two inches above the threfhold, to be twenty feet length in the crown, but the ftone fetting may be contracted to ten feet at the tail: this being the cafe, there will remain nineteen feet difference between the upper and lower reaches of the canal, to be divided into two equal rifes of nine feet and a half each, in confe-quence, the piquet at Bell Furrows lock, fet by Mr. Jeffop, being fifteen feet nine inches above the upper gate threfhold of Oxclofe, it will be thirteen feet nine inches above the lower gate threfhold of Bell Furrows, four feet three inches above the upper gate threfhold of Bell Furrows, and the fame above the lower gate threfhold of Rhode's field, that is, three inches above the furface of the water in the reach between thefe two locks; then raifing the upper gate threfhold nine feet fix inches above the lower at Rhode's field lock, the fame as at Bell Furrows, the whole will anfwer to make four feet water in the refpective reaches. It will be proper to have an over-fall alfo at Bell Furrows lock of twenty feet long, placed about three inches above gage height; thefe over-falls have been found abfolutely neceffary, even where the wafte holes have been applied, to prevent damages by overflow of banks, when the cloughs of the lock next above are left running, either by careleffnefs, or with a defign to do mifchief, the wafte holes not being fufficient to difcharge the full bores of water from the cloughs. N. B. As the bottom of the canal below Bell Furrows lock may not be quite two feet
above

above the upper threfhold of Oxclofe, it will be proper to place the lower threfhold of Bell Furrows lock, at leaft as low as the bottom of the canal below it, and if any difference, to divide the two locks accordingly, which, if not perfectly equal, is not material.

5thly. It feems as if fome dredging would be neceffary below the tail of Oxclofe lock; but of this I could not well judge, on account of a frefh in the river.

6thly. The dam at Weftwick Wath, being undertaken to be completed by contract, it is fufficient to fay, that I defired the undertaker, upon the place, as to the completing of his fcheme, the cut being very deep, and the fides fteep, it feems at prefent a good deal filled up, and I fear, being loofe fandy matter, will continue to be troublefome, even after it is cleared.

7thly. The fhoals lying below Weftwick Wath, from the frefh in the river, had fo much water upon them, it was not poffible for me to judge, whether the clearing of them, or building a fmall intermediate lock and dam, as originally propofed, will be preferable; nor can I properly judge of this matter, but when the river is in its weakeft ftate of currency, when it will be right to found it with a boat. It is certain, that if the fhoals can be effectually cleared, it will be preferable to a lock and dam, even if attended with a greater expenfe, but this I rather doubt, for in my original fcheme, which was only for three feet certain water, I propofed four feet where the water was liable to be drawn down by the mills, and, therefore, fuppofing Boroughbridge dam to pen only one foot dead water over the fhoals below Weftwick, on raifing it by a dam two feet higher, there would be three feet certain water; whereas, to deepen thofe fhoals two feet, would in effect, make but two feet water, when reduced one foot by the mills; the fhoals muft, therefore, be deepened three feet, to do the fame thing by deepening as by raifing the water two feet, and by like reafoning, to fecure four feet certain water, there ought to be five feet at a dead level within the pen of the mills; now, if this happens to be more than the general depth of the river between fhoal and fhoal, the whole will be to be deepened from whence the fhoals begin upwards; and if this be the cafe, which can only be determined with certainty by founding at a proper feafon (which I never had an opportunity of doing), the expenfe may very far exceed that of a lock and dam.

8thly. Refpecting the lock and tail cut at Milby, two things offered for confideration; firft, a double rend in the wall, near the N. E. angle of the return of the tail wing, one

being

being in the face of the wall next the lock's paffage, the other in the flank or return of the wing. They both appear to be occafioned by a fet in the foundation, probably owing to an over-weight, in extending the mafonry of the wing beyond the ground timber; at the fame time, as I underftand that the ground timber was not fupported by any piles underneath, nor had any plank piling on the down-ftream fide to keep the matter under it, and but very inconfiderable plank piling on the up-ftream fide; his, confidering that the natural bottom as is faid, was very weak and fpringy, renders it more to be wondered, that no greater derangements have happened. At prefent it has the appearance of having done fettling; and if fo, the beft thing that can be done, will be to double cramp the caping, and to cramp every other courfe downwards, and then running the joints full of fine terras mortar made liquid, doing about fifteen inches at a time, and then waiting till what preceded has fet: but, in cafe any thing further appears, to drive down clofe to the face of the ground timber a fheet of rebated plank piling of four inches thick, which may be driven down to the level of the floor by a fet, and laftly, when by the action of the cloughs, the matter appears to be driven away from below the timber, to fupport the whole by dropping in a quantity of rubble to fill the fame within fix inches of the level of the floor, and to extend as far as the ground fhall appear to be taken away, or five yards below the apron.

The other affair refpecting this lock's tail is the fhooting in of the land bank on the north fide, which being a tender fpringy foil, efpecially as is faid towards the bottom, has been in all probability occafioned by its being dug over fteep, and too great a weight laid upon the top, by the matter which came out: there feems to me at prefent no effectual remedy for this, but to wheel the extra matter fo much further back, that a flope may be formed of at leaft two to one from the level of the water's furface, upwards, and beginning the foot of the flope where the ground is whole; allowing alfo a fet-off for the road, at or about the natural furface of the ground, then clearing away the piles and timber as much as poffible, when the water is low, by dredging, or otherwife to deepen the channel, letting the loofe matter come in, and gradually forming its own flope; and if, after this, a layer of rubble of about a foot thick be depofited upon the face of this underwater flope, beginning at the bottom, and diminifhing the thicknefs upwards, in all probability the weight of this rubble will fo confolidate the ground, as to prevent its fliding in future.

J. SMEATON

Austhorpe, 12th May, 1770.

The

The REPORT of JOHN SMEATON, engineer, concerning the repair of the navigation dam at Westwick Wath, upon the river Ure.

HAVING carefully computed the probable expense of building a new dam at West-wick Wath, including all materials; and also the probable expense of making up the breach, and repairing the dam as it now stands, in order to render the same effectual, I find the difference will scarcely amount to £200, and as the old materials, I think, must be of that value in the new erection, it seems there will be no material difference in point of expense: however, as the making up and repair of the present dam may be the sooner executed, and consequently a saving of time in the opening of the navigation, and though the present dam so repaired will not subsist upon principles so regular and uniform as the proposed new one, yet, as the new dam I have computed upon is but 160 feet between the land walls, and the old dam, when made up and repaired, will be above 200. I am of opinion, upon the whole, that the repaired dam will be as safe as the new one; I am, therefore, inclined to advise the commissioners to the repair of the dam as it now stands.

The general method in which I propose to do it, is to let the timber frames stand as they now do, as a pen to the water, and to add a slope of stone on the down-stream side, extending twenty-four feet in base, and after that (that is, below the skirt) to lay in a body of rubble to the breadth of twenty-four feet more, by way of rough apron or security to the skirt. While this is doing, the breach is gradually to be filled with a body of rubble, and the top of it at last to be raised fifteen or eighteen inches higher than the crown of the dam; by this means, whenever more water than that thickness goes over the dam, it will begin to find a vent over the rubble dam likewise. When the rubble dam is made up, it will be made water-tight by throwing in upon the slope side up-stream several boat loads of gravel, which will shut up the interstices of the rubble, and a few boat loads of earth upon that will render it quite water-tight.

ESTIMATE of the quantity and sort of materials to be used in making up the dam and breach at Westwick Wath.

	£	s.	d
The slope surface of the dam, being twenty-four feet broad, twenty inches in depth, and 100 feet wide, mean length will contain 4,000 cube feet of scapelled stone, which, if brought from the quarries near Ripon, and put in place at seven-pence, will be	116	13	4

To

	£	s.	d.
Brought over - -	116	13	4
To rubble for filling under the setters, and covering the earth apron above the dam, 327 cube yards, which, if produced, and laid in place from the before mentioned quarries, at two shillings, will come to - - - - -	32	14	0
To one cube yard of earth per foot running, in order to form an apron above the dam, this getting, and properly depositing, at sixpence for 100 feet, - -	2	10	0
To rubble in the skirt apron, supposed at a medium twenty-four feet broad, and three feet thick, will, in 120 feet length, contain 320 yards, which, at two shillings, will come to	32	0	0
To 100 yards of rubble to line the shore on the north side along the apron, &c. -	10	0	0
To 120 feet running of bearing piles, stringed-piece and grooved sheeting, of three inches thick, and seven feet long, at thirteen shillings per foot running, - -	78	0	0
To facing up the south side of the south wall of the cloughs, and putting it into the form as directed, - - - - - - - -	15	0	0
	286	17	4
To contingencies on the above, at ten per cent. — — - -	28	13	0
To making good the body of the present dam, - - - -	315	10	4
To 3,300 yards of rubble for making up the breach, at two shillings, - -	330	0	0
To six boat loads of gravel and earth, at forty shillings each, - - -	12	0	0
Contingencies at ten per cent. - - - - - -	34	4	0
To making up the breach, - - — - -	376	4	0
Total, - - -	691	14	4

It is to be obferved, that though the prices are added to the above in order to fill up the blanks, yet the quantities are the only circumſtances which can properly be afcertained by me, as a thorough knowledge of the prices of materials and carriage, upon which the ſum total greatly depends, can only be had from an acquaintance with the country, and which, I apprehend, many of the commiſſioners to poſſeſs to a much greater extent than myſelf.

In regard to the rubble ſkirt or apron, it cannot be completed till after ſome conſiderable freſhes have happened, that the holes made thereby may afterwards be filled up with a proper body of rubble, which will probably be required to ſecure it, and which ought to be got down in readineſs, that after the floods have happened, it may be applied before any damage is done by the want of it.

If opportunity ferves, the breach may be begun to be filled up, as foon as the materials can be got to the place, but not fo high as to pen the water over the dam, keeping always the body of the rubble higheft and ftrongeft next the land, to prevent its widening.

The quantity contained in the above eftimate fuppofes the breach fuch as it was when I faw it, viz. about thirty yards wide, and eight feet deep, at ordinary water; proper allowance muft therefore be made for fuch alteration as has happened, or may happen, before it is put into a defenfible ftate.

The firft work for the body of the dam is to complete the piling of the fkirt, and at the fame time, filling in the rubble for the body, over which the floods will go without damage. The fkirt of the dam is fuppofed to be above water in a dry feafon, if it fhould not fo happen, after the water has been let off as much as can be, the fkirt muft be raifed, while the underfide of the ftring-piece is about level with the furface of the tail-water.

J. SMEATON.

Austhorpe, 19th January, 1771.

N. B. If the rubble cannot be put in place for 2*s.* 6*d.* per yard, or under, I believe it will be advifeable to begin a new dam in a new place.

VIEW

VIEW of the works of the navigation of the river Ure, on Tuesday, May 14th, 1771, by JOHN SMEATON.

MILBY LOCK AND CUTT.

THE tail of Milby lock, the wharfing on the south side wants lining with rubble inside the timber work.

The down-stream angle of the stone work of the lock on the south side is cracked in the same manner as the north side did the last year, when settled, it may be cramped and grouted (that is, filled with liquid mortar), as the former has been done

The current over the top of the lock having taken out some of the earth behind, the walls near the upper gate turning posts should be filled with rubble, and lagged at the top to prevent its blowing out.

A breach in the south bank near the lock should be made up and stout, and some low places where the water has run over should be raised.

A sand is said to be gathered in the cut above the bridge, and the head of the cut has a narrow bar of the original bank of the river, that has never been taken out, over which there is but three feet water at dam's height; if an opportunity offers of drawing down the river, both the above will be easily cleared by setting men into the water, but if not, the head may be cleared by a large rake, which will be useful on future occasions in clearing heads and tails of cuts, and the sand within the cut, whenever it proves obnoxious to the passage of vessels, may be taken out, by drawing a sheet of rebated piles across the cut above the same, to be well supported down-stream by stays, and then drawing off the cut's water below by the lock's cloughs.

SHOALS IN THE REACH BETWEEN MILBY AND WESTWICK CUTS.

As the clearing of these shoals has been agreed for at the meeting on this day to a stipulated width and depth, they need no further remark.

WESTWICK CUT AND DAM.

The tail of the cut being sanded up, and there being a deep part of the river below,

it

it is prefumed the fand may be blown out by the action of the lock's cloughs, if not, it muft be cleared with the rake. The wharfing at the lock's tail, it feems, will want fupporting with rubble like Milby.

The cut above the lock has a breach into the river which will want to be carefully made up of a good height and ftrength, and the matter well confolidated, and fome other places on the fide next the river will want raifing and ftrengthening where the water has run over.

The whole cut wants clearing, and the fides made more floping; the bridge wants finifhing.

The breach at the dam, though grown, it is faid, about fix yards wider than when I firft viewed it, yet, by being lefs deep, I do not apprehend to be in a materially worfe condition than before: this being contracted for upon particular plans and fections, which have been amply explained to the undertaker, needs no further mention here.

THE LONG CUT, &c.

The ford at Oxclofe, not having been completed over the river, the part unmade is blown to a confiderable depth, and unlefs repaired before any heavy frefhes come on, what is done will require to be removed, and the whole begun in another place.

The banks at the tail of Oxclofe cut will require to be guarded with rubble as the former.

The over-fall at the head of this lock will require to be conftructed as per former report, to convey the wafte water: it may be done in a floping form, with fcapelled rubble fetting, and the tail of the trench to convey off the water to be well guarded with rubble on entering the river or tail cut.

The terminations of the parapets of the bridges fhould have a heavy ftone laid on each, otherwife the bricks cannot be properly fupported, and will follow one another.

The banks for thirty feet at the tail of Bell Furrows lock fhould be wharfed up with rubble walls, inclining about one foot in four; they may be built without mortar, but fhould be covered about one foot above the cut's water, with pretty heavy ftones, a little fcapelled by way of caping.

The

The tail of the drain that brings down the water from the over-fall fhould be con-ducted through the bank, and the bottom well fet and fecured at the termination by rubble.

The head cloughs of this lock want fome amendment, and the fhutting of the lower gates againft the threfholds fhould be re-examined when the water is off.

Rhode's lock fhould be wharfed at the tail in the fame manner as mentioned for Bell Furrows, and the north clough at the head wants rectifying.

The leakages that appear in various places near thofe locks are nothing more than what always happen in gravelly foils. The water is to be conducted in fmall gutters from lodging upon the furface of the lands for the prefent feafon, and before the next, they will probably difappear. If, upon drawing off the water, any fufpicious places appear, they will be beft ftopped by being puddled with corn mould earth.

The bank near the funken tunnel on the fouth fide wants confiderably to be ftrengthened, and the mouth of the tunnel to be finifhed : it fhould be made wider at the entrance, and an upright bar or two be put down to prevent hay or bufhes that may enter from choaking it; fome fmall part of the bank on the north fide near the elbow next the turnpike road fhould be fomewhat ftrengthened.

I apprehend it will be neceffary in fome convenient place in the upper reach of the cut to have a rough over-fall, fomething like that propofed at Oxclofe; for, if the clough by which the water enters from the Skeld, fhould happen in a flood or fpeat to be left running, the banks of the cut will infallibly overflow, and probably make a breach before it is difcovered, efpecially, if happening in the night; this over-fall may be made on the fouth fide, near the faid tunnel, and its water be conveyed away through the fame.

J. SMEATON.

Austhorpe, 29th May, 1771.

The

The REPORT of JOHN SMEATON, engineer, upon the state of the navigation of the river Ure from Ripon to Boroughbridge lock.

HAVING carefully viewed and founded the river and cuts the 28th September laft, I find that when the cuts are cleaned, there will be four feet water from Ripon in to Weftwick lock.

From Weftwick lock downwards, which is now the fubject of the principal impediments, I find that in the ftate the river then was, viz. four inches water over Boroughbridge dam, there was no more than three feet nine inches and a half water over the threfhold of the lower gates of Weftwick lock, fo that when the water is reduced to dam's height, and fubfided nearly to a level, as in time of droughts muft always be the cafe, this, together with the foundings taken by Mr. Jeffop laft fummer, convinces me, that in that ftate of the river, there will not be more than three feet water over that threfhold. I found likewife that the places that had been dredged, had, at the time of my view, four, and four and a half feet water; but that in feveral places between and below the places that had been dredged, the river's bottom was no more than three feet and a half, and in one place, viz. at Rockliff fhoal (where I originally propofed building the intermediate dam) there was only three feet three inches, all below to the cut's mouth at Boroughbridge being at leaft five feet, and in general fix feet and upwards of water: hence it appears, as alfo from Mr. Jeffop's foundings taken at the time before mentioned, that in a dry feafon there will not be more than two feet fix inches on Rockliff fhoal, with a full dam at Boroughbridge, while at all other places, there will be about three feet.

From the cut's mouth to Milby lock there are fome impediments, but as it is always in the power of navigators to remove thofe, and there being four feet water over the upper threfhold at Milby, at dam's height, I confider this cut as capable of complete four feet water with a full dam.

It feems, therefore, that in cafe a three feet navigation in dry feafons, three and a half feet in common ones, and four feet in open winter weather, will fuffice, that this may be procured with a full dam at Boroughbridge by lowering Rockliff fhoal fix, eight, or nine inches; but then, in cafe the miller of Boroughbridge mills fhould draw his water within head (as in dry times he can very eafily do) the navigation, as it now ftands,

muft,

muſt, at thoſe times, be conſiderably defective, and as the original expedient of building a dam at Rockliff, or any other intermediate dam may not be perfectly agreeable to the miller at Boroughbridge, he may poſſibly ſee it to be (as it moſt certainly is) for his intereſt to enter into a firm agreement with the navigators not to draw his water within dam, and to prevent its running over and waſting in dry ſeaſons, to fix on a ſett of boards of nine or ten inches broad, which I can engage ſo to make, that they may be ſtruck and removed at any time of flood, and if thoſe are kept in repair at the joint expenſe of the navigation and mill, it will be a real improvement to the mill as well as benefit to the navigation, by giving them ſo much additional head of water when they moſt want it, and the mill will then have as good a latitude in drawing as it now has.

If the above cannot be acceded to, then it appears neceſſary either to lay a catch dam acroſs the river juſt below the head of Boroughbridge cut, or to build a lock and dam at Rockliff ſhoal as firſt intended.

The former expedient will cure the preſent inconveniences in the moſt effectual manner, becauſe it will render the navigation entirely independent of the mills; but at the ſame time, I cannot take upon me to report, that in caſe the catch dam is raiſed as high as the mill dam (which, as the works are now laid, is abſolutely neceſſary) there will not be a real injury to the mill; on the contrary, my opinion is, that there will be a real injury, which will require recompenſe, and if this recompenſe could be eaſily agreed upon it would be well, but otherwiſe, as it is of ſuch a nature as not readily to be adjuſted by a jury, it appears to me better to ſubmit to ſome extra charge and inconvenience than embark in a troubleſome piece of litigation.

The dam at Rockliff ſhoal, as originally propoſed, was intended to raiſe the water two feet, in order to clear all the ſhoals above without dredging; but, as now they are in a great degree cleared, it will be ſufficient to raiſe the water there a ſingle foot and this may be done by a rubble dam; that is to ſay, by adding to the preſent, and making an artificial ſhoal, ſo as to keep up the water a foot above its preſent ſurface, that is, much about as high as the ſhoals intermediate, between this and Weſtwick, penned the water before they were cleared away. By the ſide of the ſhoal to be made at Rockliff, there muſt be a lock to give paſſage to veſſels, but as this lock need not to be raiſed higher than till the veſſels can go over the ſhoal itſelf, I apprehend the whole of this buſineſs at Rockliff may be done for about £1200, which may be more particularly eſtimated if this mode of relief be adopted.

When

When this is done, and the Boroughbridge cut cleared fo as to be fix inches below the upper threfhold of Milby lock, together with the cut above alfo cleared, and particularly the tail cut of Oxclofe lock, there will be then complete four feet water, with fill ponds in the drieft feafons: and after the mill at Boroughbridge has drawn down one foot, there will ftill remain three feet in the Boroughbridge cut, and fhould the miller then attempt to draw and keep down his water below one foot within dam, as this cannot be with any other view than to diftrefs the navigation, I would then advife to lie in a fhoal at the head of Boroughbridge cut, which will give him the command of the pond to one foot within head, but will prevent his wantonly drawing it lower.

I would advife that the navigation fhould be furnifhed with proper dredging utenfils for making fuch clearances as can be done without taking off the water, the principal of which is a machine of known conftruction, ufed in the neighbouring river, called a hell rake: alfo a fett of grooved piles, with proper beams and braces, for readily taking off the water from any cut, fo as occafionally to clear it by hand.

For want of the former, the head of Boroughbridge cut has never been fufficiently cleared, and for want of the latter, the filt that has been lodged there in time of floods is obliged to remain: but I am apprehenfive, from the pofition of this cut's head, it will be fo liable to gather filt, that it will be worth while to have a ftaunch fixed at the head; by fhutting of which the water may be readily taken off the cut, and the foil, while light, wafhed out by a current of water.

J. SMEATON.

Austhorpe, 9th January, 1772.

P. S. In my report of May 12th, 1770, I advifed that the three overfalls from the ponds of Oxclofe, Bell Furrows, and Rhode's field locks, fhould be made twenty feet in the crown; now the two already conftructed above Oxclofe and Bell Furrow, are not above half that dimenfion; and though they may anfwer the purpofe in ordinary, yet I fhall not be furprifed if on extraordinary occafions, damages are done for want of a proper vent for the water, as well as from the pond from Rhode's field lock to Bondgate green, for want of one being conftructed on this line; fomewhat near the tunnel I apprehend will be the proper place.

J. S.

HEWICK-

HEWICK BRIDGE.

To his majesty's justices of the peace assembled in quarter sessions at Knares-borough in the West Riding of the county of York.

The report of JOHN SMEATON and JOHN GOTT, engineers.

See plate 5, fig. 5, page 142.

GENTLEMEN,

HAVING carefully viewed Hewick bridge, and the state of the river there, and having confidered the feveral points referred to our opinion by your order of Skipton feffions for the Weft Riding of Yorkfhire, of the 23d of July laft, we beg leave to report upon the points referred to us, which are as follows:

1ft. Whether any, and what damage has been done to a field adjoining to Hewick bridge, belonging to the right honourable lord Grantley, by the current of the water?

2dly. How the fame may be prevented?

3dly. From what caufe the water takes its courfe, fo as to prefs upon, and wafh away the faid field.

To the firft we anfwer, that it appears to us that confiderable damage has been done to the field next below the bridge, on the weft fide of the river; and that from the very loofe texture of the ground of this field, being a very deep rich foil, it is likely that ftill greater damages will enfue, unlefs timely prevented by an adequate re-medy; but, in order to point out the remedy with greater propriety, it will be ne-ceffary to give our opinion previoufly upon the laft point referred to us, viz.

3dly. From what caufes the water takes its courfe, fo as to prefs upon, and wafh away the faid field?

Nothing is more common than for rapid rivers, taking their courfe through grounds compofed of fine deep foil, fuch as that under confideration, to make depredations upon them fo as to change their courfe, of which very ftriking inftances are to be feen in the meadows below this and Weftwick Wath, and this kind of depredation often originates from flight, and even from unperceived caufes; they are not always manifefted to common perception, for it often happens that a change made in one part of the river, will be the caufe of depredations on another, and which, when begun, have generally a tendency to grow worfe, and not to cure themfelves, or to ftop, till fome powerful remedy is applied.

The forming of the new arches at the two ends of Hewick bridge does not appear to us to have any direct tendency to produce the bad effect complained of; becaufe the wear or wharfing annexed to the fouth-weft land ftool wall, as originally made, extending to or about fixty feet in a direct line, appears to us to ftand in a proper direction to carry the ftream of the river away from the part of the land in queftion, rather than towards it, or fo as to make it bear harder upon it than before, and the arch at the oppofite end is too remote to have any material effect as to this land; we find ourfelves therefore compelled to feek for the caufe of the difafter from fome other fource.

According to information given us, there was, formerly, a deep pool below the bridge, and at fome diftance a bed of gravel appeared in the low ftate of the river; at prefent the pool is nearly filled up, and the gravel bed much increafed in fize and breadth; which having a natural tendency to divide the current of the river, will caufe it, in confequence, to prefs hardeft upon the loweft and weakeft fide, which is the very fpot in queftion.

To this acceffion of gravel below the bridge we attribute the circumftance of the water taking its courfe fo as to prefs harder upon, and wafh away the faid field, in a greater degree than formerly; but from what caufe the gravel is depofited here, in a greater degree than it ufed to do, is a further queftion. It cannot be from the broken ground, becaufe the water's current is in fucceffion from the gravel towards the broken ground, and not from the broken ground towards the gravel bed, and the ground altered in opening the arches, at and above the bridge, is far too inconfiderable to produce the quantity; befides, the fine foil there is deep, and does not appear to produce gravel in great proportion.

We are therefore obliged to notice that about a quarter of a mile above the bridge, a work has been made by which the whole channel and current of the river has been
diverted,

diverted, and now paffes altogether through a new cut, which, in gravelly foil, and being worn confiderably wider than it was originally made, large quantities of gravel have doubtlefs been gradually brought down the river therefrom, fince its firft opening in the year 1779, and ftill larger quantities appear to be travelling therefrom, capable of increafing the evil we have pointed out.

It is to be obferved, that this cut being a work prior to the alteration of Hewick bridge, it would require a time for the gravel firft loofened from the new bank and bottom of the cut to travel progreffively on the bed of the river down to Hewick bridge, and it would be fome time before the gravel brought down and lodged in the deep p ol, below the bridge, would become perceivable. We find ourfelves, therefore, obliged to confider the depredation of the river upon the ground in queftion, in a greater degree than formerly, and the alteration of Hewick bridge, as contemporary events, and not that the latter was the caufe of the former, but that the primary caufe thereof was the making the new cut, and forcing the main current of the river to pafs through it by a dam made for that purpofe.

Laftly, we fhall now confider the fecond point, viz. how the ill effects of the water's preffing harder upon the faid field may be prevented?

Now it is to be obferved, that by way of prevention, an addition has been made to the end of the weir before mentioned in the way of a jetty, intended by contraction to remove the gravel bed, and work a channel through it, and thereby to fave and eafe the weak fide, which is the weft; but by continuing it in the fame line further out by 120 feet than the original weir, that direction has carried it too much towards the oppofite fide, fo as too much to ftraiten the river in that place, which, in of time great frefhes, gives it a tendency to fpread when it quits the jetty, and finding the greateft declivity towards the broken ground, the fet of a confiderable part of the current is almoft directly turned upon it, and as the rapidity of this part of the current has made itfelf a deep channel beyond the end of the jetty, this tends ftill to increafe its action and preffure upon the faid field.

Had Mr. Smeaton been called in before the jetty had been run out from the weir, he fhould have advifed to have fortified the curve of the broken ground by lining it with a flope of rubble ftones promifcuoufly thrown in, and not to have attempted to regain the ground by a jetty. As it is, it may be a matter of confideration depending upon the eftimates herewith delivered, whether to carry forward the jetty 420 feet further

in

in the direction fhewn in the plan, (in which direction about four yards of its extreme end is already turned) or to remove the jetty, and difpofe of its materials, fo as to form a flope againft the broken border, as far as they will go, completing the remainder with frefh materials, as either method we look upon as a likely, and the moft practicable way to preferve the ground in queftion from further depredations.

References to the plan, plate 5, fig. 5, page 142.

THE black line marked B B B, fhews the ftate of the ground before the bridge was widened.

A A, the jetty, 159 feet long.

The quantity of land wafhed away on the fouth-weft fide of the bridge, fince it was enlarged, is one rood thirty-five perches, and it is probable, if fomething be not done to prevent it, that in time the water may take the greateft part of the field away.

ESTIMATE for lining the broken bank near Hewick bridge with a flope of rubble ftones, according to the method fuggefted by Mr. SMEATON, in which cafe the ground will not be gained.

	£	s.	d.
The length being at prefent 550 feet, to form a fufficient slope, will contain at an average, three cube yards of rubble to a yard running, that is, it will contain 550 cube yards of ftone, which, at 3s. per yard, will amount to - - - - -	82	10	0
Allow for depofition and contingencies - - - - - - -	7	10	0
	90	0	0

The estimate to continue the weir from A in the plan to B, being in length 420 feet, will coft 130 0 0
By this latter method there is a poffibility the ground may be regained.

J. SMEATON.
JOHN GOTT.

Leeds, 1ft October, 1782

BIRMINGHAM

BIRMINGHAM CANAL.

QUERIES by Mr. COLEMORE, with Mr. SMEATON's answers, concerning the Birmingham canal.

QUERY 1ft.—To fee if the prefent leakage is not to be remedied, and that at a moderate expenfe?

Anfwer.—The prefent leakage is no ways alarming, and nothing more than what might be expected from the nature of the foil through which the canal paffes for about three-quarters of a mile preceding the prefent termination, when particular care is not taken to prevent it. The foil is a red fand, tolerably firm, while it lies in its own natural bed, and underneath is a red fandy rock, clofe enough in the lump, but with open joints that frequently occur. This kind of foil is commonly, in fome degree, leaky at firft, and more or lefs fo, and for a greater or lefs length of time, according as proper remedies have been at firft applied, and in proportion as the water introduced into the canal has more or lefs of the requifite quality to keep it tight. This kind of foil, however, though leaky at firft, in time grows tight of itfelf; infomuch, that it would require a greater degree of art to prevent its growing tight, were the percolation of the water through the pores of the fand for a length of time a requifite quality. I am of opinion, from what I have feen, that the moft effectual methods have not been taken at firft; and I am further of opinion, that had a proper ufe been made of the muddy water wafhed down from the marley and clayey grounds, &c. that do or might be made to fall into the canal in rainy feafons, by being carried in proper quantities through the courfe of the fame, and difcharged at or near the termination by means of a fluice, wafte, over-fall, or difcharger, fo as to keep up a conftant movement at times, when water is abundant : I fay, had this been done when the canal was opened, and at all convenient times fince, inftead of its being difcharged by an over-fall or wafte, not far from the fummit, and thereby the water towards the faid termination rendered in a great meafure ftagnant, I am of opinion that the canal would have, by this time, been rendered fo far tight as to have been without complaint.

Query 2d.—Whether it is, in its prefent ftate, confiderable enough to endanger the navigation ?

Anfwer.

Anſwer.—I do not, in the preſent ſtate of the leakage, think it conſiderable enough to endanger the navigation, or even to be a great impediment to it, as it appears to me that the leakage will be much more than balanced by the lockage water alone, which muſt be diſcharged from this loweſt ſtretch of the canal, by a waſte or diſcharger if not by leakage ; and if to this leakage we add the conſtant leakage of water through the locks, and the regulating water, which laſt Mr. Brindley has on another occaſion ſaid may be ultimately equal to the lockage itſelf; I ſay, all theſe things conſidered, it does not appear in any degree probable, that the navigation would be endangered, even ſuppoſing, what I can by no means admit, that it were always to continue as it is.

Query 3d.—To examine what probability can be aſſigned of the ſame inconvenience attending the extenſion of the canal to New Hall Ring, and if the ſuſpicion of ſuch inconvenience be a fair and ſufficient reaſon for the company to alledge in bar of their being compelled to fulfil their agreement to bring the canal there ?

To aſcertain this, Mr. Colemore ſubmits it to Mr. Smeaton to have the ground bored, as Mr. C. wiſhes to have his claim ſupported upon the cleareſt evidence, or to abandon it if he may be called upon ſo to do, upon any principle of equity and fair dealing.

Anſwer.—Not only from the general appearances, but from pits which I cauſed to be dug in Mr. Colemore's ground, in or near the courſe of the propoſed canal to New Hall Ring, in order to lay the ſtrata open to view; it appears to me that the ground from the preſent termination eaſtward, to New Hall Ring, is not leſs fit, and much of the ſame nature as that already paſſed through for about three-quarters of a mile preceding, and therefore muſt be ſubject to the like leakage, in proportion to its length, (which is under one-quarter of a mile), ſuppoſing it to be executed in the ſame manner : but as I am very well ſatisfied, that the extenſion may be made in ſuch a manner as to be ſubject to much leſs leakage at firſt, and to be more ſpeedily ſtopped, I cannot look upon the apprehenſions of increaſing the leakage, as ſufficiently grounded to be a bar to the extenſion, eſpecially when I conſider the neceſſary ſupplies above mentioned, as a ſufficient counter-balance, ſuppoſing the part to be made ſubject to a proportional leakage to that which is already made.

Query 4th.—To examine whether the termination of the canal, at New Hall Ring, be not equally convenient for the public with any other ?

Anſwer.

Anſwer.—This is not properly a queſtion for an engineer to anſwer, who cannot be ſuppoſed to enter into the particular conveniencies attending the commerce of a large trading town in a few days; but from what I have obſerved of the ſituation, and have ſeen of the town, I ſee no reaſon it ſhould not be as convenient as any other that has been pointed out to me : but I am of opinion that the greater the number of the owners of wharfs, the better it will be for that part of the trading public, who are not immediately intereſted in the navigation.

Query 5th.—To eſtimate the expenſe of finiſhing the canal from its preſent termination to New Hall Ring?

Anſwer.—According to my eſtimate, the canal may be finiſhed from its preſent ter mination to New Hall Ring, being 401 yards, according to the meaſure of Mr. Yeoman, for the ſum of £473. 10s. excluſive of the value of two acres of land, which, at twenty-five yards broad, it will occupy, and excluſive of the particular charges that may attend the conſtruction of the wharfs thereupon.

Query 6th.—To ſee if the ſhorteſt, the moſt expeditious, and the cheapeſt way the company can go from the preſent termination to their own ground, at the brick-kiln-piece, be not through Mr. Colemore's land, which lies on the eaſt ſide of the Dudley road?

Anſwer.—This, I think, can hardly admit of a doubt; for, according to Mr. Yeo-man's meaſures and the plan conjointly, the preſent termination of the canal is nearer to the ſpecific point in the brick-kiln-piece, by way of Mr. Colemore's land, than by the other tract marked through Mr. Farmer's land, by a difference of 292 yards, and the length of tunneling will be leſs than half as much. The prac-ticability of doing it either way, I think, is not to be doubted, but which may prove eaſieſt length for length, I apprehend, cannot be certainly known but by executing them both : ſince, therefore, the chance of unforeſeen difficulties is equal in the ſame length either way, the ſhorteſt runs the leaſt riſk of meeting them; and, upon equality of ground, the expenſe and time will be leſs in proportion to the diſtance. A ſhorter line may, indeed, be drawn from the preſent termination to the brick-kiln-piece, through the gardens and grounds of Mr. Baſkerville, without touching Mr. Colemore's, but, I apprehend, this will not be found eligible on account of damages; and, upon the ſup poſition that the company are to carry the canal to New Hall Ring, then the diſtance

through

through Mr. Colemore's land, will be as near as any that can be drawn through Mr. Baſkerville's.

Query 7th —Whether it is not probable the company would at preſent prefer this di-rection, if they had it not in view to prevent Mr. Colemore from making any ad-vantage by wharfs and warehouſes, and to engroſs them wholly themſelves ?

Anſwer.—What may be the reaſons why the company would avoid coming into any part of Mr Colemore's grounds, is not eaſy for me to ſay with certainty ; but ſo far as I am enabled to judge of the matter, it carries that appearance, and ſo I muſt imagine, till the company are pleaſed to explain themſelves otherways in a manner more ſatis-factory. It may, indeed, be alledged, that though this paſſage to the brick-kiln-piece is nearer, and can be done at leſs expenſe than by the other line, yet, when done, that the diſtance from the point of departure common to both is leſs by the latter than through Mr. Colemore's : and, according to Mr. Yeoman's mea-ſure, the difference will be 114 yards, reckoning both ways from the ſaid point ; but then there will be double the length of tunnel, viz. about ſix chains through Mr. Colemore's, and at leaſt twelve the other way, and I look upon a tunnel to be ſo great an impediment to the paſſage of veſſels, when they are going in contrary di-rections, eſpecially near wharfs, where they are often crowded, that the hindrance and loſs of time will, upon an average, amount to far more by the increaſe of length of tun-nel, than can be ſaved by a difference of diſtance of 114 yards : an under ground tunnel is in ſome places an uſeful expedient, by ſubſtituting a leſs evil for a greater; yet it is ſtill an evil always to be avoided, where it can.

Query 8th.—To conſider if the canal be continued to New Hall Ring, whether the ground on the lower ſide of it would be hurt for building by the leakage ?

Anſwer.—I am well ſatisfied the extenſion of the canal may be performed in ſuch a manner, that there will be no leakage hurtful for building, as what may happen may eaſily be carried off in proper drains.

Query 9th.—To eſtimate the expenſe of making a brick ſide drain to carry off ſuch leakage ?

Anſwer.—I apprehend this is impoſſible to be done, becauſe, on the ſuppoſition of leakage, it will be proper to make the drains according as the leakage happens to break out, and I do not apprehend the places can be aſſigned beforehand.

Query

Query 10th.—To confider if any rifk to the whole navigation can be incurred by extending the canal from its prefent termination to New Hall Ring, and whether any, or what fecurity, need be given the company againft fuch rifk, if Mr. Colemore were to continue the canal on his own account?

Anfwer —The extenfion appears to me to be attended with no difficulty, but what occurs in daily practice. I cannot admit of any rifk to the whole navigation in the carrying the fame into execution. I, therefore, do not fee that any fecurity can properly be expected. No human judgment is infallible, and fhould any thing of fo extraordinary a nature happen, as abfolutely to require it, the fame juftice of parliament which may oblige the company to extend the canal, will, on a fimilar application, enable the company to fhut it up again : but if works were to have a negative upon them by the profpect of chances fo remote as this, it would be impoffible to do any thing in the way of artificial navigation.

To the committee of the Birmingham canal.

The REPORT of JOHN SMEATON, engineer, upon the several matters referred to his inspection and opinion by the said committee, at a meeting held at the navigation office, at Birmingham, the 9th and 10th of October, 1782.

HAVING, in confequence of your directions and inftructions, viewed the general fcope of ground occupied not only by your prefent canal and navigation works, but alfo fuch further extenfions, additions, and improvements, as you have had in contemplation; being on this view attended by your own engineer and furveyor, Mr. Bull and Mr. Snape; after full confideration, I am now in condition to anfwer fome of the moft material queftions that you have done me the honour to refer to my enquiry.

In acquitting myfelf of this bufinefs, I find myfelf enabled to do it by taking the main body of your enquiries together, rather than by making anfwers to each feparately, for by this method I fhall be able to weave into one piece the whole bulk of matter that you have laid before me.

Your leading enquiries are, whether the courfe of the prefent undertaking, or the intended extenfion from the Wednefbury branch to the coal mines, with the feveral collateral branches, can be improved?

Having carefully viewed the feveral lines of extenfion pointed out to me, as well that leading from the New Hall branch, in the environ of the town toward Deretend and Bordfley, as that from the Wednefbury branch toward the collieries; I find the whole very judicioufly and carefully laid out, fo as to conduct the navigation in the beft manner towards the feveral points propofed; and the means pointed out of returning the water by a tunnel from this low level, into the branch of the Wolverhampton canal, that is extended to Oker Hill by means of an engine there, will be very effectual, and attended with beneficial effects upon the whole navigation, as I fhall more particularly fhew hereafter.

In going over the feveral lines of this bufinefs, it forcibly ftrikes me that the principal leading hinge upon which the whole, both primarily and ultimately, muft turn, is the

procuring

procuring a fufficiency of water, not only for fupplying the extenfions you have propofed, but for upholding and fubfifting the navigation as it now ftands, and in refpeƈt
whereof the feveral claims made upon you by the millers, for reftitution of the water intercepted for the fupply of the canal, appears to me a very ferious bufinefs; for as the
value of all mills and ftreams of water capable of turning mills, which can be employed
in the very extenfive manufaƈtures of this country, are very great in themfelves, and
alfo neceffary powers for the very exiftence of the manufaƈturers; it therefore is not very
eafy to make a compenfation in money that will be adequate to the thing taken away:
befides, as the water taken away from the firft, or any given mills upon a ftream fuccef-
fively, turns the fucceeding, the whole ftring of mills in fucceffion have their refpeƈtive
claims upon you, fo that the making the due and adequate fatisfaƈtion to them all, in
their proper and proportionable degree, would be in a manner endlefs.

This, in a great meafure, appears from the memorial of the gentlemen concerned in
the mills of Afton brook, delivered to you gentlemen of the committee of the Birmingham canal, and bearing date the 28th of March, 1782, the matter of which memorial,
though defired by you to be confidered by me in the light of a fecondary bufinefs, yet
naturally leads to matter that appears to me the primary; and more efpecially as I underftand that a fimilar demand has been urged by the gentlemen concerned in thofe mills,
whofe waters fall into the river Tame from the weft fide of your fummit; and which
claims in proportion to their value, that is, diftance from the feats of manufaƈture, are
undoubtedly capable of being fupported with equal juftice and propriety.

It appears then to me, and is plainly fo underftood by the gentlemen who delivered
the memorial above referred to, that you have no right to any water for navigation
without recompenfe to the mills, but either fuch water as you can colleƈt into your
refervoirs in fuch rainy feafons, when the mill ponds are overflowing, and the mills
fully fupplied, or fuch water as is drawn out of the bowels of the earth from the mines,
which had not previoufly made its appearance in fprings upon the furface; and, upon
this footing, the fecuring of an adequate fupply of water at all feafons, becomes a matter of the firft import, not only to the fuccefs of any extenfions that may be thought
proper, but for fecuring a continuance and permanency to the prefent lines of navigation.

I will, therefore, now endeavour to inveftigate and lay before you,

1ft. The fupplies of water that your prefent canal fhould take, according to calculation;

tion , 2dly. The quantities that were actually confumed during the dry months of the fummer 1781, and from thence deduce the quantity of regulating water that was actually confumed, with other ufeful deductions, fhewing how the neceffary quantity of water, for the maintenance of the whole, is to be maintained and fecured.

1ft. The length of your prefent lines of canal, taken together, is $27\frac{1}{4}$ miles; this, at the mean breadth of thirty feet, gives a furface of 4,316,400 fquare feet. The foakage and evaporations in dry feafons, I have ufually computed at the lofs of one-fifth of an inch per day in depth, which, upon the whole furface as above, amounts to 71,940 cube feet: but as we have not fo diftinct an idea of high numbers, as thofe of a fmaller denomination, it may be more commodious to reduce our meafures to that of the ordinary meafures of the canal locks; which, according to my computation, at fix feet deep, contain 3,388 cube feet, which, for a round number, call 3,400; then 71,940 will contain $21\frac{16}{100}$, which per week will amount to $148\frac{12}{100}$; fay, for foakage and evaporation of the canal per week, 150 locks-full.

According to the account of the four fummer months of the prefent year, viz. June, July, Auguft, and September, there have paft the fummit about 250 boats per week, loaded and empty; thofe would generally require two locks-full to a boat: fometimes, indeed, they would pafs alternate, fo that two boats would pafs with a lock-full; that is, reckoned for each fide the fummit, the amount would be a lock-full to a boat; but as this only fometimes happens, to allow for unavoidable wafte, we may fairly reckon two locks-full to a boat: the expenfe, therefore, of lockage through the fummit may be called 500 locks-full per week.

1ft. The lockage of the Wolverhampton branch into the main trunk of the Staffordfhire canal, is eftimated at twelve boats per day loaded, and as many empty, that is, twenty-four locks-full; but the loweft lock at Autherly being a ten feet rife, will expend the fame water as if they were all ten feet locks, and which will be equivalent to forty locks-full of fix feet, fuch as the reft are, that is, 240 per week.

The account, therefore, with other intervening locks, will ftand thus :—

		Locks-full.
To lockage of veffels through the fummit per week, - - - -		500
To ditto by the Autherly locks into the main trunk, - -		240
Total lockage per week, - - - -		740

The

						Locks-full.
			Brought over			710
The soakage and evaporation estimated at	-	-	-	-	150	
Leakage of locks-full per day, in each string, viz. one on each side the summit (one of which will supply the Wednesbury branch also), that is, eight per day,				-	56	
Leakage of Autherly string, four ten feet locks will be two and two thirds more than common locks-full per day, that is, per week, eighteen and two thirds, say				-	19	
		Dead stock, or regulating water,	-	-	-	226
		Total supply per estimate,	-	-	-	965

2dly. The supplies that furnished this business in the dry months of the summer, 1781, were as follows:

					Locks-full per day.
The company's new engine at Smethwick,	-	-	-	-	565
Ditto engine at Spon lane,	-	-	-	-	415
					980

			Locks-full.	Butts.	
The spring near the lock office, called Hadly's spring,	-	-	1	1	
The springs, called G. Smith moor springs,	-	-	-	7	3	
The Rood End springs,	-	-	-	-	46	0	
Crosswell springs that feed Hill's mill,	-	-	-	115	6	
					170
					1150

1. Lord Dudley's old engine,	-	-	-	-	-	127
2. Brown and Co. Coastly moor,	-	-	-	-	54
3. Bloomfield, the patent engine,	-	-	-	-	74
4. Wilkinson's, at Bradley moor,		-	-	-	26
5. Capon field new engine,	-	-	-	-	-	15
Penn and Co's.	-	-	-	-	-	75
Lane's,	-	-	-	-	-	33
Tomkey's,	-	-	-	-	-	53
Catchem moor,	-	-	-	-	-	19
						478
						1628

But in this period, the canal had lost so much in depth, that the vessels could only go with burthens of seventeen tons, instead of twenty-four; so that having lost full twelve inches in depth, it would be expending from itself at the rate of, per week,

73

Total of the supply, — 1701

N. B. The

N. B. The Thimble mill water and springs that used to be taken in to supply the great reservoir during this time were turned down to the mills; and hence it appears, that of the above supply, amounting to 1700 locks-full per week, only 170 of it was natural water supplied by springs, which is but one-tenth part of the whole, had the others been taken in, viz.

	Locks-full per week.
The Thimble mill springs,	45
Hanson's pool springs,	23
	68
To which add the western springs,	170 as before
The whole of the springs	238 would make

but about one-seventh of the whole supply.

3dly. From the two preceding statements we may draw the following:

	Locks-full per week.	
Water furnished by springs,	170	
Ditto by colliery engines,	478	
by loss of depth upon the whole surface of the canal,	73	
Total of new water supplied to the canal per week,		721
From which deduct the known consumption that never returned, viz. by lockage to Autherly,	240	
By lockage of the Wednesbury branch, at one boat per day each way, will be	12	
The known consumption that never returned,		252
The real weekly consumption of regulating water expended in the several articles of evaporation, soakage, leakage of locks, and waste,	469	
The regulating water, as by computation preceding,	225	
Weekly expense of regulating water more than computation,	244	

Hence it appears, 1st, That the weekly waste of dead stock, or of regulating water, is full double to the computed quantity. 2d, That the real quantity, 469, is almost equal to 500. The quantity computed for lockage of the summit. 3dly, That the quantity of spring water actually taken in, viz. 170 locks-full, was not one quarter of 721, the quantity of new water wanted, beside the return of the company's engine, and had the whole of the springs been taken in, it would not have amounted to one-third of the whole supply of new water.

The

The quantity therefore produced by nature appears very inadequate to the supply of this canal in dry seasons; and therefore it is not to be wondered, that before the company's engines were built, the navigation was totally stopped in dry seasons for want of waters.

For with respect to the two reservoirs, they could afford no assistance in the case, as they would be exhausted before the pinch of the season came on, as we shall see must be the case, on comparing their contents with the weekly wants above stated.

The great reservoir at Smethwick, is computed to contain 1514 locks-full when full, but as I am informed, if full by the winter's rains, it will have lost eight feet of its top water, by the time it is wanted; but supposing it to lose only six feet, it will then contain but 621 locks-full: the rest of the reservoirs I understand hold water pretty well, and their capacities are computed as follows:

		Locks-full.
4th. Great reservoir at Smethwick, after losing six feet at top	-	621
Little ditto, near Mr. Hansom's	-	100
Titford reservoir	-	500
	Whole content of all the reservoirs	1221

Hence, if the company's engines were laid aside, and the summit to be supplied from the reservoirs at the rate of 500 locks-full per week, the whole, with allowance for leakage, &c. arising in this part (viz. fifty-six per week, as per first statement) would be consumed in fifteen days nearly; and with the assistance of all the springs coming in during the time in aid, the whole would be gone in three weeks, and then the canal would be left to subsist upon the springs alone in the manner before stated, and such casual showers as do not commonly happen in the four dry months of the year, to be of much service to such an extent of navigation.

From what precedes, it plainly appears how very inadequate not only all the springs but all the reservoirs are, towards an effectual supply of the Birmingham canal with water: As it therefore must in a great measure depend upon artificial means, and as there do not appear to me any, but the same artificial means of restoring the water of the springs to those mills from which it is intercepted, it seems by far the clearest way, to dismiss all the waters of the springs, and suffer them again to take their natural courses: for though some advantage might be made in using those spring waters for lockage from the summit each way, and a method may be pointed out, whereby

precisely

precifely the fame quantity of water may be let out from the level canals, extending from the foot of the Smethwick locks each way for the ufe of the mills, that was taken in at the fummit; yet as I know no means of effecting this, but by a continual attention to the water gages, that muft be conftructed for this purpofe, fo that in all the variation of the influx, that of the efflux may be regulated conformably thereto; yet as a mifapplication will always be in the power of the perfon attending, fo as to give the mills either lefs or more than their due, this of courfe will often prove un-fatisfactory; and as it appears, that the whole of the fprings are fcarcely one-third of the water wanted, and not one fourth of the water raifed by the company's engines, what fupplies the two-thirds may juft as well fupply the other part; for when an engine or engines are once eftablifhed, with proper perfons to attend them at all times, as appears to me indifpenfably neceffary, there is only the trifling difference in the quantity of coals neceffary to do the whole work, or two-third parts of it.

We are now led to remark, that though engines can return the water from the lower level to the higher, fo as to anfwer the purpofe of lockage, and leakage of locks, yet fince the returning engines cannot create water, what is loft by evaporation and foakage muft be a fheer lofs, that muft be made good by the introduction of new adventitious water, to fupply the place of what is gone: but towards the fund of fupply, the water of the fprings cannot be applied, becaufe what finks into the earth, and mixes with the atmofphere is loft to the mills, as well as to the navigation, and if to be fupplied elfewhere, or otherways, what will fupply the mills will fupply the navigation.

It has been already ftated, that the quantity of regulating water, or wafte of the dead ftock weekly, amounts to 469 locks-full, which, together with what is neceffary for the lockage to Autherly, and down into Wednefbury branch, viz. 252 locks-full more, make the whole amount of new water into the Wolverhampton canal, to be 721 locks-full per week.

Now it has been remarked, that the 469 locks-full of regulating water, actually loft, being full double of what might be expected according to my ordinary computation thereof, this is a reafon why the more fufpicious parts of the canal fhould be fearched, in order to try to ftop the leaks: yet as the quantity of foakage depends in a great meafure upon the nature of the foil through which it paffes, and though in making a computation for the fupplies of a canal, as the lofs by foakage will always be fomething, and in fome foils may be a very great quantity, and therefore fome kind of allowance muft be made for it; yet as it is out of all human power to forefee

or

or fay before hand, what it will be in the particular cafes, in this refpect when a canal is made, calculation muft fubmit to the experience of the particular fituation, and therefore when the fact greatly exceeds the computation, it is, as has already been faid, a good reafon for looking out and examining: yet after all, any difference that we may find, may be owing to the nature of the foil; and indeed, when we confider that many parts of the Wolverhampton canal, in paffing through the colliery grounds for confiderable fpaces together, pafs over thofe from under which the coal has been worked, and in feveral places the canal actually let down, fo that the fides need raifing to an additional height, on that account we are not to wonder at extraordinary leakages; nor even if it fhall prove, that thefe extraordinary leakages are incapable of remedy; fo that a canal in fuch a fituation, being fubject to thefe derangements, I cannot hold it in a fafe fituation, unlefs it can command a confiderable redundancy of water beyond all calculation.

For thefe reafons we cannot reckon a lefs quantity of new water neceffary to uphold the navigation in its prefent ftate, than what we now find, viz.

Locks-full per week.

5th. The nine colliery engines supply as before stated	478
And if you take the springs amounting to	238
You have nearly the quantity	716

But as there are feveral colliery engines that raife, and are likely to raife, confiderable quantities of water that might go into the canal, that now do not, as it will be their real intereft fo to do, fo it will be your intereft and advantage to lay fuch reafonable inducements before the owners of thofe collieries, as fhall make them ready to fupport the canal, with all the water in their power.

Account of engines that actually raife water that would go into the canal, but which is turned another way.

Locks-full per week

6th. Tomkey's executors (late Wilkinson's)	107
Ditto next beyond	120
Barber's upper engine	56
Water that may be had	283
To which add the nine engines that now supply the canal	478
Here then is plainly a surplusage beyond what is found to be wanted of forty locks-full	761

Brought over   761

In further aid, to stand against such accidents as may happen amongst the engines, it seems proper that the company should raise the water into the canal, that is now delivered by Bickley's new engine upon the surface, which, though considerably below the level of the canal, may be done by an engine of the smallest size; the water of this engine amounts to   -   -   -   -   -   -   122

Total recommended for present service   883
Total wanted for ditto   721

Overplus   162

Besides the above, the waters of the three following working engines might be lifted up into the canal, if occasion required, that now being delivered below it, take their course a different way.

Locks-full per week.

7th. Aston and Co.   -   -   -   -   -   -   -   12
Dall's hole   -   -   -   -   -   -   -   90
Barber's low engine   -   -   -   -   -   -   -   14

Further supply   116

It is very natural for the mind to suggest, that though the whole number of working engines may be at present fully sufficient to keep up the dead stock of regulating water for the canal, and also the Autherly lockage, yet, as the collieries are daily in a state of change, and successively working out, it may happen in time, that being wrought out, and ceasing to draw water, the canal may fail of supply: but in answer to this it must be observed, that so long as Birmingham, Wolverhampton, &c. are wanting coals, new collieries will be opened for their supply, which will afford new quantities of water; and for this reason, I have brought to account only those colliery-engines that I found actually at work, reserving those that were building; and those that may probably work in a short time, to balance those that may not long continue at work, on account of the fields of coals being worked out, for which they were built. The following is therefore a list of such engines as probably will work in a short time, with the quantity of water they may be expected to afford at twelve hours work per day, viz.

Hancock and Co.'s new engine   -   -   -   -   -   63
Lord Dudley's new engine   -   -   -   -   -   79
Fownes and Co. the engine stopped on account of the coal being on fire, but water putting in to extinguish it   -   -   -   -   -   50

Expected supplies   192

N. B. Hancock and Co.'s feems intended to deliver into the canal, the other two will deliver below it, but being at no great diftance therefrom, can be lifted into it by the fmalleft engines.

The following contains an account of fuch other waters as have been pointed out to me as a fupply, viz.

	Locks-full per week.
9th. Manmoor Green, a colliery formerly worked here, whose out-burst - - -	31
Broadwater, a colliery formerly worked, whose out-burst here - - - 57	
Ditto Sparrow's Forge - - - 166	
Total of Broadwater's out-bursts - - —— 223	
Total supply of collieries out-burst water -	254

Now it is to be remarked, that it is the greateft part of twenty years fince the two collieries above mentioned ceafed working, and that the whole of the out-burft water goes to the fupply of Gafbrook mill, near Autherly, and in dry feafons is a material part of the fupply of that mill. The out-burft, near Broadwater, goes to Willingfworth Hall mill, and, in dry feafons, is a confiderable part of its fupply; and the out-burft that goes to Sparrow's forge near to Wednefbury (which is by far the moft confiderable) is the whole fupply of that work.

If the water of Broadwater colliery wafte were drawn by an engine to a proper depth, I doubt not but it would bring to it the greateft part, if not the whole of the water of the out-burfts, amounting to 223 locks-full per week, and drawn at a greater depth, doubtlefs the feeder would be increafed; but we may be fure it would not be much increafed, becaufe from the remains of the old engine at Broadwater, its fize and powers may be gathered, and from the reputed depth from whence it drew, the quantity may be nearly eftimated that it could do, which only moderately exceeds the quantity now given by the out-burfts.

The Manmoor Green water may alfo be drawn by an engine put down to a greater depth than where it iffues, and its feeder might alfo be probably increafed, in what degree is uncertain; but as the quantity now given is a mere trifle when applied to the purpofes of the canal, and the engine, by the accounts of it, was a fmall one, no quantity of water of confideration can be expected here; and indeed, as both the waters of Manmoor Green and Broadwater have been fo long occupied by mills, though it may not amount to a legal prefcription againft the inheritors of the freehold, yet being in poffeffion thereof at the time of paffing the Birmingham Canal Act, I

muft

muſt ſubmit it to the gentlemen of the law, whether, as reſpecting the canal, they would not ſtand upon the ſame grounds as other mills, as to a reſtitution or recompenſe for loſs of water: and in caſe of application to parliament for any new powers, they would (I ſpeak with all due ſubmiſſion) doubtleſs, on petition, prevent the water being taken without a recompenſe; and therefore would then be eſtabliſhed upon the ſame grounds as other mills poſſeſſing their water by a more ancient preſcription.

But ſuppoſing the whole of the natural waters that I have viewed, excluſive of the colliery engines, were collected for ſervice, they would ſtand thus.

	Locks-full per week.
10th. The ſprings - - - - - - - - 238	
Manmoor Green and Broadwater out-burſts - - - 254	
	492
The company's reſervoirs, ſuppoſing 1221 locks-full, equally expended during the four ſummer months, they will amount to per week - - 70	
And ſuppoſing the canal itſelf to be lowered ſix inches, which I ſuppoſe it might be without conſiderable prejudice, this will amount per week to - - 36	
	106
Total collection, excluſive of colliery-engines -	598
Quantity wanted -	721
Deficient -	123

It has been ſuggeſted, that each of thoſe colliery waſtes may be capable, excluſive of their natural feeders, as comprehending a quantity of hollow ground from whence the coal has been worked, of containing a great quantity of water as reſervoirs, to be filled yearly by the winter's rains, which, being drawn out into the canal during the ſummer months, without defalcation from evaporation, or ſoakage, will afford a material ſupply of water to the canal, when it is moſt needed.

How far at this time theſe hollows may exiſt, is altogether problematical; for the coal wrought at Manmoor Green, was what is in theſe parts called a thin ſeam, that is, ſix or ſeven feet thick; and though there is not much appearance of falls from the ſurface, yet the under ſtrata may be ſo broken, and ſhivered, as to fill the bulk of the cavity: beſides, if this coal reſts upon that kind of ſtratum of clayey ſtone, which is uſual for the coal to reſt upon in other parts of the kingdom, and called by various names,

names, in this length of time it is probable, in like manner as in other places, that the pillars, by finking into the foal, fill, or thill, the roof and foal may be fo nearly come together, that but little cavity remains.

At Broadwater colliery, being upon the thick feam, the furface itfelf appears in many places fo broken, that I have no doubt, but that a great part, ifi not the greateft part, of the cavity is already filled up: but if it were to happen that the Birmingham canal were to be much preffed for want of water, then it might be worth while to try thofe expedients; and if we fuppofe each of thofe waftes of equal capacity with the whole fum of the refervoirs the company now have, viz. to furnifh feventy locks-full per week each, that is, one hundred and forty locks-full the two, then this will exceed the deficiencies laft-mentioned, by feventeen locks-full.

11th. The whole of the supplies of new water taken together that it seems at present practicable to procure, is as follows:

		Locks-full per week
By springs into the summit		238
By nine colliery-engines, now delivering into the canal	478	
By three ditto, that may deliver into the canal, but do not	283	
By one ditto, that cannot deliver, but recommended to be made	122	
The supply recommended		883
By three engines that cannot deliver into the canal, but might be made to do it, if occasion required	116	
By Manmoor Green, and Broadwater out-bursts	254	
By the company's reservoirs	70	
By six inches of the canal as a reservoir	36	
Visible waters that may be procured		476
By problematical waters that may possibly be procured from Manmoor Green, and Broadwater wastes		140
Total of waters that appeared to me capable of being collected		1737

Befides all which, as the fprings into the fummit amount only to 238, whereas the quantity wanted there (as before ftated) is 556; the difference, 318 locks-full, would be required to be raifed by the company's engines, and would be near upon the work of Spoon-lane engine.

The whole of the above ftatements being carefully confidered, it will clearly appear, that though it may be laid down as a practicable thing to fupply the prefent demands of the canal with water, yet it cannot be done otherwife than by taking the greateft

part

part of it from the mills that were in poffeffion of it before the Birmingham canal was made; befides the fupply of a confiderable part by the company's returning engines. Now, as there appears a fufficiency capable of anfwering all probable contingencies, by means of the full ufe of the company's engines, and a judicious application of the colliery-engines' water, it cannot be doubted, but that the fupply of the canal by the latter means will be the moft eligible.

It may be faid, indeed, fuppofe the owners of the collieries fhould refufe to deliver their water into the canal, but on fuch terms as might make it dearer to the company even than paying the millers their full value? I beg leave to fuggeft what on other occafions I have been affured of; viz. that no man can acquire any property in water, otherwife than in regard to the ufe he makes of it; and therefore I conclude, that though the colliery owners may, at their own choice, deliver their engine water into the canal, or into a water-courfe that will naturally lead it under, or away from the canal, yet that the canal company may, in preference of all others, take it up again, and raife it into the canal, before any third perfon has acquired any property therein.

Refpecting the propofed extenfions of the canal.

1ft With refpect to the extenfion from New Hall Ring to Deretend and Bordfley, I have already remarked, that it appears to me very judicioufly laid out to conduct the navigation to the points referred to; but it will appear from the preceding ftatement, that it can have no water for lockage, but what is the refult of a rainy feafon, which may probably be, *communibus annis*, about four or five months; that is, when all the mills in general have a fuperfluity of water, and therefore, that a fire-engine in fome proper place near New Hall Ring, with a tunnel to conduct the water thereto, will be neceffary to be eftablifhed along with the canal itfelf; for if the whole of the natural fupply at the fummit were to be ufed, it is probable it would not do more than fupply the lockage of this branch, and then it will be wholly taken from the mills; for though this water were ufed to lock from the fummit down to the prefent level canal to Birmingham, yet, if its equivalent be let out by gages to the mills, then this water would be wholly taken away from the lockage of this extenfion; but there is no doubt of its being effectually fupplied, by returning the water by an engine.

2dly With refpect to the extenfion of the Wednefbury branch to the coal grounds lying upon a lower level towards Bilfton and Darlafton, the ground alfo appears to me judicioufly chofen to conduct the canal through this diftrict to the points required, and
except

except some deep cutting to the north-westward of Broadwater old engine, there does not appear any part of the work likely to be atteded with any particular extra expense: but here again, as the lockage water from the summit is returned from the Wolverhampton canal back again to the summit, here will be no water to lock down with, into the lower level, except the surplusage of the Wolverhampton canal supply above its ordinary confumption of regulating water, that is,

	Locks-full per week.
12th. The supply being stated - - - - - - - -	883
The water wanted for other stated purposes - - -	721
There will remain for lockage to the lower level	162

but as this will only lock eighty-one veffels, it is plain this is not to be fecurely done without a returning engine, to lift the water back into the Wolverhampton level; and for this purpose nothing can be better adapted than the propofed fituation at Oker Hill, which will in reality be attended with feveral material advantages.

This propofed extenfion, with its branches, will meafure about feven miles, which being about one quarter part of the prefent canal, will require for its evaporation and foakage waters about one quarter part of what the prefent canal requires, which being computed to be 469 locks-full per week, one-fourth of this will be 117 locks-full per week; but the overplus being 162, it appears that there will be a fupply of regulating water for the canal, in this firft inftance, and fomething to fpare for lockage. And when collieries are opened upon it, being intended for their immediate fervice, they can think it no hardfhip to fubmit to a claufe, obliging them to turn all fuch engine water into the canal as is delivered above its level, and that the company be at full liberty to lift all fuch into the canal that they fhall judge proper, and in confequence it is not to be doubted but that a quantity will be procured, not only fully adequate to the regulating water of this extenfion, but which, by means of the engine at Oker Hill, may be brought in aid of all fuch deficiencies as may happen upon the prefent lines by the working out of the collieries now there; and thus the whole work be permanently eftablifhed.

Refpecting the expediency of a new canal.

A new canal being faid to be at this time in agitation, propofing to take up coals from the fame diftrict of country, between Wednefbury and Oker Hill, where the

Wednefbury

Wednefbury extenfion of the Birmingham canal is propofed to pafs, and to conduct them by a different paffage to the fkirts of the town of Birmingham, near Deretend; and from thence away to Fazely, to join the line of the propofed canal, from the main trunk of the Staffordfhire canal, at Fradley Heath, to Coventry, and fo to Oxford: As a perfon confulted by you, it will be proper to fubjoin fome remarks upon the operation and expediency thereof.

And in the firft place I muft obferve, that what may be the neceffity or expediency of conveying the Wednefbury coals to Fazely, that is, to the line of the Coventry canal, or the profits attending fuch conveyance, I pretend not to judge; but, whatever they might be, confidered in this fingle point of view, they will undoubtedly turn out to be the greater, by the work itfelf being done upon the moft direct and practicable line, and at the leaft expenfe, this being the obvious way to make a productive fcheme.

What the line may be that I am told is in contemplation, I cannot completely judge of till the propofers have publifhed their fcheme, but this I obferve, that the country lying between Wednefbury and Birmingham, after quitting the valley of the river Tame, in order to get thither by way of Afton, &c. is very uneven, not naturally adapted to the courfe of a canal, nor yet fuch as that a canal can be adapted to it without incurring very large extra expenfes; whereas, upon fuppofition that the principal object is to make a navigation from Wednefbury to Fazely, nothing can be better adapted in point of ground, through fo long a courfe of country, than to purfue the courfe of the river Tame, which might be done either by conducting a canal by the courfe of the river's valley down to Fazely (in the whole courfe of which there feems nothing material to hinder) or by making the river itfelf navigable; or perhaps by what will be found the moft eligible, practicable, and cheapeft mode of all, to do it partly by one and partly by the other; that is, to make a canal where the river does not fuit, and to take the river where it does. To do this by way of Salford bridge (about three miles from Birmingham) there is no natural impediment, on fuppofition of a competency of water.

But in lieu of a propofition fo plain and evident, to take the canal out of the valley of the Tame, through every impediment, and by a much longer courfe, to the fkirt of the town of Birmingham, and then from thence to return back into the valley of the river Tame to go down to Fazely, can certainly be intended for no other purpofe than that of fharing the profits with you, that arife from the carriage of coals from

the

the diſtrict of Wedneſbury, for the upply of Birmingham, in which article I would pre-
ſume that the greateſt part, if not the whole of the profit of your undertaking conſiſts.

When it is conſidered how many canal ſchemes have been abortive, how many more
have fallen ſhort of paying common intereſt of the vaſt ſums of money expended upon
them, in proportion to thoſe few that have ſucceeded, ſo as to become profitable ; and
when it is further conſidered, that the few which have ſucceeded, have been ſucceſsful
in conſequence of ſome favourable circumſtances occurring, either wholly unforeſeen, or
not ſeen in the light of the capital articles of profit, which they have afterwards turned
out, we muſt conclude, that were it not that ſome have proved ſucceſsful, the ſpirit of
canal making that a few years ſince even raged amongſt us, would be wholly extinct,
and the public be deprived of the advantages further to be derived from this kind of
enterprize, by which it is already found to be ſo largely benefited. For if in thoſe
ſucceſsful caſes, the proprietors are by an act of legiſlature defeated in the profit
they have ſo dearly earned by a partition, the whole of the real motive for this kind
of enterprize will be totally aboliſhed ; for in caſe it were poſſible in reality to turn out
a ſufficient inducing profit for two, a third has an equal right to thruſt in, and after
that a fourth upon the ſame principle.

A canal has been granted running parallel to a former eſtabliſhed navigation, but
then it was where expedition and cheapneſs of carriage were not only offered, but in-
ſured on terms not even in any degree met by the old eſtabliſhment. A canal, in a
ſimilar ſituation, has alſo been refuſed, though propoſing greater expedition and lower
terms of carriage, but the practicability of materially lowering the carriage, con-
ſiſtently with the neceſſary profit to keep the works alive, not being fully proved ; and
the whole propoſition being met by the old proprietors on reaſonable grounds, it
was judged better fully to eſtabliſh one, than put them both to hazard : and
in both theſe caſes, the water poſſeſſed by the former navigation was held ſacred,
though each of them had a large river for their ſupply. Much more might be ſaid
upon the above heads, and painted in ſtronger colours ; but it comes more within my
province to ſtate the mechanical difficulties that ſeem to me likely to attend the eſtabliſhment of a double navigation in this place.

It is principally to be remarked, what has already been clearly ſhewn, viz. that the
Birmingham canal, not only is now, but of neceſſity muſt be in future, ſupported by
artificial means ; and that this canal, independently of any one article of buſineſs done
upon it, requires ſupporting with a dead ſtock of water, whoſe weekly conſumption, or
waſte, amounts to no leſs than 469 locks-full per week ; and that to do their buſineſs

upon the prefent eftablifhment, they neceffarily require (befides the return of almoft 1000 by the company's engines) 252 more, making in the whole, 721 locks-full of new water; that the whole quantity the Birmingham canal can readily command, does not exceed 883, and that is no more than neceffary to have a r afonable fecurity of having enough : yet, fuppofing us to calculate to a fingle lock-full (a tafk I can by no means engage to perform)

<table>
<tr><td></td><td>Locks-full per week.</td></tr>
<tr><td>The exceedence here will be only - - - - - -</td><td>162</td></tr>
<tr><td>To which, if we add the only water that would turn off from the engines, that need not be taken up at prefent for the fupply of the Birmingham canal, viz.</td><td></td></tr>
<tr><td>Afton and Co., Dall's hole, and Barber's lower engine - - - -</td><td>116</td></tr>
<tr><td>The whole amount of overplus would be no more than ——</td><td>278</td></tr>
</table>

Now, if a canal of equal extent with that of Birmingham, or greater, were executed, we muft expect it to require as much, or more regulating water, and lockage water, in proportion to its bufinefs. It therefore appears, that if after the Birmingham canal is barely fupplied, there is not much above half as much regulating water as it muft reafonably be expected to want; and if the new canal finds means to procure a part of that water which would have been drawn out of the earth, and fed the prefent canal, then there will be two companies depending upon the fame joint ftock of water, and there not being enough for both, the confequence would be, that both would greatly fuffer: for if a new canal be fupplied with 278 locks-full for regulating water, that fhould have 469, there will be a deficiency of 191 locks-full; and if only half of this deficiency were deducted from the old canal, fo as to create a deficiency of ninety-five locks-full per week, the very ill effect that muft neceffarily attend, this will appear from what was experienced in the fummer of 1781, for the want only of feventy locks-full more than they were then fupplied with, had reduced them a foot in depth upon the whole canal, and obliged the boats to go with fuch fhort loadings, that three boats were employed to do the bufinefs ufually done by two; and had they been deficient ninety locks-full per week, the navigation muft have been wholly ftopped for want of depth of water in the canal, notwithftanding the power of the company's engines to return the lockage water.

It further appears, that the new canal muft be in the fame predicament refpecting the being fupported with artificial water, as the prefent one; for it cannot be fuppofed to acquire any more right over the mill waters, without reftitution, than the other; and therefore the out-burft waters of the Broadwater colliery wafte cannot be applied in aid of the new canal without the fame inequality and injuftice as to the prefent one.

one. It may be fuggefted, that as new collieries will be opened in confequence of a
new canal, new waters will be raifed by thofe new collieries towards its fupply ; but as
the confumption of coals at Birmingham, can only be of a limited quantity, the queftion
will be, whether they can be more effectually carried by two canals than by one, that
is, when they are both to fubfift upon the fame original ftock of water ?

A certain produce of coals, or, if I may fo fay, a certain number of collieries, are
fufficient for the fupply of Birmingham : if new collieries are opened upon the new
canal, the fame number lefs will be opened afrefh, or continued upon the old; the
water therefore drawn, that can be brought to fupply the old canal, will be diminifhed
by the quantity brought into the new, and fince the regulating water of the two canals,
will be doubled, and without carrying a coal, will be nearly equal to the quantity that
will effectually ferve the old one in the higheft ftate of bufinefs in fhort water feafons ;
it therefore follows, however paradoxical it may at firft fight appear, that one canal
may do a great quantity of bufinefs with the fame water that would barely fupply
the wafte of two.

It perhaps may be urged by thofe who would favour the new fcheme, from a flight
confideration of my eleventh ftatement, which is intended to fhew, that 1737 locks-
full per week, is the utmoft that can be obtained by an entire fweep and command
of all the waters in the country which can be brought to bear, that is, without paying any regard to the rights and ufes of mills now eftablifhed ; and therefore that
this quantity divided between the two canals, that is, 868 to each, will be almoft as
much as I have recommended at prefent to be taken in, as a full fupply for the now
fubfifting canal, viz. 883, as per fixth and eleventh ftatements, and more than 721,
which I have afcertained to be the prefent neceffary fupply of it ; yet when all the waters
are deducted out of that account, that are either perfectly problematical, or already appropriated to working mills (which I apprehend to be fully as ufeful to the manufactures
of the country as the canal itfelf) the overplus remaining will not exceed 278 locks-
full, as before ftated, even when the Birmingham canal is reduced to the bare fubfift-
ence of 721 locks-full per week.

How very hazardous it muft be to the prefent fubfifting undertaking to be circum-
fcribed in fuch a narrow manner, which, going over many tracts of hollow grounds,
that now contribute to make it take double the quantity of regulating water that it
might be expected to do, and which will every day be increafing, muft appear to
every judicious perfon; and if it be further confidered, that befides the 721 locks-full

afcertained

afcertained as above, and alfo, that there is a neceffity for the company's re-turning engines, which raife no lefs then 980 locks-full, making in the whole 1700 locks-full per week, equal or nearly upon to the full amount of the whole country's produce, it will clearly appear, that if even this whole produce were divided between the two undertakings, that the new one would have an equal neceffity to apply to returning engines, and as thefe would be neceffary in the firft inftance, they would very greatly embarrafs an infant project.

Having gone this length, I conclude, that it muft appear to every judicious perfon who confiders and weighs the above ftatements and premifes, that the part of the country in queftion does not in reality afford the quantity of water whereon to fubfift two canals, even after all artificial means are ufed, with the beft management of the water that nature produces; and furthermore, that which is likely to be fully fufficient for one, will be an apparent ruin to two; nay, even if the fcheme were confined to the fingle purpofe of carrying coals to Fazely, as above pointed out, which, though it would no ways interfere with the prefent company in point of trade, yet as the regulating water that it muft fubfift upon being drawn from the fame fources, would evidently put the recourfes of the prefent canal in obvious and imminent hazard. I muft therefore conclude with fuggefting, that if a canal from the Wednefbury collieries to Fazely be in reality neceffary, it can be done to the beft advantage by carrying forward the new propofed extenfion of the Wednefbury canal, by way of Walfal and Shenton to Fazely, which being a courfe as I apprehend fhorter by more than one half, the quantity of regulating water will be very greatly lefs, and I ap-prehend the additional collieries for this fervice would be more than adequate to the leakage, but till this courfe is furveyed and levelled, I forbear to fpeak further of it.

J. SMEATON.

Austhorpe, 16th December, 1782.

———————————

BUDE HAVEN CANAL.

The REPORT of JOHN SMEATON, engineer, upon the practicability of a canal proposed from Bude Haven to the river Tamar, at Calstock, in the county of Cornwall.

HAVING viewed the country through which it has been propofed to carry this canal, and remarked the leading circumftances relative thereto, I have no doubt but that a canal is, phyfically fpeaking, practicable in the courfe that has been chalked out, and by means of fuch kind of engines and machines as have been contrived and pro-pofed by the ingenious Mr. Edyvean ; but how far it may be eligible and ufeful in point of expenfe to execute the whole, or any part thereof, or in any other different line or mode, which is the fubject of the prefent queftion, can only be known by a careful in-veftigation of the expenfe, when turned in every practical point of view.

The county of Cornwall, in general, feems but ill adapted for the making of canals acrofs the country, being fo very frequently interfected with valleys, that to preferve a level for any confiderable fpace between two given points, it becomes neceffary to go through a vaft meandering courfe, inafmuch that in the prefent cafe, where the diftance between Bude Haven and Calftock is under thirty miles, the courfe chalked out for the canal, fuppofing it to afcend the fmalleft perpendicular between the two feas, at a place, on this account, called the Pack-faddle, will be above ninety miles in length.

The time allotted to look over this extenfive courfe, with its collateral circumftances, being but feven days upon the ground, did not admit of the taking particular furveys, levels, and dimenfions ; and indeed to have done that fully and fatisfactorily, would have taken as many months as I fpent days upon it ; but this, in fact, would have been a wafte of time, becaufe to have gone into accurate admeafurements of fuch fchemes, that muft in all probability be rejected, when at the fame time the merits of the quef-tion are capable of being brought into a narrow compafs, would indeed be fpending time to no purpofe. For this reafon I have contented myfelf with viewing the general line and face of the country, which, affifted by the actual meafures taken, and obferva-tions made by Mr. Edmund Leach, who was defired to attend me upon this view, and who has fince communicated to me, not only the meafures of lengths and propofed di-menfions of the canal, but many other judicious obfervations on thefe grounds, affifted

by

by my own experience, of the probability of what is likely to happen in the execution of such a kind of work in a country so constituted, I have been able to bring together into one view comparative estimates of the diff rent modes that appear to me the most likely to take place in the present case, from whence it may be judged by the well-wishers to, and promoters of this scheme, which is the most likely to be adapted to the capital that is practicable to raise, and which is most likely to pay interest for the capital so advanced, or what is, as I apprehen , the same thing, likely to be, upon the whole, the greatest advantage to the country; and if it shall appear that any of the five modes that I have particularly investigated, shall be eligible to be put in execution, then I would recommend the tract so chosen to be more particularly and carefully surveyed, levelled, and dimensions taken, from whence an estimate may be made of particulars that may be expected to come somewhat near the real expense which will be incurred, and as I apprehend, that, in any of the ways, an amendment of the act of parliament will be required, I have, therefore, no ways confined my views to the limits thereof.

The materials that I have chiefly made use of from the communication of Mr. Leach, are as follows:

1st. The measures of lengths of the different districts or stages, upon the line of the canal, as he measured them, being in the whole, according to his original survey from Bude Haven to Kelly Rock, upon the Tamar, in the parish of Calstock, ninety-one miles.

2d. The dimensions of the canal proposed, viz. twelve feet bottom and four feet deep, digging from three-pence to sixpence, but at an average four-pence per cube yard.

3d. The value of the lands, which he observed to be at various prices, worth from five shillings to forty shillings annual rent, and averaging the quantity of each sort, with the quality, he makes the value of the whole at fifteen shillings per annum; which, at thirty years purchase, will be twenty-two pounds ten shillings per acre.

4th. The quantity of sand and lime likely to be carried annually upon the canal, wherein he observes, that the distance between Bude Haven and Kelly Rock, in a right line, according to Martin's map, is twenty-eight miles, and he supposes that an equivalent surface to that of five miles on each side this direct line, that is, a surface of twenty-eight miles long, and ten miles broad, will receive benefit from this canal; in

consequence,

confequence, 179,200 acres of land will be concerned therein, of which he fuppofes one-twentieth part, that is, 8960 acres yearly, to be broken up for tillage.

He fuppofes further, that one half of this, viz. 4480 acres, to be manured with fand and the other half with lime, and that the value of the carriage of each may be efti-mated upon what it is at Launcefton, which is nearly half-way between the two ex-tremes.

That the ufual quantity of Bude or Widemouth fand, to an acre, is fixty horfe feams, and weighs about eight tons, which cofts in land carriage to Launcefton three pounds, but he allows ten tons to an acre and one-third, which is nearly the fame proportion.

The ufual quantity of lime laid upon an acre is 100 Winchefter bufhels, weighing about $3\frac{1}{3}$ tons, the land-carriage of which to or near Launcefton is one pound fixteen fhillings, this will be ten tons to three acres.

Upon thefe foundations, which are fufficiently diftinct, and of which every gentleman will be enabled to judge, perhaps better than myfelf, I have proceeded to form the comparative eftimates before mentioned, according to five different propofitions for executing a canal, and which are contained in five fchedules, accompanying this report; and as the merit of the queftion is chiefly contained in what arifes from thefe fchedules, what I have further to remark will moft properly come in by way of explanations, ob-fervations, and abridgements of thefe fchedules.

Schedule firft contains the eftimates and refult of the original propofitions for making a canal from Bude Haven to Kelly Rock, by way of the Pack-faddle, and by means of five engines, as propofed by Mr. Edyvean, for transferring the cargoes from the level of one canal to that of another, without the ufe of locks. The whole length from Bude to Tamar, at Kelly Rock, being ninety-one miles, and from Bude to Launcefton forty-nine miles, by the courfe of the canal.

The eftimated capital for the execution of this work is £119,210, and the capital that is likely to be fupported by the tolls upon fand and lime is £145,192, which exceeds the former by no lefs a fum than £25,990; and hence, upon the firft face of it, it would appear to be a practicable fcheme; but it muft be confidered and remarked, that the eftimate of the expenfe, as it ftands, is the leaft that, from the nature of the coun-try, can happen; and that there is no allowance for fome deep cutting through confi-derable

derable rising grounds that occur, notwithstanding so very long a course to avoid them; but, exclusive of these considerations, the exceeding sum being but twenty-two per cent. upon the sum estimated (at the very lowest) to be wanted; this seems inadequate to the risk of a greater cost, or less profits, or both, and also that of keeping the whole in repair; for though the articles of the carriage of coals, culm, wood, timber, iron, lime for building, corn, merchandize, &c. are not taken into the account, yet, if the profit of all these articles put together, will pay for their overseers of the repairs of the works, their clerks, toll-gatherers, &c. it will probably be as much as the tolls upon these articles will produce. As the plan thus circumstanced does not therefore appear to me eligible to the adventurers as it stands, we will now attend to what is stated in the next schedule.

Schedule, No. 2, contains the estimates and result of the same propositions shortened, in point of length, between Bude and Launceston, by carrying the canal over Greena Moor, as pointed out by Mr. Call; which course having viewed, I found the same practicable, and by which means a very long meandering course, up the valley of the river Altry, and several included valleys, will be avoided; and, by which means, according to Mr. Leach, there will be cut off fifteen miles from the length.

In this tract, as Greena Moor lies considerably higher than the ground at the Pack-saddle, an additional lift will be required into the canal of Partition, and another engine to drop down into the former level, or works to avoid them, in all probability, of equivalent charge. I have, therefore, computed the expense on this supposition, for as the lowest land between Treneglofs and Lancast rises still much higher than Greena Moor, I see no probability of a passage that way upon the same level, a circumstance recommended by Mr. Call to be adverted to.

The total length, therefore, by Greena Moor, being reduced to seventy-six miles, and the distance between Bude and Launceston to thirty-four miles.

The estimated capital is reduced to	£102,087 and.
The capital supposed to be supported by the tolls,	146,971
The difference by which the estimated capital is exceeded by the capital supposed to be supported,	44,884

that is, forty-four per cent. upon the estimated cepital.

This

This being double per cent. upon the former, looks much more like bufinefs, but yet, in affairs of fuch great uncertainty, it may be advifeable to look further.

Schedule, No. 3.—Suppofe the northern part of the canal only to be executed, that is from Bude to Launcefton, upon Mr. Leach's original plan,

The total length then being forty-nine miles, the estimated capital will be reduced to	£65,228
And the capital supposed to be supported by tolls, - - - -	83,440
The difference, or excess of the capital supposed to be supported above that estimated, will be - - - - - - - -	£18,212

that is, twenty-eight per cent upon the estimated capital.

Hence it appears that, compared with fchedule, No. 1, the execution of the northern part of the canal, from Bude Haven to Launcefton, is a fafer adventure than that of the whole, and confequently a much better adventure than the execution of the fouthern part from Launcefton to Calftock, but yet it appears that the execution of the whole, upon Mr. Call's plan, will be a better adventure, but a much greater capital, than this; we will, therefore, next fee what will come out from a partial execution of the canal upon Mr. Call's plan.

Schedule, No. 4, contains the ftate of the cafe on fuppofition of executing the northern part of the canal from Bude Haven to Launcefton, by way of Greena Moor, according to Mr. Call's plan.

The total length thereof is thirty-four miles, and the estimated capital thereon will be reduced to - - - - - - - -	£48,114
The capital supposed to be supported by tolls, - - - -	84,490
The difference or excess of the capital supposed to be supported above that estimated, will be	36,376

that is, the exceedings are seventy-six per cent. upon the estimated capital.

We are now arrived at not only a moderate adventure in point of capital, but a great probability of tolls anfwerable to the rifk. Yet, as the great object appeared to be the bringing of Bude fand into the heart of the country, whether to Launcefton or elfe-where; when I came into the valley of the river Tamar from the Pack-faddle, and obferved how gentle its decline was, how eafily practicable for a canal by locks, as far as the eye could reach, in comparifon with the very long meandering courfe that refulted from

VOL. II.   F f   keeping

keeping the level; I was tempted to take a further view of the valley, and found that from the point oppofite the Pack-faddle, down to Grefton bridge, it continues quite open, and very practicable for fuch a canal, and its decline is fo eafy, that for that length, as it appeared to me, a very moderate number of locks would make good the defcent, and carry the veffels three or four miles further fouth than Launcefton, to which place, if thought neceffary, a branch canal may be carried up the valley of the Kenzie river to the bridge between Launcefton and St. Stephen's. Finding this length down to Grefton bridge fo practicable, I was induced to examine the whole of the valley down to Calftock, to fee whether there was any, and what, natural impediment to the mak-ing a navigation to the tides-way at Calftock, in cafe the fame fhould hereafter be found expedient; but I found none to the making of a navigation upon the bed of the river, in the ordinary way of making rivers navigable by means of locks and dams, there being every where flat ground on one fide or the other, one place only ex-cepted, under Hingfton Downs, which continues for fo fhort a length, and the fall fo eafy, that a dam erected below this place would pond the water above it, fo as to con-tinue the navigation through it. Mr. Call alfo pointed out the eligibility or making the gently declining valley of the Bude river navigable by locks, to the foot of the fteep rifing hills; now, this being done from Bude Haven to the foot of Longford Hill, then afcending by engines and canals to a canal of partition at the Pack-faddle, as per No. 1, and then down to the locking canal of the Tamar valley to Grefton bridge; this not only appeared the fhorteft way of bringing the Bude fand into the heart of the country, but feemed to be attended with far lefs rifk and uncertainty of expenfe, as well as likely to be executed for the leaft original capital; and this has induced me to lay before the promoters of a navigable communication from Bude, a ftatement alfo of this matter, follows :—

Schedule 5, contains the refult of a fcheme for executing a canal, partly by locks and partly by engines, up the valley of the Bude river, and from thence acrofs the Pack-faddle into the valley of the river Tamar, and down that valley to Grefton bridge.

The estimated capital, - - - - - -	£46,108
The capital supposed to be supported by tolls, - - - -	94,150
The difference by which the estimated capital is exceeded by the capital supposed to be supported, - - - - - - -	48,042

that is, exceeding by one hundred and four per cent.

Here

Here the capital fuppofed to be fupported, being more than double the capital efti-mated, affords every reafonable affurance of a fuccefsful enterprize, fo that were the exceeding coft and the falling fhort of the tolls, to be very great, yet it is likely that on this principle there will be a fufficiency to pay the intereft. Indeed there is one article of the eftimate of the expenfe, upon this fcheme, that wants fupporting by actual meafures; for, with regard to the level between the point of the Tamar, oppo-fite the Pack-faddle, and the furface of the river at Grefton bridge, I can only judge of it by eftimation; fo that it is as poffible that it may need more than ten locks, as that it may be done with lefs; yet, when it is confidered that this canal takes in about four miles in length more of the interior of the country, to compenfate for four miles next Bude Haven, whofe fand will fcarcely come upon the canal, except to thofe who lie immediately thereon; I fay, this being confidered, and alfo that by going four miles nearer the lime country, a quantity of that article will be brought to the canal by land carriage to go north, and that the expenfe of even three or four locks more, would make no very material enlargement of the eftimated capital; this appears to me to be the moft fecure and eligible fcheme, and if fo thought by the promoters of the canal, the actual taking of this level would put the affair out of doubt.

In regard to the expenfe of continuing the navigation by the bed of the river from Grefton bridge to the tides way at Calftock; for want of levels I can fcarcely guefs at it, but I am fatisfied, that it may be done that way, from Grefton bridge, for lefs money than it can be carried in any other way from Launcefton to Calftock.

If this mode of execution fhould be adopted, I expect that on a re-furvey it may appear practicable to pierce the hill at the Pack-faddle, and to go upon a level from the Tamer into the valley of the Bude, till by one fingle lift or engine this canal may be joined with the locking canal in the Bude valley, and thereby all transferring of the cargoes avoided, except only one; but this muft depend upon a more minute infpection of particulars, the eftimate now made depending upon nothing but what I know the ground will really admit of.

It has been a queftion, whether machines in the manner propofed by Mr. Edyvean, or of any other fimilar conftruction, would do the bufinefs with fufficient difpatch, that is, whether one machine would do at each lift; but on computation from what I have already done in the way of drawing coals by water, I have no doubt but that the cargoes may be changed without the leaft hurry, at the rate of one ton in four minutes; fo that a cargo of twenty tons will be changed in one hour and twenty minutes; and if 1680 cargoes are

carried

carried annually in forty weeks, fix days to the week, that will be feven cargoes per day, which will be changed in nine hours and twenty minutes.

I have only further to remark, that if £2. 16s. 0½d. be allowed to the confumer for carrying a cargo fideways from the canal, if Bude fand is carried from Bude to Launcefton, fourteen miles according to the direct line, exclufive of the crookednefs and inequalities of roads, for £7. 10s. as has been ftated in the fchedules; this £2. 16s. 0½d. would carry it 5¼ miles in a direct line fideways, which makes good the original fuppofition of taking in a fpace of five miles broad on each fide, whereas any lefs allowance to the confumer would not do it.

J. SMEATON.

Austhorpe, 8th Jan. 1778.

SCHEDULE, No. 1.

Probable ESTIMATE for making a canal from Bude Haven to the river Tamar, according to the meafures defigned by Mr. Edmund Leach.

The bottom width twelve feet, and four feet depth, length ninety-one miles.

ESTIMATE of work per mile.

	£	s.	d.
The bottom width being twelve feet, the slopes should be as three to five on each side, this, upon four feet depth, will contain, per mile, 14,600 cube yards, which, at fourpence per yard, will amount to - - - - -	243	6	8
Allow for banking up hollow places, and extra cutting through uneven grounds, hard matter, &c. one half the above, - - - - -	121	13	4
The common width of ground over all, including cut and cover on both sides, for a canal of the above dimensions, will be about seventy-six feet, which will contain 921 acres per mile, which, valued at the mean price of fifteen shillings per acre annual rent, and thirty years purchase, according to Mr. Leach, will be £22 10s. per acre, purchase, and per mile, - - - - -	207	5	0
Allow half the above for extra widths in high banking and deep cutting, passing places, turning places, docks, basons, cranes, engines, temporary damages, and spoil of ground, - - - - -	103	12	6
To six small tunnels at £10. each, and two larger ditto, at £20. each, making in the whole eight subterraneous passages, per mile, - - - -	100	0	0
To one road bridge per mile at £60., and two more for communication between the lands severed, at £30. each, making three passages over the canal, per mile, -	120	0	0
To making towing-paths, back drains, fences, gates, &c. at two shillings and sixpence per yard running, - - - - - -	220	0	0
Neat estimate per mile, - -	1115	17	6

ESTIMATE of the works of Schedule No. 1, applicable to the whole length of the canal.

	£	s.	d.
To ninety-one miles of canal, at £1115 17s. 6d. per mile, - -	101,544	12	6
To five engines for lifting and lowering the cargoes from the level of one canal to another, which, supposing the incidental expense of sinking shafts, and making tunnels, cannot be stated at less than £1000. each, - - - -	5000	0	0
To building a dam, or other equivalent work, across the Bude river, for communicating the canal across the valley and river, - - - -	100	0	0
To three principal aqueducts, tunnels for crossing the rivers Attry, Kenzie, and Inny, at £100. each, - - - - - - - -	300	0	0
To making six overfalls for discharging the waste water in time of great rains, at £40. each, - - - - - - - -	240	0	0
To six sluices for emptying the several stretches of canal, in order to clean, &c. at £30.	180	0	0
To six small shuttles and aqueducts for taking in water from the rivulets, at £10. each,	60	0	0
To six stop-gates for preventing the whole of the water from flowing off in the long reaches, in case of accident, or occasion for a partial cleansing and repair, -	120	0	0
To cutting an aqueduct from Shernock moor weir to the canal of partition near the Pack-saddle, being in length four miles, supposed a five feet bottom, three feet deep, and slopes as one to one, this will contain 14$\frac{2}{3}$ cube yards per rod, say 15, which, at 1$\frac{1}{4}$d. per yard, will be £30. per mile, and for four miles, - - -	120	0	0
Allow half of this for extra cutting and banking, - - - -	60	0	0
The ground occupied by the aqueduct and banks will be half a chain broad, which, in four miles, will contain sixteen acres, which, if valued at the average price of £22 10s. will come to - - - - - -	360	0	0
To a new weir at Shernock moor, and stop-gates for taking water into the aqueduct,	80	0	0
In dry seasons the navigation will be likely to take all the water of the Tamar into the Shernock aqueduct, and thereby deprive the mills of Bridge Rule and Langerton thereof; if we suppose the average damage done to each at £5. per annum, this, at twenty years purchase, will make a capital of - - - -	200	0	0
Neat estimate, -	108,364	12	6
Allow for contingencies, and unforeseen expenses upon the above, at £10. per cent.	10,836	9	3
	119,201	1	9

ESTIMATE

ESTIMATE of the expenfe of freight in bringing twenty tons of fand from Bude Haven to Launcefton by the faid canal, being in length forty-nine miles by its courfe.

	£	s.	d.
To filling the sand into carts or boats in the naven, and carrying it to the foot of the first engine, at 1½d. per ton, - - - - - -	0	2	6
To delivering the sand into the buckets, hoisting the same, and loading and trimming the boat in the canal above, at 1½d. per ton, - - - -	0	2	6
Trackage upon the canal from the first and second engine, at three-pence per mile loaded going, and returning light, at two-pence, that is, five-pence per mile, upon a length of 6½ miles, by the course of the canal, - - - - -	0	2	8½
A trip will easily be made in this reach per day, allowing for stoppages, the boatman at one shilling and sixpence, and the boat or boats to carry twenty tons, at three shillings, together per day four shillings and sixpence, - - - -	0	4	6
Expense of first reach, -	0	12	2½
In the second reach, changing the cargo, as in the former, at 1½d. per ton, -	0	2	6
Trackage in this reach, four miles, at five-pence, - - - -	0	1	8
Two trips may be made in this reach per day, therefore, the man and boat hire will be half a day, at four shillings and sixpence, - - - -	0	2	3
Expense of second reach, - -	0	6	5
In the third reach, changing the cargo, - - - - -	0	2	6
Trackage, 38½ miles, at five-pence, - - - - -	0	16	0½
In this reach it will take three days to make a trip backwards and forwards, reckoning twelve hours to the day, at four shillings and sixpence, - - -	0	13	6
Expense of third reach, - -	1	12	0½
Ditto, of second reach, - -	0	6	5
Ditto, first reach, - -	0	12	2½
Total expense of twenty tons from Bude to Launceston, -	2	10	8
The land-carriage from Bude to Launceston being stated by Mr. Leach at three pounds per eight tons, this, for twenty tons, will cost - - - -	7	10	0
Saved upon one cargo, - -	4	19	4
And if upon this saving we allow half to the consumer for land-carriage from the canal to his situation, and the other half to pay the interest of the capital, there will be for each purpose, - - - - - - - -	2	9	8

Mr.

Mr. Leach suppofes that 8960 acres of land will be manured yearly within the reach of the canal by fand and lime jointly; and suppofing one half, that is, 4480 acres, be manured with fand, and the other with lime, then, as he ftates it, that ten tons of fand is at an average laid upon $1\frac{1}{3}$ acre, there will be wanted for 4480 acres, 33,600 tons annually, amounting to 1680 cargoes of fand annually of twenty tons each.

But 1680 cargoes, upon a profit of £2 9s. 8d. per cargo, amounts to the annual fum of £4172.

Again.—It is faid, that 100 bufhels of lime, weight $3\frac{1}{2}$ tons, does an acre of land at an average; therefore, 4480 acres will require 14,933 tons, equal to $746\frac{2}{3}$, fay 747 cargoes of twenty tons each, the land-carriage of which to Launcefton is ftated at £1 16s. per 100 bufhels, that is, per cargo, £10 16s. Now, if the neat expenfe of freight is ftated to Launcefton at the fame rate as from Bude, viz. £2 10s. 8d., which, though the diftance is lefs, it reafonably may on account of greater rifk in carrying lime than fand, then

	£	s.	d.
The land-carriage being	10	16	0
The water carriage,	2	10	8
There will be a faving per cargo,	8	5	0

And if this be also equally divided between the consumer for land carriage, and greater waste and loss in the commodity, and the other half to the proprietors for payment of interest, there will be a profit, per cargo, of

	£	s.	d.
	4	2	8
And this upon 747 cargoes will amount to per annum	3087	12	0
The profit upon sand, as before stated,	4172	0	0
Total	7259	12	0

	£	s.	d.
Now, if £7259 12s. will pay the interest of a capital of	145,192	0	0
And the estimated capital be	119,201	1	9
The capital maintained will exceed the capital to be raised,	25,990	18	3

that is, the estimated capital is to the exceeding, as one hundred to twenty-two, nearly.

SCHEDULE,

SCHEDULE, No. 2.

State of the cafe on fuppofition of the execution of Mr. Call's propofition of carrying the canal by Greena Moor, by which the courfe of the canal from Bude Haven to Launcefton will be fhortened fifteen miles.

	£	s.	d.
Saved by fifteen miles of canal, at £1115 17s. 6d. per mile,	16,738	2	6
Ditto, by omission of the Shernwick moor aqueduct,	620	0	0
And by damages to Bridge Rule and Langerton mills,	200	0	0
	17,558	2	6
But as by this scheme two more engines or lifts will be wanted, or what will amount to an equivalent expense, from this must be deducted	2000	0	0
Neat savings,	15,558	2	6
Neat estimate per schedule, No. 1.	108,364	12	6
Neat expense per Mr. Call's plan,	92,806	10	0
To which add ten per cent. contingencies, as before	9,230	13	0
The estimated capital,	102,087	3	0

ESTIMATE of the expenfe and freight in bringing twenty tons of fand from Bude Haven to Launcefton, by the canal fo fhortened, being in length thirty-four miles.

		£	s.	d.
Expense of the first reach, as before,	6¼ miles,	0	12	2¼
Ditto, second, as before	4	0	6	5
Ditto, third and fourth, supposed together,	8 as per last,	0	12	10
In the fifth reach, changing the cargo,		0	2	6
Trackage, 15¼ miles, at five-pence,		0	6	5¼
A trip made in two days, man and boat, four shillings and sixpence,		0	9	0
Neat expense of the freight of a cargo,		2	9	5
Land carriage from Bude to Launceston, as before,		7	10	0
Saved upon one cargo,		5	0	7
This being divided equally between the consumer and proprietors will leave to pay interest per cargo,		2	10	3¼

Now,

	£	s.	d.
Now, 1680 cargoes of sand, at £2 10s. 3½d. amount to per annum, - -	4,224	10	0
And 747 cargoes of lime, as before, at £4 2s. 8d. amount to - -	3,087	12	0
Total per annum, - -	7,312	2	0
But £7312 2s. will pay the interest of - - - -	146,971	4	0
The estimated capital, - - - - -	102,087	3	0
The capital maintained will exceed the capital raised, - - -	44,884	1	0

that is, the estimated capital will be to the exceeding as one hundred to forty-four.

SCHEDULE, No. 3.

State of the case upon supposition of executing the northern part of the canal, that is, from Bude Haven to Launceston, upon Mr. Leach's plan.

ESTIMATE of the expense.

	£	s.	d.
To forty-nine miles of canal, at £1115 17s. 6d. per mile, - - -	54,677	17	6
To three engines for three lifts, at £1000. each, - - -	3,000	0	0
To building a dam across Bude river, as before, - - - -	100	0	0
To two large aqueduct tunnels for crossing the Attry and Kenzie waters, at £100. each, - - - - - - - -	200	0	0
To five over-falls, sluices, stop-gates, and shuttles for taking in water, -	500	0	0
To Shernick moor aqueduct, and damage to bridge rule and Langeston mills, -	820	0	0
Neat estimate, - - - -	59,297	7	6
Add ten per cent. contingencies, as before,	5,929	15	9
Estimated capital, - -	65,227	13	3

Now it is supposed, that in this case, the carriage of lime will be lost, but the quantity of sand before specified in schedule, No. 1, will be carried from Bude to Launceston, viz. 1680 cargoes, leaving a profit of £2 9s. 8d. per cargo, amounting to, per annum, - - - - - - -

	£	s.	d.
	4,172	0	0
But £4172. will pay the interest of - - - - -	83,440	0	0
The estimated capital, - - - - -	65,227	13	3
The capital maintained will exceed the capital raised, - - -	18,212	6	9

that is, the estimated capital will be to the exceeding as one hundred to twenty-eight.

VOL. II.   G g   Consequently,

Confequently, the executing of the north end of the canal upon Mr. Leach's plan from Bude to Launcefton, will be a better adventure than executing the whole, and there-fore much better than executing the fouth part from Launcefton to Calftock only.

SCHEDULE, No. 4.

State of the cafe on fuppofition of executing the northern part of the canal from Bude Haven to Launcefton, upon Mr. Call's plan.

ESTIMATE of the expenfe.

	£	s.	d.
To thirty four miles of canal, at £1115 17s. 6d. per mile, - - -	37,939	15	0
To five engines, at £1000. each, - - - - -	5,000	0	0
To dams, aqueducts, over-falls, and sluices, as per estimate, No. 3. - -	800	0	0
Neat estimate, - - - -	43,739	15	0
Add ten per cent. contingencies, as before, -	4,473	19	6
Estimated capital, - - - -	48,113	14	6

Here again it must be supposed that the carriage of lime will be lost, but that the 1680 cargoes of sand will remain to be carried from Bude to Launceston, as spe-cified schedule, No. 1, but which leaving a profit of £2 10s. 3¼d. per cargo, as per schedule, No. 2, this will yield per annum, - - - 4,224 10 0

But £4224 10s. will support a capital of - - - - 84,490 0 0

The estimated capital, - - - - - - 48,113 14 6

The capital maintained will exceed the capital raised - - - 36,376 5 6

that is, the estimated capital will be to the exceeding as one hundred to seventy-six.

Hence it appears alfo, that not only the execution of the north part of the canal will be more advantageous than the whole, but that the fhortening of the length will be a great advantage alfo.

SCHEDULE,

SCHEDULE, No. 5.

Scheme for executing a canal partly by locks, and partly by engines, to go up the valley of the Bude river by locks from the Haven to the propofed engine at Longford hill, from thence acrofs the Pack-faddle by canals and engines, as per Mr. Leach's plan, and then to fall down by an engine into the valley of the Tamar, and down that valley by locks and canals to Grefton bridge.

IT is neceffary to premife, that the cutting a canal in the eafy declining vallies of rivers, being a matter of different confideration from that of cutting canals upon the floping and winding fides of hills; to the latter of which the eftimate for a mile of canal in Schedule, No. 1, is accommodated; a particular eftimate has been made upon the fame dimenfions for the former; and that the advantages and difadvantages fo nearly compenfate one another, that as the difference turned out but a few fhillings per mile, the fame price is taken in the following, as in the former eftimates, that uniformity may be preferved.

ESTIMATE of the expenfe.

	£	s.	d.
To 3½ miles from Bude Haven to Longford hill, at £1115 7s. 6d.	3,905	11	3
To six locks, at £600 per lock,	3,600	0	0
To an engine at Longford hill, according to Mr. Leach's plan,	1,000	0	0
To four miles of canal to the next engine, at £1115 7s. 6d. per mile,	4,463	10	0
To an engine for lifting the goods to the canal of partition, crossing the Pack-saddle,	1,000	0	0
To two miles of canal from the second engine across the Pack-saddle beyond Thorn, where a third engine is to be placed,	2,231	15	0
To the third engine to transfer the cargoes to the level of the Tamar's valley,	1,000	0	0
To 15½ miles of canal from that engine to Greston bridge,	17,295	1	3
To ten locks, at £600 each,	6,000	0	0
To two capital aqueduct bridges for crossing the Attry and Kenzie rivers, at £300 each,	600	0	0
To the Shernick moor aqueduct, and damage to Bridge Rule and Langeston mills,	820	0	0
Neat estimate,	41,916	17	6
Add ten per cent. contingencies as the former,	4,191	13	9
Estimated capital,	46,108	11	3

ESTIMATE

ESTIMATE of the expenfe of the freight of twenty tons of fand.

	£	s.	d.
Bringing the sand to the beginning of the canal, and lading the same on board the vessels, taking it as much as bringing and hoisting by the first machine, as before stated, at three-pence per ton, - - - - - -	0	5	0
Trackage from Bude Haven to Longford hill, 3½ miles, at three-pence, - -	0	1	5½
This distance will allow two trips per day, therefore man and boats, - -	0	2	3
First district from Bude to Longford, - - -	0	8	8½
The second reach as per schedule, No. 1, - -	0	6	5
In the third reach, or canal of partition, changing the cargo into the same, -	0	2	6
Trackage of the canal of partition, two miles, at five-pence, - - -	0	0	10
Changing the cargo at the third machine, down to the valley of Tamar, - -	0	2	6
In this reach three trips may be made per day, which, at four shillings and sixpence boat and man, - - - - - - - -	0	1	6
Third reach, being in the canal of partition, -	0	7	4
In the district of canal from the third machine to Greston bridge, being about 15½ miles, at five-pence, - - - - - - - -	0	6	5½
This will take two days to a trip, therefore boat and man, at four shillings and sixpence, - - - - - - - - -	0	9	0
Expense of the fourth stretch, from the third machine to Greston bridge,	0	15	5½
Canal of partition, or third reach, - - - -	0	7	4
The second reach, - - - - -	0	6	5
The first stretch, - - - - -	0	8	8½
Total expense of freight, - -	1	17	11
Suppose the sand at Greston bridge worth no more than at Launceston, viz. as before,	7	10	0
Saved per cargo, - - -	5	12	1
And suppose this divided as before between the consumer and proprietors, there will be to each, - - - - - - -	2	16	0½
Suppose also the number of cargoes to Greston bridge no greater than if stopped at Launceston, that is, 1680 cargoes, at £2 16s. 0½d. - - -	4707	10	0

But

	£	s.	d
But £4707 10s. will support a capital of - - - - -	94,150	0	0
The estimated capital, - -	46,108	11	3
The capital maintained will exceed the capital raised, - - -	48,041	8	9

that is, the estimated capital will be to the exceeding as one hundred to one hundred and four.

This way, therefore, appears to be the safest adventure of all.

J. SMEATON.

Austhorpe, 8th January, 1778.

KINGSTON

KINGSTON AND EWELL CANAL.

The REPORT of JOHN SMEATON, engineer, concerning the effect that the execution of a canal from the river Thames at or near Kingston to Ewell, in the county of Surry, will have upon the several mills worked by the stream in its course.

FOR effecting a navigable canal through this valley, two schemes have been submitted to my consideration, one of them drawn up by Messrs. Nichols and Broughton, which, in general, keeps the flat ground in the bottom of the valley, the other by Mr. Whitworth, who, in general, keeps the rising grounds in preference to those of the valley; and having viewed the two different lines proposed, and considered in what manner each of them is circumstanced, that we may see how the mills will be affected by the execution of either plan, it is in the first place necessary to ascertain the quantity of current water in the river Ewell, particularly at those times of the year, when the mills are capable of using it all. The time of my view was in the middle of August last, and from my observations then taken, it appears that the river Ewell afforded, according to my calculation, at or about 700 cube feet of water per minute, that is, at the rate of 4987, or nearly upon 5000 hogsheads per hour; but I was informed, that though this was a very dry season, yet the river depending almost wholly upon springs, it was not the time of the shortest water, and judging by the average difference of effect reported respecting the working of the mills in those short-water times, I conclude, that at those times the river does afford full 500 cube feet per minute, or 3562 hogsheads per hour; for though at some very particular times there is still less, yet these happening but seldom, I lay them out of the account.

To determine what part of the water ascertained as above will be needed for the navigation, it is necessary to know the cost that will attend the execution of either plan; that from judging from the quantity of business which must be done to support that capital, we shall be enabled to ascertain what water must be taken to do that business, and that both these plans may be put upon the same grounds, I have made an estimate of the probable expense likely to attend each, in such a way as I should myself direct the execution, both of which I have hereto subjoined.

The probable expense of Mr. Whitworth's line, so estimated, amounts to the sum of £20,236, which is for the performance of the works and purchase of lands, exclusive

clusive of all expenses preceding the application, and in procuring an act of parliament, as well as interest of money, committees, meetings, and law charges, during the procedure of the works. Therefore, to answer not only those contingencies, but the repairs of the navigation after it is opened, the salaries of lock-keepers, clerks, and other necessary agents to be employed therein, it appears to me necessary, that the gross tolls which may be expected to arise therefrom, should be equivalent to the interest of five per cent. of a sum half as large again as the estimated sum, and calling it £20,000, there will be needed tolls to the amount of the interest of £30,000, which is £1500 per annum.

Mr. Whitworth states, that a chaldron of coals is carried from Kingston to Ewell by land-carriage for six shillings, but to give a preference to the water-carriage, and to extend the present sphere of business, this ought to be done for four shillings, which must include freight and tolls.

Now the freight and incidental charges of loading and unloading must be first paid, whether more or less remains for the tolls; I have therefore particularly considered the srticle of freight, in the fort of vessel that seems best accommodated to such a trade, that is, such a vessel as will not only navigate this canal, but take up from, and carry her cargo to London, a vessel of thirteen feet breadth, fifty feet in the tread of the keel, and drawing loaded about three feet four inches water, will carry thirty-six tons neat weight, which, according to my computation, will be rather better than, but say, twenty-five chaldron of coals. And though I would suppose, when bargemen are brought to work one against another, it will be done for considerably less, yet in a case of this kind, I cannot compute upon a less freight and incidental charges from Kingston to Ewell than twenty shillings upon a cargo of twenty-five chaldrons; but twenty-five chaldrons of coals, at four shillings, is five pounds, so that there will remain four pounds per cargo that may be taken for tolls, that is, at the rate of 3s. 2 $\frac{10}{13}d$. per chaldron.

Again, I suppose, that though coals may be the principal article with which barges will go up with full loads, yet of corn, timber, and the produce of the mills, I would suppose them at an average to carry half loads down, that is, eighteen tons of goods, which, if paying the same tonnage as coals, will pay downwards half as much as up, that is to say, forty shillings, which will be as. 2 $\frac{3}{4}d$. per ton, tolls, and with freight at the same rate as coals, amounting to 6 $\frac{1}{2}d$, in the whole, will amount

to

will amount to 2s. 9 ¼d. which will be at the rate of 6 ½d. per quarter upon corn, freight and tolls, to the Thames at Kingston.

A barge paying therefore a toll at an average of £4 up, and £2 down, will pay per trip £6. But £1500, the annual sum to be produced, divided by £6, will make 250 trips of such vessel per annum, to quit the outlay and pay the interest.

Now 250 trips requiring a lock-full up, and another down, there will be annually consumed 500 locks-full, for in the simple case of navigable locks, supposing them all of the same rise, what will work one will work all the rest; these 500 locks-full, averaged upon 365 days, will be as follows:

						Cube feet.
Lockage per day -	-			-	-	8,968
Leakage of locks (supposed two locks-full per day)	-	-	-	-		12,194
Evaporation from the surface of the canal	-	-	-	-		6 314
Soakage into the ground, allow the same	-	-	-	-		6,314
Water expended per day by the navigation	-	-	-	-		33,790

Now seven times this quantity, viz. 236,530 cube-feet, will be the average consumption for a week, whereas the river in its lowest state affording 500 cube feet per minute, will supply in one day 720,000 cube feet, that is, the river in its lowest state supplies above three times as much water in one day, as the navigation will consume in a week; if therefore the mills are supposed to lie still on Sundays, and the Sundays' water be turned into the canal, this will be much more than sufficient for its supply for a week, and upon this construction of a canal I have made my estimate; the lockage and leakage out of the reservoir or pond of the first mill excepted, which amounts to no more than one thirty-fourth part of the river's water in short-water times, and about one forty-eighth part only of the water in summer, such as I found it, which would be recompensed by five pence a trip upon each barge, amounting to £5 4s. 2d. say £5 5 per year, as I shall shew more particularly hereafter, if occasion shall require.

My estimate upon the same prices and co-incident proportions, for the line of Messrs. Nichol and Broughton, amounts to £18,029, calling this £18,000, the sum for which the tolls must pay interest will be £27,000; the interest of which, at five per cent. will be £1350, so that, supposing the number of cargoes to be the same,

as

as by the former, and the tolls the fame alfo, then here will be an exceeding of tolls, above the eftimated out-goings and intereft, of £150 per annum, to make good further contingencies; or it will allow the tolls upon a chaldron of coals to be lowered five-pence, fo that they will then be, tolls and freight, three fhillings and feven-pence per chaldron, inftead of fix fhillings land carriage, and this will occafion a proportionable extenfion of trade and confumption; or otherwife, if the tolls are kept at the fame price, the annual amount thereof will be raifed by five cargoes lefs.

But whatever may be thought proper as to prices, it is certain that this will occafion a lefs confumption of water; for the canal being dug in the bottom of the valley, will be more likely to raife fprings, than take in water by foakage, and therefore that article, amounting to 6314 cube feet per day, which is, in fact, the moft uncertain of the whole, may be laid out of the account.

In cafe the navigation is brought only to the tail of the paper mill, as has been fug-gefted, there will be faved nearly upon either fcheme as follows :

	£	s.	d.
In making the wharf, about	200	0	0
In digging	133	3	8
In land	225	0	0
By twenty feet of lockage, at £80	1600	0	0
By two over-falls and tunnels	40	0	0
By one quarter a mile of towing-path, &c.	55	0	0
Neat estimate	2253	3	8
By ten per cent. contingencies upon the above	225	6	4
To be deducted	2478	10	0

which, deducted from the estimate of Mr. Whitmore's line, leaves — — 17,758 7 8
and, deducted from the estimate of Messrs. Nichol and Broughton, leaves — 15,550 12 4
But then a road to the wharf must be considered.

It has been fuggefted that the navigation may be fupplied by the Sunday's water, the uppermoft lock excepted, and, if this mode be adopted, it will behove the proprietors of mills to have a claufe inferted in the act, to prevent any water from being unneceffarily drawn; and for this purpofe it will be proper, that a couple of men be employed to attend the boats up and down through the locks to fee that no water be mif-fpent; their wages to be paid by the navigation company, but their appointment and difmiffal to

depend upon the occupiers of mills for the time being. It appears practicable, there-
ore, in nature, to supply the navigation without lofs to the mills; but, as every human
purpofe is capable of being defeated, let us fee what damage the mills can fuffer, in cafe
of a different application.

If the lockage only be taken conftantly from the river, which indeed would be convenient, it will deprive the mills of one ninety-fifth part only of their water in the ordinary
fummer's drought, and not more than one fixty-ninth part thereof in the fhort-water
feafon, which lofs may very well be compenfated by the claufes common in this kind
of acts; the millers being allowed to bring up mill-ftones, timber, and materials for the
re-building and repairs of their mills, toll-free. But if the whole expenditure of water,
viz. 33,790 cube feet be immediately to be taken from the river at all times promifcuoufly as wanted, it will take one-thirtieth part of the water in common fummer feafons, and one twenty-firft part of the water in fhort-water times, which, though capable of a recompenfe, yet as even then it would be neceffary to keep a couple of men
to attend the paffage of the veffels through the locks, to fee that a much greater mifapplication of water by negligence of bargemen, &c. than has been calculated did not
enfue; the evil may be almoft as eafily prevented in the whole, (except as to lockage
only), as in part.

It appears to me, therefore, upon a full view of the matter, that under the proper reftrictions, it will be the intereft of the proprietors of mills to encourage this fcheme of a
navigable canal, on fuppofition that each fcheme is equally eligible to the land-owners,
the keeping the low grounds not in the exact line pointed out by Meffrs. Nichols and
Broughton, but fomewhat near the fame, will be more eligible to the proprietors of
mills, not only as coming nearer to them, but by being liable to lefs rifk and mifapplication, as being attended with no foakage water.

The great general queftion, therefore, is reduced to this, refpecting the eligibility of
a navigation, that fince 250 trips, or fomewhat toward the fame, appear to be requifite
to fupport the capital neceffary to make it, whether the confumption of the country will
be likely to require 6000 chaldrons of coals, or their equivalent in other goods, to go up
the river annually, at four fhillings per chaldron, freight and tolls, and 22,000 quarters
of corn at 6½d., freight and tolls, to go down, or their equivalent weight in timber, or
other commodities, to the amount of 4500 tons? This can only be judged of by thofe
well acquainted with the extent of country moft likely to be fupplied from, or bring
commodities to Ewell, to whom I muft refer this point. It being evident that a lefs
quantity

quantity upon the whole will not quit the coft; but if a much greater is fuppofed, then the tolls can be laid lower.

J. SMEATON.

Austhorpe, 29th Jan. 1778.

N. B. In the above calculations, I fuppofed that for one-half of the year, the wafte water of the river, more than what fupplies the mills, will work the navigation, in ordinary courfe, without any particular attention to the lofs of water; that one quarter more the river is affording at the average rate of 700 cube feet per minute, in ordinary droughts in fummer, and that the other quarter may be called the fhort-water feafon, when the river is fuppofed to afford 500 cube feet per minute. It appears alfo to me, that in the two laft-mentioned ftates, the quantity of water will be nearly the fame at Kingfton as at Ewell, having remarked no branch of fupply of any confideration, and that the fmall fprings taken in during this courfe, nearly balance the evaporation. As, therefore, in all probability the barges will not need attending above half the year, and in the whole year there being only 250 trips, this will not be at the average of a trip per day. Two men will, therefore, eafily attend, one to each barge, both up and down, but I do not propofe to fix it to two men; but that, in proportion to the trade, every barge fhall be attended by a man both up and down, and indeed this will, in fact, be much for the company's intereft, as the beft means of preventing damage by the mifufe of their works.

The great exceeding of my eftimate, upon the line of Meffrs. Nichols and Broughton, above their own, will, doubtlefs, attract attention, wherein it is to be noted, that they have not taken the purchafe of lands into their eftimate, and the digging feems not to be valued at above $2\frac{1}{2}d$. per yard. No allowance for hauling tracts, &c. or wharf at the head.

J. S.

ESTIMATE for the Ewell canal, upon the line of Mr. Whitworth.

	£	s.	d.
The bottom being sixteen feet, slopes three to five, five feet deep, will contain 23,793 cube yards per mile, which, at four-pence, will come to £396 11s. per mile, and for five miles one furlong to - - -	2,032	6	$4\frac{1}{2}$
Allow for banking up and extra cutting through uneven grounds, hard matter, puddling, candy, and channelly grounds, making passing places, &c. one-third of the above. - - - - - -	677	8	$9\frac{1}{2}$
Carried forward -	2,709	15	2

According

	£	s.	d.
Brought over - -	2,709	15	2
According to the above dimensions, the ground wanted in the plain cutting will be one and a half chain broad, that is, twelve acres per mile, which, if valued at the average price of £60. an acre, as supposed by Mr. Whitworth, will come to £720. per mile, and for 5⅛ miles to - - - -	3,690	0	9
Allow for extra widths in high banking and deep cutting, passing places, turning places, landing places, temporary damages, and spoil of ground, one quarter of the above, - - - - - - -	922	0	0
To lockage, eighty-six feet, at £80. - - - - - -	6,880	0	0
To making an over-fall and tunnel at each lock, supposed No. 12, at £20. each,	240	0	0
To three turnpike road bridges, at £150. each, - - - -	450	0	0
To seven smaller road bridges, at £90. each, - - - -	630	0	0
To five communication bridges between lands, to make in the whole at the average of three passages per mile, at £50., - - - - -	250	0	0
To making a wharf at the head, as per Mr. Whitworth, - - -	420	0	0
To making an aqueduct over the river below the paper mill, with banking, -	200	0	0
To ditto, near Malden powder mills, with ditto, - -	300	0	0
To extra cutting and impediments, between No. 63, and No. 67, inclusive, -	252	8	2
To extra charge in the Thames lock, lock pit, drainage, foundation, and building,	250	0	0
To making small tunnels, supposed at the average of one per mile, that is, No. 5, at £15., - - - - - - - -	75	0	0
To making towing-paths, back-drains, fences, gates, stiles, &c. at two shillings and six pence per yard running, that is, per mile, £220., and for 5⅛ miles,	1,127	10	0
Neat estimate, - -	18,397	3	4
Add ten per cent. contingencies upon the above, exclusive of all expenses preceding the application, and procuring an act of parliament, as also interest of money, and law charges, &c. during the proceedure of the work, - -	1,839	14	4
Total estimate for works and land, - -	20,336	17	8

ESTIMATE for the Ewell canal, upon the line of Messrs. Nichols and Broughton.

	£	s.	d.
The bottom being sixteen feet, slopes three to five, and five feet deep, will contain 23,793 cube yards per mile, which, at four-pence per yard, will come to £396 11s. per mile, and for five miles one furlong, - - -	2,032	6	4½
Allow for banking up and extra cutting through uneven grounds, hard matter, making passing places, &c. one-eighth of the above, - - -	254	0	9
Carried forward -	2,286	7	1½

Brought

	£	s.	d.
Brought over	2,286	7	$1\frac{1}{2}$

According to the above dimensions, the ground wanted in the plain cutting will be one and a half chain broad, as the other, that is, twelve acres per mile, which, if valued at the average price of £60 per acre, as before, will come to £720. per mile, and for $5\frac{1}{8}$ miles, to   -   -   -   -   -   **3,690 0 0**

Allow for extra widths in high banking and deep cutting, passing places, turning places, landing places, temporary damages, and spoil of ground, one-eighth of the above,   -   -   -   -   -   -   -   **461 5 0**

To lockage, eighty-six feet, at £80.,   -   -   -   -   -   **6,880 0 0**

To making an over-fall and tunnel to each lock, supposed No. 12, at £20.,   -   **240 0 0**

To one turnpike road bridge, and altering Clatton bridge, at £150. each,   -   **300 0 0**

To two road bridges of a lesser kind, at £120. each,   -   -   -   **240 0 0**

To three small road bridges, at £90. each,   -   -   -   -   **270 0 0**

To five communication bridges between lands, at £50. each,   -   -   **250 0 0**

To making a wharf at the head, as before,   -   -   -   -   **420 0 0**

To expenses in crossing the river, supposed three times,   -   -   -   **150 0 0**

To making small tunnels, supposed at an average one per mile, at £15.,   -   **75 0 0**

To making towing-paths, back-drains, fences, gates, stiles, &c. at two shillings and sixpence per yard running, that is, £220. per mile, for $5\frac{1}{8}$ miles   -   -   **1,127 10 0**

| Neat estimate,   -   - | 16,390 | 2 | $1\frac{1}{2}$ |

Add ten per cent. contingencies upon the above, exclusive of all expenses preceding and procuring an act of parliament, as also interest of money, and law charges, &c. during the proceedure of the work,   -   -   -   -   **1,639 0 $2\frac{1}{2}$**

| Total estimate for works and land,   - | 18,029 | 2 | 4 |

J. SMEATON.

Austhorpe, 27th January, 1778.

the

RIVER TYNE CANAL.

The REPORT of JOHN SMEATON, engineer, upon the practicability and probable expense likely to attend the extension of the navigation of the river Tyne, by a canal on the south side of the river, from Stella, by Ryton, towards the grounds opposite Wylam.

HAVING taken a view of the ground through which a canal was propofed to be carried, by Captain Bainbridge, and in ccmpany with him who pointed out the fame, the fcheme appears to me very feafible, without any natural impediment, fave that the natural banks lie fteep upon the river in paffing by Ryton; neverthelefs, by an extra expenfe, which will appear no ways formidable in proportion to the whole, this difficulty may be overcome. This report being founded upon a mere view, without having either meafures or levels, which muft, neverthelefs, be fuppofed, in order to bring the matter to a calculation, the eftimate cannot be thought to be very correct, but yet may ferve, by giving an idea of the probable expenfe that is likely to be incurred, to fhew whether the fcheme is likely to bear; becaufe, if it will bear an expenfe of £6000 or upwards, exclufive of the coft of procuring an act of parliament, and of all plans and furveys, with other charges, previous and preparatory thereto, as well as clear of intereft of money advanced during the performance, then it will be advifeable that an accurate plan and level be taken, from which an eftimate may be made that will be likely to come as near the matter as the nature of this kind of fubjects can poffibly admit; but as this cannot be done without time and accuracy, and, in confequence, at a confiderable expenfe, that will be avoided by this previous eftimate, in cafe the fubject is not likely to bear the expenfe thus afcertained.

The following is an eftimate for a canal of the fmalleft dimenfions I can poffibly recommend to be executed, yet it will carry veffels of twelve feet wide, and drawing $2\frac{1}{2}$ feet water, which, at forty-two feet in the tread of the keel, lighter built, will carry twenty-one tons neat weight, that is, nearly eight chaldrons, Newcaftle meafure: veffels of this kind would, doubtlefs, be able to carry coals down to the harbour of Shields, in moderate weather, fo far as regards the performance of the navigation, but how far they might anfwer in point of ftrength, I cannot properly judge, but apprehend the keels are built of fuch bulk and ftrength as they are, not only on account of fecurity of navigation in bad weather, but to refift the blows they receive againft the fides of the fhips, as I apprehend they are often liable to be fqueezed between them; on this account they are very particularly guarded in point of ftrength, by the breadth of their gunwales.

I am

I am not very certain of the length of canal computed upon, not having had a good opportunity of confulting the beft county map, but if the length be greater or lefs than three miles, as I have fuppofed it, the coft of the work eftimated by the mile, will be in proportion to the length, and the other charges thereon will remain the fame.

Austhorpe, 26th Feb. 1778.   J. SMEATON.

ESTIMATE for making three miles of canal from Stella, by Ryton, towards the grounds oppofite Wylam, on the fouth fide of the river Tyne.

	£	s.	d.
To be three feet deep of water, twelve feet wide at bottom, with sufficient slopes on each side.			
The ordinary digging per mile,	283	17	9
Purchase of lands, at £40 per acre	460	10	0
The chance of bridges, tunnels, towing-paths, back drains, fencing, &c. in that situation	430	0	0
Neat estimate per mile	1174	7	9
To three miles of canal, at £1174, 7s. 9d. per mile	3523	3	3
To three locks, estimated at	1910	0	0
To probable expenses attending the taking in water	100	0	0
The extra expense likely to attend the passing of the steep banks at Ryton, supposing the length one quarter of a mile	176	0	0
Neat estimate	5709	3	3
Allow for contingent expenses, temporary damages upon the above, at the rate of ten per cent.	570	18	4
	6280	1	7

J. SMEATON.

Austhorpe, 26th Feb. 1778.

KANQUARRY CANAL.

The REPORT of JOHN SMEATON, engineer, upon the practicability and utility of making a navigable canal from Kanquarry to or near the new bridge over the river Plym, in the county of Devon.

HAVING, in the month of September laſt, by the deſire of John Parker, eſq. taken a view of the traćt of ground between the valuable ſlate quarry, called Kanquarry and new bridge, to which place the tide reaches, and admits of barges to go up from Plymouth, there is no doubt of the praćticability of the propoſition, both reſpećting the ſituation of the ground and a ſupply of water, nor is there any doubt but that by a canal ſo effećted, the produce of this quarry might be brought down much cheaper than by the preſent mode of land carriage ; the principal or previous queſtion then is whether the buſineſs that is likely to be done upon this canal, be likely to quit the outlay upon it. For this purpoſe I have carefully computed the probable expenſe, without particularly entering into the mode of it, that muſt naturally attend ſuch a ſcheme, and to make a canal of twelve feet bottom, with ſuitable locks to carry down the veſſels into the tide's way, the length being about $2\frac{1}{4}$ miles, and the perpendicular deſcent, about thirty feet, will coſt the eſtimated ſum of £2476, which we may call £2500.

The land carriage I eſtimate at ſixpence per ton per mile, therefore, for $2\frac{1}{4}$ miles, the expenſe will be $13\frac{1}{2}d$. per ton.

The freight of the ſlates per ton, by water, including loading and unloading, I compute at $6\frac{1}{4}d$. per ton, the ſavings therefore on land carriage, ſuppoſing the navigation could be made and upheld for nothing, would be $7\frac{1}{4}d$. per ton.

The intereſt of £2500, at five per cent. per annum, is £125, but I eſtimate that to keep all in repair, and ſupport the principal for ever ; the navigation, beſides paying the above ſum at intereſt, ought to raiſe ſixty-five pounds per annum more, in the whole £190 per annum ; to raiſe which ſum upon a profit of $7\frac{1}{4}d$. per ton, there muſt be navigated 6290 tons annually, all carried more than this will yield a profit of $7\frac{1}{4}d$. per ton, and all leſs will be a loſs to the proprietor of $7\frac{1}{4}d$. for every ton deficient of the above quantity.

Apprehending

Apprehending the above tonnage of flates to be confiderably greater than is, or is likely to be vended from the faid quarry, I have turned my thoughts to what may be done by means of a rail road, fuch as is ufed in coal countries for the land-carriage of coals to the navigable rivers, &c. and I find that to conftruct a waggon rail-road, proper for this fervice, would coft the eftimated fum of £1089, which we may fairly call £1100, the intereft whereof, at five per cent. per annum, is fifty-five pounds, but to keep this in perpetual repair in like manner as the navigation, it ought to raife fifty-five pounds more, that is £110 per annum.

I compute further, that the expenfe of loading down the flates by this road, will coft, loading and unloading included, feven-pence per ton, which, deducted from 13½d. fuppofed the prefent land-carriage, leaves 6½d. per ton to raife £110 per annum, which will require an out-put of 4060 tons; all over will, in like manner as in the navigation, be a profit of 6½d. per ton, and all under fo much annual lofs.

It is further to be obferved, that in cafe the prefent land-carriage of a ton of flates amounts to more than 13½d. then the difference will be in favour of either fcheme; becaufe the difference between the eftimated price, and that of the prefent land-carriage being greater, an out-put of a fmaller number of tons will compenfate the out-lay, and the profit be fooner arrived at.

I have mentioned that a canal of a twelve-feet bottom is as fmall as in my opinion can be executed with advantage; I do not mean that a fmaller canal may not be made for lefs money, but as the extra bankings, cuttings, bridges, and locks, will be nearly the fame, and nothing faved but in the plain cutting in the middle of the canal, when this compared with the want of freedom, and facility of paffage, or the neceffity of ufing fmaller boats, the freights being thereby increafed, more than the proportion of the out-lay is diminifhed, the probability of profit upon the whole will be lefs.

Nor can engines for lowering the materials be ufed to advantage in lieu of locks, in this fituation, unlefs the barges could be brought under the engines, fo that the materials could be lowered into the very veffel, (which, indeed, is very practicable), but the addition of expenfe incurred hereby, would be as great or greater than could be faved from the conftruction of locks, which are certainly the moft commodious mode of navigating yet found out.

J. SMEATON.

Austhorpe, 15th May, 1778.

KNOTTINGLEY LOCK.

The REPORT of JOHN SMEATON, engineer, upon the situation proper to be adopted for a new lock at Knottingley.

THE making of a new cut through the marſh is doubtleſs the moſt eligible mode of doing the buſineſs, but as I underſtand that this is attended with difficulties ſcarcely to be ſurmounted, then it appears to me that the moſt eligible ſcheme, and which, when finiſhed, will nearly amount to the ſame thing, will be to form the cut by artificial means within the trough of the river, by placing the lock lower down upon the peninſula, ſomething near the place propoſed by Mr. Jeſſop; for in regard to the rebuilding the lock in the preſent ſituation, the going into it, and out of it, at the tail below, is ſo extremely crooked, awkward, and inconvenient, that I think nothing but an abſolute neceſſity ought to induce the re-building of it in the ſame place. That it will coſt more in this new ſituation than in the old one is very certain, and to do it completely, conſiderably more than Mr. Jeſſop has eſtimated, but then it will render the navigation more complete; this is a circumſtance, which, with the preſent trade upon the rivers, and particularly at this critical paſſage, ſhould outweigh all ſmall conſiderations: beſides, the compariſon, in point of expenſe, ought not to be made between building the lock in the old, and in the new ſituation, but between building the lock upon the new cut through the marſh with all convenient expenſes, and rent charges thereto incident, and in the new ſituation; and when theſe matters are duly weighed, it would ſeem, that judging upon the ſurface of the thing, the buſineſs may be done as effectually, and upon the whole, at as little expenſe, by adopting the new ſituation as by making a new cut through the marſh. The principal difference will be, that a new lock upon a new cut will not obſtruct the navigation at all, becauſe it may be opened while the other is paſſable, but then, though the obſtruction to the navigation might exceed a week which Mr. Jeſſop has propoſed, yet there is an equal probability of the re-building of the locks exceeding four weeks, ſo that we may fairly lay the account of the lock's re-building where it is, to produce obſtruction of full three weeks more than the preſent, than in the new ſituation, and the former when done, a much worſe lock, becauſe built more in a hurry, and without the neceſſary time for the maſonry to ſettle and harden.

I would therefore propoſe, in order to make the beſt of the new ſituation, to place the lock further down, and nearer the natural deep water than propoſed by Mr. Jeſſop

by

by the full length of a lock, and inſtead of a double wall with a rib of earth between, to build a ſolid aiſler wall, well founded upon piles at the level of the tail-water, of about eight feet thick at baſe, and ſix feet at top, with a ſolid caping of ſingle ſtones croſs and croſs, and well cramped together. This wall, reaching from the old lock-tail to the new lock-head, I ſuppoſe will be the only extra expenſe attending the new ſituation, which, as near as I can compute, will amount to about   per rood, running meaſure, and though we reckon ninety yards length inſtead of fifty, which, calling thirteen roods, the whole will amount to no more than about £   to balance which, on one hand we ſave the value of three weeks ſtoppage, and have the advantage of a baſon that will hold a conſiderable number of veſſels, and on the other we ſave the purchaſe or annual rent of land, part of the digging of the cut, and poſ-ſibly the expenſe of flood-gates at the head of the cut, which, if needed, may here be eaſily applied at the old lock.

As a further improvement and ſafety to the entry of the preſent lock, being too near he dam, I would propoſe to drive about ſix large piles, reaching about three feet tabove common water, and caped with a rail, ſo as to lead in the veſſels in the ſame manner as if the cut had been ſo much longer, theſe being at twenty feet diſtance, will make in the whole one hundred feet extenſion.

It has been ſuggeſted by Mr. Martin, that the having two locks, or ſeparate paſ-ſages, would be very advantageous, particularly below the junction of the two rivers, not only for facilitating the paſſage of the numerous veſſels, but that one lock might be open, while the other was cloſed for repairs; and this circumſtance would doubtleſs ſtill give the preference to a cut through the marſh, if it could be obtained, but as the chief difficulty as I underſtand has laid amongſt the land owners, having right of common upon the marſh, it appears to me, that provided the flood-wheel of Brother-ton's mills were taken away, there would then be room for making a new and ſeparate additional lock and paſſage, parallel to the preſent one, ſtill leaving the other two wheels of Brotherton's mill, that are in common uſed, ſtanding for uſe where they are; and in caſe this ſupplemental lock were firſt conſtructed, then the alteration and re-moval of the preſent lock would be done at leiſure without any ſtoppage, or any other inconvenience than the expenſe of building the ſupplemental lock.

J. SMEATON.

Auſthorpe, 29th Auguſt, 1778.

SOWERBY

SOWERBY BRIDGE CANAL.

The OPINION of John Smeaton, engineer, respecting the best way of supplying the Sowerby Bridge cut with water from Holling's mill.

MY idea always was to bring the water through the high grounds, by means of a tunnel made in the way of an adit or fough, fuch as thofe made for draining collieries, which I would recommend to be done as follows; beginning from fome proper place near the head of the canal, and in the ftraighteft direction, that one or more ftaples or fough pits can be got down, and the matter coming thereout can be difpofed of, to carry on the fame fo as to come out at the day into the holme, that begins on the north fide of the river juft above Sowerby bridge, then to go on by an aqueduct cut along the fkirt of the high ground, till you begin to interfere with the tail race of Holling's mill, then you fet in an adit again, and go under or round the north end of the mill, and fo into the head. This is the general outline of the project which I have formerly communicated; what is now particularly to be mentioned is, that I would begin the tail of the tunnel or adit, fo low that its bottom fhall be at leaft as low as the bottom of the canal, which the defcent to the river there will eafily admit of a trench to clear any water that fhall be raifed in carrying on the tunnel; the bottom of the tunnel to be carried on upon as dead a level as poffible, or to allow for a better current you may begin upon a lower level; but as it will take too great a length of time to carry it on wholly from the tail upwards, I would advife the tunnel to be begun at both ends, and meet in the intermediate way.

The weft end of the tunnel may be begun about three or four feet below the level of Sowerby bridge dam's water, and being furnifhed with a fhort pump to raife the water about fix feet, the drainage water will probably be overcome by pumping half an hour at the beginning of each day, and now and then when wanted, to carry this upon the deadeft level they can, and at the meeting to reconcile the two drifts together, by taking away all fudden turns, humps, and elbows; it is poffible that a confiderable quantity of water may be raifed in this tunnel, which will in part fupply the navigation, and which they can call their own; and it is alfo poffible, that the tunnel may prove fo full of fprings, as to make pumping for the head of it two expenfive; if this fhould be the cafe, its bottom may be carried upon a level about fix inches above Sowerby bridge dam's water, fo as to clear itfelf, and the bottom may eafily be cut down after

the

the meeting, so that the whole drainage will go off by a natural descent at the tail. The adit or tunnel at Holling's mill I would also advise to be begun by another set of men at the same time, beginning from the tail, about a foot above the common tail water of Holling's mill, and running in upon a dead level; in the face of the bank, just above the mill, to put down a staple or sough pit, and to carry up the drift to this pit; in this pit you must establish a good clough or sluice of about two feet wide, and when the whole is completed and secured, cut away from the pit into the mill head.

While the above is doing, another set of hands may be making the opencast, which will be best if its bottom is inclined, so as to reconcile itself with the tail of the mill tunnel, and the head of Sowerby bridge tunnel, but if the ground will not readily admit of this, its bottom may be kept upon a higher level; however, no part of it should be upon a higher level than four feet below Holling's mill dam-head. When the tunnel and aqueduct are completed, they may be joined by banking up the tail drain of the mill tunnel, and the water, in rising to its level, will find its way down the aqueduct, though its bottom should be higher than that of the tunnel; and a small shuttle or sluice being established upon the tail drain of Sowerby bridge tunnel, the water will be forced into the canal by a proper aqueduct, either open or covered, that must be prepared to give it passage, and by having a shuttle upon this last passage, the water may be let out of the tunnel at any time, in order to repair or alter any thing without letting off the canal.

I covet to lay the bottom of the tunnels so much lower than might seem necessary, because, as it can be done at the same expense, the pumping excepted, it will avoid all extra expense in making them larger than they must unavoidably make them to work therein; for by this means the tunnels being in a manner filled with water, they will vent a great quantity upon a small difference of level from first to last, the water always tending to rise to its own level; whereas, were the bottom made even with the surface of the canal, the water running then upon the bottom will require not only a considerable width to carry a quantity, but there will be a good deal of level lost to make it run, insomuch, that the whole would scarcely be sufficient. I also covet the water ways to be thus ample, not that I suppose any considerable quantity will be wanted in ordinary from Holling's mill, but that the canal on any occasion of emptying it may be again filled in a short time upon an emergence.

A sluice will be proper at the head of the bridge tunnel to prevent the floods from making a passage through it, and the aqueduct having its bottom made about four feet
wide

wide, with proper flopes, may afterwards be walled at three feet, or covered at two and a half feet wide, but I think it will be beft to leave the aqueduct open at firft, till it is feen how it acts, and that if any amendment is neceffary (which can eafily be done while open) it will be beft done after trial.

If any part of the tunnels want walling, I would by all means have them made fufficient for a man to get through, and the ftaples, one at leaft in the middle, to be walled, and properly covered, fo as to be ufed on occafions.

J. SMEATON.

Austhorpe, 14th January, 1772.

QUERIES

DUBLIN GRAND CANAL.

QUERIES from the company of undertakers of the Grand Canal, as amended and approved the 3d day of August, 1773, and ordered to be laid before JOHN SMEATON, esq. engineer, and F. R. S.

YOUR approved integrity and abilities as an engineer, induced the company to requeſt you to come to Ireland and inſpect the line and tracts propoſed for the intended navigation of the Grand Canal from the city of Dublin to the river Shannon, and alſo for the collateral canals to the Boyne and Barrow, with ſo much of the river Barrow as lies between Portarlington and Athy, and perſonally to view the country adjoining to theſe tracts, and to report in writing the lines you would recommend, and the plans according to which you would adviſe the undertakers to proceed in making them navigable, and to adviſe what you think neceſſary, not only with regard to the conſtruction and dimenſions of the locks and navigation, but alſo the mode of conducting and carrying the ſame into execution, in the moſt effectual and expeditious manner, for the common good of the kingdom, and the intereſt of the undertakers; and though we are confident, that in ſo doing, nothing material to our intereſt, or for the general improvement of this kingdom, relative to this ſubject, that can occur, will eſcape your notice, or paſs without being fully repreſented to us, yet we are deſirous (among other matters which vou may think neceſſary) of having the following queſtions particularly conſidered, and as fully anſwered as the nature of them will admit.

QUERIES.

1ſt. How and where would you adviſe that the canal ſhould terminate next the city of Dublin? Here you are to take into conſideration the different kinds of merchandiſe and manufactures which are likely to paſs by the canal, and, whether for ſuch goods it will be neceſſary for the company to erect any, and what kind of wharfs, warehouſes, toll-houſes, or other offices, contiguous to the termination, and where, and what avenues and paſſages may conveniently be opened from thence to the different parts of the city?

2d. Do you think it would be of general advantage to make a navigable communication between the canal and the harbour of Dublin, ſo that goods paſſing by the canal (particularly ſuch as are imported, or to be exported) may be ſubjected to as

little

little land carriage as poffible ? if fo, you are to fay where, and of what dimenfions, and in what manner you would recommend fuch a communication to be, and to confider the quantity of water that will be neceffary for its fupply. Several tracts have been propofed for the communication, between the canal and the harbour of Dublin; in particular, one, nearly in the fame direction as the new environ road, one to pafs the city bafon, and to crofs Thomas-ftreet, near Crane-lane, and to terminate near lord Moira's houfe, and one to crofs James's-ftreet, to enter the Liffey oppofite the barracks; the others will on enquiry be pointed out to you.

3d. Is there any part of the prefent cut for the canal (between the city of Dublin, and Sallin's-bridge, near the Liffey) from which you would now advife a variation ?

4th. There are three places pointed out as proper for croffing the river Liffey, the firft near Chain-bridge, by which means it is faid, the deep finking on the high ground, or hill of Downings, would be avoided, and a very confiderable expenfe faved thereby to the undertakers. The fecond is at Waterftown Ford. The third the prefent line. You are therefore not only to examine them, but alfo the river up and down ftream, and to recommend the place which fhall appear to you moft for the company's advantage. You are likewife to fay how the canal fhould pafs that river, whether by a dam or penlock, or by an aqueduct bridge; if by the latter, give the dimenfions of it, and what additional works may be requifite there.

5th. There are three ways propofed for avoiding the high ground of Downings, befide that mentioned in the foregoing queftion: Firft, by carrying the canal from Sallin's to Carrah-bridge, and there croffing the river. Secondly, by continuing the canal further up the eaft fide of the river, to crofs at Gammonftown, and through the bog of Donore, to the Blackwood river. Thirdly, by departing from the prefent line at Aughpadin, and paffing by Killebegges and Blackwood. You are requefted to be particular in examining thofe three paffages, and to give your opinion whether it would be moft for the advantage of the company to adopt one of them, or to carry the canal through Landenftown, Downings, and Graig, in its prefent direction. You are to obferve that the hill of Downings is a clayey lime-ftone gravel, and to fay whether any, and what ufe may be made of it on the bog of Allen ?

6th. If you think it expedient that the canal fhould be continued in the prefent line to the Blackwood river, would you advife finking it through Landenftown, Downings, and Graig, to the fame level as at Sallin's-bridge, as by that means fome lockage

would

would be faved, the canal of partition made much longer, and the Morrell, with part of the river Liffey, brought to the point of partage?

7th. Whether you would recommend carrying the canal in a tunnel or arch through the hill of Downings, or making it an open navigation?

8th. You are to confider the nature of the bog of Allen, and fay, whether you think a permanent canal can be made through it, and the methods you would recommend to effect it?

9th. You are to examine the line (as marked out) from Downings to Lullymore, and fay, whether the canal can be carried in that tract on one level quite through the bog, or how many different levels will be neceffary?

10th. You are to examine whether the low part of the bog, between the Cufh and the Figuile rivers may not be avoided, and the canal of partition carried feveral miles farther than in the prefent line, by paffing from the wood of Allen towards Edenderry, and north of river Lyons, to or near Philipftown, and alfo, whether a more favourable paffage to the river Shannon cannot be found, than through the high grounds of Knockvalleybay, and by the Maiden and Brufna rivers, and whether you would advife this junction by a canal, or clearing the Brufna where it is not navigable?

11th. There are three tracts propofed for the cut of communication between the canal and the river Barrow. Firft, from Cardiffstown or Sallins by Naas, in nearly a right line to Athy. Secondly, along the Blackwood river to enter the Barrow at Monftereven. Thirdly, by a cut from the canal between the Cufh and Figuile rivers, to enter the Barrow near Portarlington, you are therefore to fay, which of thefe you would recommend, or what other, and mention the depth and dimenfions you would recommend for this collateral canal and its locks, and for boats of what burthen?

12th. There are two tracts propofed for the collateral branch to the river Boyne, one from the canal near Lullymore to Edenderry, the other by the Black Water, give your opinion which of thefe the company ought to adopt, or what other, and mention the depth and dimenfions you would recommend for this collateral canal, and its locks, and for boats of what burthen?

13th. What harbours, wharfs, and warehouses do you think neceffary at the great roads where they are interfected by the canal, and where it is probable, goods may be loaded or unloaded ?

14th. Do you think lock-houfes are neceffary, mention their ufes ?

15th. What probable quantity of water will be neceffary for the fupply of the canal, what depth and dimenfions would you advife the canal to be made, and what burthened boats would you recommend ?

16th. Do you think the Blackwood river fufficient for the fupply of the canal, if not, you are to fay from whence you would propofe drawing the greateft neceffary quantity, and at what probable expenfe ?

17th. If you think part of the river Liffey neceffary, would you advife taking it up at Gammonftown, and to enter the canal at Sallins, or to draw it from New Bridge round Rafberry and the ifland of Allen to the canal of partition, at the Blackwood river, or where elfe ?

18th. If you think the Morell river neceffary, pleafe to fay how, and where you would propofe that it fhould enter the canal ?

19th. Will there be refervoirs neceffary, if fo, where, and of what dimenfions ?

20th. Do you think it the intereft of the company to begin the works at, or weft of Downings, before thofe from Dublin to Aughpadin fhall be finifhed, and be fo good to advife the particular parts of the navigation you would recommend to be firft executed ?

21ft. Is it neceffary to have convenient places on the canal during its execution wherein implements, &c. may be fafely depofited, or do you think the company's general ftores in Dublin entirely fufficient and convenient for that purpofe; may not the Dublin ftores be fixed on the line?

22d. After receiving full information and fatisfaction, relative to the materials which the company can conveniently procure for the locks and other works, you are requefted to give your opinion of them, and in particular, to fay, whether building them of

ftone

ftone fquared or dreffed with the hammer (copeing quoins, cove and other particular ftones excepted) would not prove fufficiently permanent, and more to the advantage of the company, than by building entirely with cut or chifeled ftone.

23d. You are to give your opinion of the permanency and conftruction of the locks now building for the company, near the city bafon, by Mr. Trail, as well as thofe built by Mr. Omer, at Clonaughlefs, Ballykealy, and Clonburrowes.

24th. In giving your opinion upon the above queries, you will confider and declare your fentiments, whether the parts of the line already cut, and the works done thereon, are of fuch ufe and benefit, that it will be moft proper and advantageous to go on and continue them, or in any and what places, and how deviate from them?

25th. If there occur to you any convenient places on the line for mills, bleach-yards, or any other, and what benefits that may arife to the company from the canal water banks, land, &c. be fo good to mention them, and alfo if you would recommend a turnpike road to be made on the banks of the canal, and in what manner? note, the company by their act of incorporation have a right to erect turnpikes.

26th. There is a contract between the city of Dublin and the company, that the city bafon fhould be amply fupplied with water for the ufe of the inhabitants: how and in what manner would you recommend the company to provide for that fupply of pipe water?

27th. It is fubmitted to you, whether for the reafons that will be mentioned to you, inftead of endeavouring to avoid the bog of Allen, it will or will not be much for the advantage of the company to carry the line as far through it as can be done with pro- priety, and whether the draining and improving that great wafte by means of the canal running through it, will not alfo be a matter of great national utility?

The

The REPORT of JOHN SMEATON, engineer, upon his view of the country through which the Grand Canal is proposed to pass, and in answer to several matters contained in the queries of the company of undertakers of the Grand Canal, as agreed to on the 3d of August, 1773.

(1.) HAVING carefully perused the matter contained in the queries, which the company of undertakers of the Grand Canal have done me the honour to lay before me, and having examined the face of the country through which the lines of this very extensive scheme of navigation are proposed to pass; a scheme so extensive, that it has taken fourteen days merely to go over the ground; and though it cannot possibly be expected, that in so small a space of time I can give a positive and distinct answer, to all the several matters contained in the said queries, which, in fact, comprehend the modelling of the whole of this great project, which I could scarcely undertake to perform with accuracy, and upon my own knowledge, in twelve months of uninterrupted attention; yet from the view that I have taken, and the assistance that I have received, from the several plans, sections and papers that have been put into my hands, relative to the affair, as well as from the personal attendance and information of Mr. Trail, the company's engineer, I find myself enabled to pass some general judgment upon the outlines and leading marks of the great design before me; and as I doubt not, but that the company will be desirous of being acquainted with my sentiments as early as possible, I will endeavour to throw together the principal matters that have occurred to me, in as short a manner as possible; and as I find I can do it in a shorter way, by a consideration of the whole matter before me, than by pursuing the queries in the particular order in which they are stated, I beg leave to do it in this way, for the present, reserving a more particular discussion of the matter contained in the queries, to a future report, to be made more at leisure, and upon such further surveys and enquiries, as may be thought necessary by the company, in consequence of what is suggested in this present report.

(2.) In the first place I beg leave to observe, that the execution of a scheme of inland navigation, is of all others the most embarrassed with obstacles and difficulties, arising from a difference of interest, property, and opinion, which increase in proportion as the design is more extensive. I perhaps shall not be wrong in saying, that no very extensive enterprise of this kind has yet been brought to a completion, even where opposition of interest and opinion may be supposed in a great measure out of

the

the queſtion.—The Czar Peter, after a very great expenſe, and carrying on his works to a great length, was obliged to lay aſide the execution of his great project of a canal from the Volga to the Don, and which has never ſince been reſumed: and the canal of Languedoc, though driven tnrough by Louis XIV. at the expenſe of above £600,000 ſterling, was at laſt terminated, though a canal of ſix feet water, in a river that in dry ſeaſons in ſummer for a great extent has not above three feet of water; and to remove which defect, it has been computed by the French engineers, it will coſt a further ſum amounting to £50,000, ſo that even this canal may be ſaid to be incomplete as to execution; and it is very certain, that as to revenue over and above repairs, it is ſo ſmall that it bears no proportion to the intereſt of the capital expended.

(3.) The greateſt fault of the plan now before me, is, that (as it appears to me) it is too extenſive; I cannot help therefore moſt earneſtly adviſing the company, to con-fine their propoſition for the preſent, to ſome one point of view; and to take that which is moſt immediately and moſt certainly likely to turn to advantage; provided that it be carried on in ſuch a way as that the firſt plan may be capable of extenſion; for one propoſition carried through to ſatisfaction, is the moſt likely way to produce a further extenſion, and gradually to get the whole into execution.

(4.) From every thing I have ſeen, heard, and read, concerning the trade that is likely to be carried on upon this canal, I am of opinion, that flat-bottomed veſſels from thirteen to fourteen feet wide, fifty-ſix feet extreme length, ſtem and ſtern poſt, and drawing from three and a half to four feet water will anſwer every purpoſe:—ſuch veſſels, even if decked, and not made unneceſſarily heavy, will carry thirty-three tons neat weight.—I chooſe veſſels of this conſtruction rather than thoſe more narrow and long, becauſe they are better adapted to ſail in the tide ways, and in broad lakes, whenever the canal ſhall be brought down to the Liffey at Dublin, or extended to the Shannon, which I think are points always to be kept in view:—open boats, lighter built, of the above dimenſions will carry above forty tons; and it ſeems to me, that veſſels not exceeding theſe ſizes will anſwer the purpoſes of the trade of the country better than larger, becauſe they will be leſs detained in getting a loading to or from any particular place, will make more quick returns, and be more eaſily managed: it is indeed not poſſible for me in a ſmall ſpace of time to have a comprehenſive view of the trade, manufactures, and connections of a country; but this I know, that the artificial navigation, which of all thoſe in England has turned out the greateſt profit to the undertakers, and does the moſt buſineſs, (which is that of the Air and Calder in Yorkſhire) is carried on in veſſels not exceeding the above dimenſions.

(5.) The

(5.) The fize of the largeft veffels being fixed, the fize of the canal and locks will be determined thereby, viz. the locks to be fourteen feet wide, fixty feet long in the clear pool, and four feet one inch of water over the threfholds.—A canal for fuch veffels where there is no extra digging or hardnefs in the foil; in order to enable the veffels to pafs one another, and move freely, fhould be dug to a twenty-four feet bottom, with flopes or batters as three to four, that is, to fall back four feet, in three feet perpendicular; and to be dug to four feet nine inches depth of water, when there is four feet one inch over the lock threfholds : but where there is extra cutting, hard matter, or rock, the bottom may be reduced to a lefs width, that is, to twenty, eighteen, or even fixteen feet; but then if thofe contractions are extenfive, paffing places fhould be made in the moft favourable parts; and the depth may in rock be reduced to a little more than the lock threfholds.

(6.) The grand canal, as it has already been dug, is in general not wider than necef-fary for veffels of the dimenfions fpecified; but on the contrary, in many places is more contracted than I fhould have laid it out. I do not find, however, any place too narrow to admit of the paffage of fuch veffels; and though it may be proper to widen fome of them, yet I would recommend to make as few alterations as poffible in the prefent works. This reduction propofed in the draft of water, from what was originally propofed, viz. five feet three inches over the lock threfholds, and fix feet in other places, will, I apprehend, make a confiderable difference in the expenfe, not only of finifhing the works that have been begun, but thofe that are to be done; as I have found that more expenfe attends the making and maintaining depth of water than width.

(7.) In the next place I beg leave to acquaint the company, that I have viewed all the courfes pointed out in the feveral queries, for obtaining a paffage from the eaft country into the bog of Allen; but as I obferve that a continued chain of high ground, ftretches itfelf on the weft fide the Liffey from north to fouth, they differ only in being more or lefs high, more or lefs broad, or more or lefs round about;—had the courfe of the canal been undetermined, no money laid out, or lands purchafed upon any of thefe paffages, I might have hefitated in determining my choice upon ocular infpec-tion; and might have caufed furveys and fections to be made of others, as well as the prefent courfe through the high grounds at Downings; but when I confider that what cutting has been made there will be in aid of what is to be done, together with the grounds purchafed and paid for; I fee no way of avoiding this paffage now, but what will in effect be the changing a lefs difficulty for a greater: it feems therefore moft eligible, to make the beft of this paffage: but if any gentleman that has

pointed

pointed out any of the several ways mentioned, to avoid the deep cutting at the hih of Downings is unsatisfied by this opinion, which is the best I can form upon ocular inspection only, I am now from this view enabled to direct such surveys and sections to be made, as are necessary to determine the affair upon scientific principles.

(8.) Finding, previous to my view, the passage at Downings every where mentioned as a hill, it seemed extraordinary that any one should think of going through a hill, if they could go round it; but as I have already mentioned, here is a chain of rising grounds, and this is the lowest neck between two of the more elevated rises, that is most directly in the line, and shortest; and I am not sure that it is not the very lowest and most eligible, even if nothing had been done. The deep sinking in the rocks at Gollerstown, I think might have been in a great measure avoided; but as this difficulty is in great part overcome, it does not seem now proper to depart from it; I cannot, therefore, now advise any part of the line to be varied from the city of Dublin to the entrance of the bog of Allen. As I look upon bogs to be an uncertain soil to form canals in, and to be avoided as much as possible, this, together with the great extent of it, between the hill of Downings and Philipstown, and the irregular surface thereof, greatly enhances the difficulty; and adding to this the low bog between the Cushaling and Figuile rivers, which will either occasion two summits of partition, or a great increase of deep sinking at Nockballyboy (being a summit west of Philipstown, and which I see no adequate means of avoiding), I say that those difficulties put together, are so great objections to the present line from the Togher of Graig to Philipstown, that I can by no means recommend a canal to be executed thereupon.

(9.) In consequence of what is suggested in the 10th query, I have reconnoitered the country north of the present proposed line from the Togher of Graig to Philipstown, and as closely as the season of the year, the wetness of the bogs and time would permit; and it seems to me that a line more favorable than the present proposed one may be found by traversing from point to point upon the skirt of the high ground, called the Wood of Allen, and from thence passing that part of the bog near the head of the Boyne towards Edenderry, falling in with the island of firm ground called Derryfacod; and from thence either through the hollow on the north side of the hill of Ballykillin; or, if that shall occasion too much deep cutting round the skirt on the south and south-west side of the said hill, to the north side of river Lyon's house, and from thence keeping nearly parallel to the Figuile river to Philipstown, by which means the very soft bog on the south of river Lyon's house will be avoided. This course it will be proper to have particularly examined by the level, and as it will take up some time, it should fhould be at a favorable feafon of the year, when the bogs are more dry; a fection and plan to be made thereof, which, from the view that I have had of the country, I fhall not only be enabled to judge of when done, but to give fome more particular directions for the doing thereof when thought proper by the company. According to this idea, the canal of partition, if found favorable this way, will extend from the eaft fide of the deep cutting at Downings to the weft fide of the deep cutting at Nockballyboy, which will probably be a courfe of about twenty miles; and in this diftrict all the difficulty lies; for, after getting through the neck of the laft-mentioned fummit, the ground lies as favorably as poffible to Tullimore, and from thence by the valley in which the Maiden and Brufna rivers have their courfe to the Shannon. Thofe rivers at the time of my view overflowed the valley in many places, fo as to make it impracticable to purfue them clofely; but as the valleys in which they lie are in general open, with flat grounds on each fide, there cannot occur any confiderable difficulty; the principal will be in cutting through a fummit a little above Firbane to avoid the mill-race and eel weirs about that place. After the Maiden joins the Brufna, it appears by the plan and foundings of the river taken by Mr. Trail, that long ftretches of deep water occur therein; it feems, therefore, a faving of expenfe where thefe occur, to make ufe of the river; but where mills, weirs, fhoals, and other impediments to navigation occur, that cannot eafily, or with certainty, be removed, there the beft and often the only practicable way is to cut a canal with locks upon it, by them; and if the prefent propofition of reducing the navigation to boats drawing no more than four feet water, be adopted, it will occur, that the length of feveral of the canals may be fhortened, and thereby expenfes faved.

(10.) Refpecting the junction of the river Barrow, after the canal is carried to the Togher of Graig, as I have already advifed a new line to be furveyed more to the north than the prefent, the branch for the Barrow will moft naturally here divide, and go by the tract of the river Blackwood to the Barrow, in which fpace there feems not the leaft impediment to the paffage of a canal, or any extra expenfe likely to attend it, except in paffing the town, and fome ground a little too high at Rathangan; and as the whole courfe of the Barrow from Portarlington to Athy lies very flat, the whole may be connected by navigable paffages at a very moderate expenfe, in proportion to the extent of country, whofe produce will be hereby brought to communicate; fo that if it be true that as great a trade will be carried on with Dublin from the Barrow as from the Shannon, I cannot hefitate in advifing the company firft to complete a navigation to the Barrow before they begin to work from the Togher of Graig towards the Shannon, and that for thefe reafons:

(11.) As

(11.) As the junction with the Barrow will at any rate coft much lefs than the junction with the Shannon, the company will be able to make a much better dividend of profits upon their outlay than if they go to the Shannon.

(12.) If the dividend is good, this will of all others be the greateft inducement to go afterwards to the Shannon, if bad, it will prevent the company from embarking in a fcheme that would turn out to their detriment.

(13.) If the company have money to execute one, and not both, that which will turn out moft profitable to themfelves, will be fo to the public.

(14.) If the company have money to execute the lefs and not the greater; the executed fcheme will make a good return, and be of great ufe; but the unexecuted fcheme will make very little, if any, return, though it may be of fome fmall ufe.

(15.) I doubt not but that with the fum of money that the company can rely on, they may complete their works to Athy; but yet I would advife the ftricteft economy that a large company can poffibly ufe, to be put in practice; and then they may have fomething to fpare towards the Shannon; but I cannot undertake to fay as much, if they proceed firft for the Shannon, even if the line now propofed fhould turn out favorable.

(16.) On the head of economy I cannot help taking notice, that there is nothing fo conducive thereto as unanimity. I have had occafion to obferve, that the worft way of doing a thing, if carried on with unanimity, will often be attended with lefs expenfes than the beft, if the execution is diftracted by a variety of opinions; perfect unanimity in a company, confifting of many members, is not to be expected long together; but it would have nearly the fame effect, if it were poffible for the minority to think themfelves bound by the opinion of the majority. I afk pardon for launching out upon a fubject not fo immediately a part of my profeffion, but as the company defire me in general terms, to point out whatever may be for their benefit and advantage, as well as thofe of the public, in carrying this grand undertaking into execution, I could not help feizing the opportunity of pointing out what is of the greateft confequence of all others.

(17.) It feems to me, that the firft thing to be aimed at is to complete the navigation from its prefent termination, near the city bafon, to Sallin's bridge; and a warehoufe of a moderate fize being built at each end, it may be expected that the greateft part of the goods that now go by land from the fouth-weftern parts of the kingdom, would come upon the canal, and bring fomething in ; and probably fome fpecies would come that do not now come at all, of which turf may be one. A confiderable warehoufe will be wanted in time at the Dublin termination, but warehoufes may be added as they are wanted.

(18.) I have viewed all the ways propofed for joining the canal with the Liffey at Dublin, and all are practicable, of which I doubt not but that by the environ road will be the moft expenfive ; but were it leaft fo, I think it moft liable to objection, becaufe the navigation by the other routs being carried by the Liffey through the heart of the city, thereby all the warehoufes, upon the very extenfive quays on each fide of the river, would become fo many warehoufes for the navigation, fo that none would be neceffary on the company's account. Of the others ; that by croffing James's-ftreet, to enter the Liffey oppofite the barracks, feems, at prefent, the leaft encumbered and the moft direct courfe : but as the going down to the Liffey will, at any rate, be attended with a great expenfe, and as the feveral avenues to the town will be connected by the environ road, and the prefent termination of the canal, I am of opinion that the joining of the Liffey fhould be the laft thing done, and not till the circumftances of the trade of the Grand Canal fhall, from actual experience, fhew it to be neceffary.

(19.) I fhould hardly propofe the moving weftward, till the navigation to Sallin's bridge was completed ; but as the cutting the neck of the hill of Downings, (or the making good any other paffage through that chain of hills), will take up a confiderable time to do it with proper economy, I would advife to recommence the work there; and I believe that not only boats at each end of the deep cut, but machinery, might be employed in the middle to good advantage. I believe it would be right to keep a feparate account of the work weft of Liffey, by which means the company will be enabled to fee what expenfe it cofts to complete the works, delivered over to them between the city bafon and Salin's bridge.

(20.) The mafonry of the locks now building by Mr. Trail, near the city of Dublin, appears to me to be done in the moft permanent and fubftantial manner ; I cannot fay the fame thing of thofe at Clonaughlefs, Ballykealy, and Clonboroughs ; but I do not think the neceffary degree of firmnefs in the laft-mentioned locks, fo much deficient on account of the fort of mafonry made ufe of, as the want of proper method and proportion. The fort of marble wherewith all the locks are built, I perceive is very expenfive to

chifel

chifel fair; otherwife I fhould not recommend the ufe of any thing but chifeled ftone in the face of the work: I perceive, however, that the beft fpecies of hammer dreffed work will ftand very well, efpecially if fet with good cement, which, indeed, is the very effence of good mafonry. I faw a lock upon the Shannon, about three miles and a half below Banagher, which has been in ufe eighteen years, is built with the fame kind of ftones, hammer dreffed, and has ftood very well; I underftand it was built by Mr. Omer; it is in the fame ftile of work as thofe at Ballykealy, &c. but is only eighteen feet and a half wide, whereas thofe at Ballykealy, &c. are twenty. The lime of this country feems very good, particularly for dry work; the lime ufed in the Shannon lock, which, I think, has ftood better than that at Ballykealy, &c. has not wafhed or decayed by the water, but had not acquired a ftony hardnefs; for this reafon I would recommend the ufe of pozzelana, which has been imported from Italy to feveral of the works that I have been concerned for in Great Britain, at the price of forty-two and forty-three fhillings per ton; as it comes much cheaper than terras, it can be made a more liberal ufe of, and as it is in many refpects preferable, it will greatly contribute to the eftablifhment of good mafonry: I therefore think, that, confidering the great difference here between hammered and chifeled work, and that it feems requifite to reduce the expenditure upon the locks, whenever it can be confiftent with that duration and permanency which they ought to have; that good large hammered work, interfperfed with good bond-ftones in every courfe, jointed and pointed with pozzelana mortar, the quoins, fell-arches, cove-ftones, and caping being chifeled, will produce work fufficiently durable, and more to the profit of the undertakers, than expending large fums of money in conftructions, that though more perfect, yet fuch as in point of utility the difference will hardly appear in fifty years.

(21.) Refpecting the communication with the Boyne, as I have recommended the line to go nearly to the head thereof, and as the Boyne runs in a very flat courfe till it is paft Edenderry, it is plain that a junction therewith, in the line propofed, will be practicable; and this, confidering the fcope of my advice to the company, to have only one immediate point in view, feems to be all that it is now neceffary to fay about it.

(22.) I have viewed two paffages from Sallin's bridge to Athy, without going through the Downings, and befide the appearance of a want of water at the fummits, will coft more than the paffage at Downings, and as this paffage would leave the extenfion to the Shannon quite out of the queftion, on all thefe accounts, I look upon the propofition

of

of a direct paſſage from Sallin's bridge to Athy, if not practicable, at leaſt as very ineligible.

J. Smeaton.

Dublin, 6th October, 1773.

(23.) P. S. There is no doubt of the practicability of bringing a ſufficiency of water to the point of partition.

J. S.

Navigation-houſe, October 6th, 1773.

At a meeting of the company of undertakers of the Grand Canal,

Redmond Morres, eſq. in the chair,

Reſolved—That a committee be appointed to take Mr. Smeaton's report into conſideration; any five of whom to be a quorum.

And a committee was accordingly appointed and ordered to meet to-morrow morning, at ten o'clock, for that purpoſe.

Navigation-houſe, 7th October, 1773.

At a meeting of the committee appointed to conſider Mr. Smeaton's report,

Mr. Edward Strettell, in the chair,

THE committee appointed to take into conſideration the report of John Smeaton, eſq. engineer, who attended in compliance with the deſire of the company, on the 6th inſtant, having requeſted his opinion of the beſt line wherein to carry the Grand Canal from the hill of Downings to the river Shannon, the junction with that river being the great object which they have moſt particularly at heart, he recommended a ſurvey to be immediately taken of the ground, as mentioned in his report, in anſwer to the tenth query,

query, from the Togher of Graig, in the bog of Allen, to Philipstown, as he expects that the said new line will be executed with greater certainty, much sooner, and at less expense than that formerly proposed through the bog of Allen; and Mr. Trail having declared that he fully agreed in the same sentiment, the tenth query having been for that reason suggested by him, it was resolved to be the opinion of the committee, that Mr. Trail and Mr. Jessop should be furnished with instructions to proceed immediately upon the said survey.

The FURTHER REPORT of JOHN SMEATON, engineer, in answer to the queries proposed by the company of undertakers of the Grand Canal, the 3d of August, 1773.

To the court of directors of the Grand Canal, Ireland.

GENTLEMEN,

I NOW fit down to give you a more particular anfwer to the queries that you did me the honour to put into my hands in the month of September, 1773, than from the want of time, and fome circumftances neceffary to be afcertained for the forming my opinion at large, after a general view of the premifes (which was taken in fourteen days) I was then enabled to do ; but yet, upon which view I took the opportunity of fuggefting fome of the great outlines relative to this bufinefs in the report then made to you, bearing date 6th October following : and though a confiderable length of time has elapfed fince the date of that report, and feveral fteps have been taken by the company in confequence thereof, which in fome refpects alter the face of the bufinefs, yet, as the grand point then in view, of carrying the navigation from Dublin to the Shannon, ftill ftands an undetermined queftion, both as to the abfolute practicability thereof, and, as to the moft practicable and advifeable road of carrying the fame into execution, I flatter myfelf that what I now have to fay will not come out of feafon, or be lefs to the purpofe on account of the length of time that has elapfed, which has, in fact, put it in my power more maturely to weigh and confider the circumftances attending, and likely to attend, this bufinefs. And as my faid report of the 6th of October, 1773, contains a confiderable quantity of matter in anfwer to the fubject of thofe queries, and as this report has been printed, together with the faid queries, which I conclude is in every one's hands any ways interefted in this matter, to avoid unneceffary repetitions, I fhall, in this further report or anfwer to the faid queries, occafionally refer to the faid printed copies of both ; and that we may go on more fmoothly together, I muft defire that the faid queries and report may be laid before you, and that you will take the trouble of numbering the paragraphs of the report from one to twenty-two (exclufive of the P. S.) to which, by this means, I can more eafily and readily refer, and though I have now feveral matters to touch upon, which will not come properly in direct anfwer to any of the queries, yet, as in the preface to the company's queries I am defired, in general words, to report, in writing, the lines I would recommend, and the plans according to which I would advife the undertakers to pro-

ceed

ceed in making them navigable, and to advise what I think neceffary, not only with regard to the conftruction and dimenfions of the locks and navigation, but alfo the mode of conducting and carrying the fame into execution in the moft effectual and expeditious manner for the common good of the kingdom and intereft of the undertakers. Under thefe general directions I fhall beg leave (after anfwering the queries, one by one, with references as aforefaid) to fubjoin what I have further to offer upon this fubject.

1ft. Anfwer to query 1ft.

The feveral objects of this query are local circumftances, which it requires to be well acquainted with the different kinds of merchandize and manufactures of the kingdom of Ireland, that are likely to pafs upon this canal, and the fituation of trade in the city of Dublin, to anfwer properly, which the fhort ftay that I had it in my power to make did not enable me to become ; and at laft it is rather a queftion to be anfwered by merchants, traders, and manufacturers, than by engineers. It does not, however, to me appear, that on fuppofition of the canal not entering the city itfelf, its prefent termination is ill chofen, and, at any event, a large wharf and warehoufes will be wanted at the prefent termination. I would advife to begin with a very moderate one at firft, to be added to as occafion fhall require, but to be pofitively ready as foon as the navigation fhall be completed from Sallin's bridge, to the prefent termination at Dublin. For I find nothing fo common, after much impatience in getting a ftretch of navigation completed, in a great meafure to lofe the ufe of it for fome time for want of warehoufes at both terminations; but a very moderate one will always do at Sallin's bridge, as the navigation is propofed to go further. See report, paragraphs or articles 17th and 18th.

To the 2d.

This is almoft fully anfwered in article 18th, fo that it is only neceffary to add, that the water that brings veffels from Sallin's bridge will lock them down to the Liffey, and as many veffels, notwithftanding fuch communications, will yet deliver at the prefent termination, and return weft, there will be a redundancy of water at the prefent termination.

To the 3d.

This is fully anfwered in article 8th.

To

To the 4th and 5th.

See what is faid in article 7th, and firft part of the 8th, to which I muft add, that the only queftion remaining in query 4th, is, where to crofs the Liffey? whether at Waterftown ford, or at the prefent line, and how to be croffed? refpecting the latter I have at prefent no doubt, but that it will be beft croffed upon an aqueduct bridge; nor do I look upon the reafon given as valid for preferring Waterftown ford to the prefent line; viz. becaufe upon the prefent, or Mr. Omer's line, the Liffey is ten feet deep in the loweft ftate of the river, and at Waterftown ford only three feet deep; doubt-lefs the cafe of building the bridge, fimply confidered, is greater at the fhallow than the deep water; but as the river Liffey muft have at all times a fufficient paffage, and on the fame width, the deeper the water the greater the fection; as well as relative to the fame depth, the greater the width the greater the fection; the determination of the preference fhould rather follow from the quality of the foundation than the fection of the river: at either place, rock, firm clay, or ftrong gravel is preferable to bog, foft clay, or fand; and if the prefent fection is not adequate to the water that is to pafs, it can at either place be made fo. The weather being rainy, and the Liffey fwelled fo as not to admit of croffing it any where, except on bridges, prevented my making any actual examination of the foundation. An accurate fection being therefore made of the valley and river, defcribing by borings or diggings, the qualities of foil, then the proper conftruction of the bridge may be afcertained, and advifed upon, and it is now time to enquire into this matter, fo that the fame may be afcertained.

In further anfwer to the 5th. I obferve, that the hill or rife of the Downings, where it is at prefent opened, is a very firm lime-ftone gravel, which feems to ftand very firm at an angle more elevated than 45°. And I am told that this lime-ftone gravel laid upon the bog, produces an excellent meadow foil; but it is of little ufe for me to expatiate to you on a piece of hufbandry, which I firft there learned myfelf; however, as the whole country, where not covered with bog, feems chiefly of the fame kind of gravel, it is, however intrinfically valuable, by no means fo fcarce as to be fetched from Downings, in preference to an eafier place.

To the 6th.

I can by no means advife fuch deep finkings as are here fuggefted; it will be found a very fufficient undertaking to get it down to the level of the Blackwood river, which

as

as fhewn in Mr. Trail's fection, appears to be the leaft finking that can be difpenfed with to obtain water to the canal of partition.

To the 7th.

The hill of Downings is fo firm a gravel, and appears to be capable of ftanding fo much fteeper than any kind of earth, clay, or gravel foil, that I have feen before, that I look upon arching to be altogether unneceffary.

To the 8th.

Having confidered the nature of the bog of Allen, with all the attention I am able, I can by no means retract what I have faid in my former report, the latter part of article 8th. I do not, however, mean to be underftood, that I think it a confiderable degree of difficulty to form canals in bog where you can choofe your level, fo as never to have any extra cutting or banking of more than three or four feet.

When a perfon takes a ride upon the Downings, or any eminence on the border of the bog of Allen, and cafts his eye upon that very extenfive flat, he would not imagine any difficulty in carrying a canal over it, or in croffing it in any direction; I therefore do not wonder that when Lord Strafford firft caft his penetrating eye over this bog, and perceived that it gave rife to feveral of your principal rivers, which run from it in different directions, he would readily form a defign for their communication through this bog as their centre, and I can as readily conceive that thofe who have gone after him, making ufe of ocular infpection only, have as readily conceived, that the practicability of this fcheme was exceedingly eafy. This, on ocular infpection, has in general been taken for granted; it has only been by the late actual levels taken, that the difficulty has in any degree appeared: thirty or forty feet of flope, in ten or twelve miles, looks to the eye like a perfect plain; but when this comes to be reduced to a water level, it will create a terrible finking or banking. It appears by thofe levels, that the middle part of the bog is the loweft of all, above thirty feet lower than the fame bog at the eaft and weft ends, while thofe are lower by above twenty feet than fome of the intermediate parts, and are further barred up by very confiderable rifes that cannot be avoided; that is, by the Downings at the eaft end, rifing no lefs than forty-one feet above the level of Blackwood river, and at the weft end, at Knockballyboy, by an afcent of thirty-one, fo that the natural drainage of this bog forming the heads of rivers, is by the middle taking their courfes north and fouth, in a direction quite con-

VOL. II.   M m   trary

trary to the defirable one of going from the eaft to the weft. Mr. Trail has very judicioufly endeavoured to avoid this difficulty upon the line firft chalked out, by making
two canals of partition, and thereby finking the middle part of the canal; but ftill,
notwithftanding thofe different levels, there frequently occur finkings of the canal in
the bog from eighteen to thirty-three feet deep for confiderable lengths together, exclufive of the deep finkings at the Downings and Knockballyboy, which, though I do
not fay they cannot be done, yet cannot be done by any method that is known
to me, or that has been fuggefted or pointed out to me, in any given time, or fubject
to any given eftimate of expenfe. It is a kind of undertaking that has, in this degree
of it, never been executed, to my knowledge, or attempted. Every thing that I
have done and feen in bog, even in moderate deep finkings, has been attended with difficulty, and uncertain expenfes, while, at the fame time, where the level of a canal can
be carried fuperficially upon bog, it is a thing perfectly eafy.

The piece of canal that has already been attempted in the bog, by no means convinces me of the practicability of the thing; what has been done is by no means a fpe-
cimen of what is to be done; it is not funk to the depth nor width, the bottom is upon a
confiderable defcent capable of carrying off the drainage, and has been open for that
purpofe feveral years, being begun by Mr. Omer.

The fact is; wherever I have feen it tried, that till you get four, five, or fix feet
deep (according to the firmnefs of the bog) you go on very expeditiously, and without
any difficulty; but in going deeper, though the bog may be very firm at top, yet underneath retaining a great quantity of water, like a fponge, it becomes very heavy, and
not readily parting with its water, the upper part begins to prefs out the lower, or
flide itfelf into the cavity, and fo keeps coming in as you take it out, till the whole
furface is lowered to a confiderable diftance. The finking a little at a time, as for inftance, one fpit of the fpade annually, giving in the mean time leave for the water to
drain off, and the adjacent bog to confolidate, is the beft method I know of, and think,
with much labour and patience, you may get ten or twelve feet deep. Drain the bog,
fay they that think it eafy; true, this is a very effectual method, but then it is in this
that all the difficulty confifts. No drain can operate below the level of the water it
contains, and by the fame rule that you can make a drain of a foot wide, and keep it
open, you may make a canal of fifty feet wide and keep it open to the fame depth.
I do not mean to limit other men's capacities, or to preclude the company from being
advifed by thofe artifts who may fee clearly what I do not, or whofe particular experience may have led them further into the profecution of this particular fubject, than

I have

I have had occasion for; but I must say, that my particular experience, so far as it has gone in this particular subject, suggests this maxim, avoid a bog if you can, but by all means possible, the going deep into it; nor will any consideration prevail, on me to pronounce that practicable, that I can foresee no adequate means of effecting, and for this reason I must answer the 8th query in the negative.

To the 9th.

From what is said in the preceding answer; it will easily be inferred, that if the carrying the canal upon different levels in the present line will be attended with extreme difficulty, nad uncertain expense, the carrying it on one level will be respecting men's strength and abilities altogether impracticable.

To the 10th.

In answer to this see the 9th article of former report, in consequence of which, and of the resolution of the committee thereon, Mr. Trail and Mr. Jessop proceeded upon the level of the proposed new line from the Wood of Allen to Philipstown, of which levels Mr. Jessop made his report to me, and, as I understand that Mr. Jessop left a duplicate thereof with the company's secretary, it is unnecessary to state them here. From these levels it appears that the ground in this line is much better adapted to carry the canal of partition upon one level, than the former line upon several levels, as laid down in Mr. Trail's section; and in case more than one canal of partition is introduced, it is very immaterial into how many different levels, each descent from its respective canal of partition, is broken; because, by the known principle in artificial navigation, whatever number of locks there are in one descent, they are all worked by the same water that will work the largest; and in regard to increasing the number of locks, as they are capable of being reduced to an estimate, I think little of building a few additional locks, when put in contra distinction to inestimable difficulties. I therefore, upon the whole, cannot avoid coming to the same conclusion upon the former part of this query as upon the 8th.

The latter part of the 10th query is answered at the end of the 8th article, to which I have only to add, that from Philipstown I see no likelihood of finding a better passage than through the high grounds of Knockballyboy, and by the Maiden and Brusna rivers.

To

To the 11th.

Of the three tracks here mentioned, having reconnoitered the country between Naas and Athy, fee article 22d. now, in cafe I thought that a paffage by the prefent line, through the bog of Allen, as propofed by feveral levels, an executable fcheme, that is, a fcheme attended with fuch a kind or degree of practicability that I would recommend the execution of it, then a branch canal from the loweft part of the main canal, between the Cufh and the Fuguile rivers, will be far the moft proper, as being not only fhorteft, but there would be very little difference of level but what might be furmounted by one, or at moft two locks between the low level of the canal and the Barrow at Portarlington; and as there would be water coming down from both the canals of partition into this low level, they could not fail of plenty of water to work them down to the Barrow; but, as I cannot recommend this fcheme, as being fufficiently executable, it will remain that the Blackwood river will be the advifeable paffage to the Barrow. See article 10th, and the reafons given in the 11th, 12th, 13th, and 14th. I would not, however, recommend the paffage by the Blackwood to fall into the Barrow at Monaftereven, for reafons that will hereafter be given. I would recommend the canal, veffels, and locks, of the fame fize, as recommended in my report, articles 4th and 5th.

To the 12th.

If either the prefent line, or the new one traced, as per query 10th, could be recommended as executable fchemes, then a branch from the canal to fall in with the Boyne near Edenderry would be beft to be adopted. See article 21. But, on a contrary fuppofition, I fhould prefer the joining of the Blackwater, with the Blackwood river, for reafons that will hereafter be fubjoined: its locks and boats ftill of the fame dimenfions.

To the 13th.

This being a local circumftance, I am not enabled to give a general anfwer.

To the 14th.

Lock houfes at particular places are abfolutely neceffary, and every lock fhould have a lock houfe, were it not for the expenfe. At particular places they are neceffary for

examining

examining the cargoes, and afcertaining the tolls. At particular places they are neceſſary for the proper regulation and diftribution of the water, and at every lock they
would be uſeful in being a check upon the boatmen and idle and evil minded perſons,
in improperly and wantonly ſpending the water and damaging the works and machinery.

To the 15th.

The dimenſions of boats, locks, and canals, are amply ſet forth in articles 4th, 5th,
and 6th, and to locks of this ſize leſs than a third of the water will be ſufficient that
will be neceſſary to work the locks, ſuch as have been built by Mr. Omer at Clonaugh-
leſs, &c.

To the 16th.

As the ſeaſon was rainy when I ſaw the Blackwood river, I have not the means of
judging of the ſpecific quantity of water it affords in dry ſeaſons; but for the locks, as
now propoſed, I apprehend the Blackwood will be ſufficient to carry on a navigation to
a conſiderable extent of buſineſs, which ſtock of water may be conſiderably increaſed
by forming reſervoirs for treaſuring up the waters of the Blackwood, &c. in rainy
ſeaſons. To which end ſeveral hollows in the bog, lying above the level of the canal
of partition, will be ſubſervient, into which the flood waters of Blackwood can be introduced, dams being put acroſs the outlets by which theſe hollows are naturally
drained.

To the 17th.

The quantity of water that will be neceſſary, will, in a great meaſure, depend on the
quantity of trade that will be carried on; it ſeems that the Blackwood, eſpecially with
reſervoirs, will be ſufficient for the firſt outſet, and as the trade increaſes, the finances
will alſo, and then additional ſupplies of water can be afforded to be brought at
greater diſtances and expenſe. Mr. Trail informs me, that by a level actually taken,
there is a fall from new bridge upon the Liffey of twenty-three feet to the canal of
partition at the Downings, which, in a ſpace of about ſix miles, is very ſufficient to
carry any quantity of water thither that can be required. On viewing the country it
appears very practicable, after erecting a dam acroſs the Liffey, a ſmall diſtance below
new bridge, to take water into an aqueduct on the weſt ſide of the Liffey, and paſſing

by

by Rofeberry or Rafpberry, to follow a level upon the afcent of the ground, from the Liffey toward the weft, and leaving the bridge of Canah, and the mill of Cabbertealy on the right, to fall in with the canal of partition at Langonftown.—This will avoid the deep cutting that feems otherways unavoidable to come at the canal of partition by the bog of Donore; the fummit or neck of that bog being proved higher than the paffage by the Downings. If the Liffey's water be not neceffary at the canal of partition, where it will be wanted in double quantity, that is, to diftribute a lock-full each way, on the paffage of each veffel, it can hardly be wanted on the Dublin fide only, where with great eafe the Morell river can be taken in.

To the 18th.

I look upon it as neceffary to take in the Morell river, which can very conveniently be done, by an aqueduct from John's town to near Sallin's bridge, which will not only be of great affiftance in fupplying the partial navigation that will always obtain upon this diftrict of the canal, but be abfolutely neceffary for the fupplying it before it is joined by the canal of partition. From obfervations upon the quantity of water ufed by the mill at John's town, upon the Morell river, it is capable of fupplying thirty-nine locks-full per day in the drieft feafon, of fuch locks as have been by me propofed, fuppofing them to pen twelve feet difference; and if notwithftanding water fhould be fcarce, it will be proper not to let veffels pafs the large locks but in pairs, and if poffible to bring together veffels going both ways; fo that four veffels may pafs the great locks with one lock-full of water.

To the 19th.

This is already in a great meafure anfwered; if wanted at all, they cannot be too large. I look upon a refervoir of fifty acres capable of holding at a mean of the area fix feet depth of water as a moderate fize; fuch a refervoir will hold about forty-two locks-full upon an acre, and upon fifty acres 2100, fay 2000 locks-full, which at two locks-full to a boat, and ten boats per day, will laft, exclufive of leakages, 100 days, and this, or refervoirs to the amount of this, being formed in the bog itfelf, will be made at little expenfe, where the inheritance of the ground is valued at fo fmall a price.

To the 20th.

My former report recommends article 17th, the firft finifhing to Sallin's bridge, which advice I am glad to find you have taken, by contracting for the works with

Mr.

Mr. Trail, upon the dimensions and depth of water advised by article 5th. The next thing I would advise to be undertaken, is the deep cutting of the neck at Downings, for the reasons specified in article 19th.

To the 21st.

I apprehend that places upon the canal for the deposition of implements and utensils are indispensible; that the company's general stores in Dublin are by no means sufficient and convenient, without others nearer to the centre of the works *pro tempore;* and yet that a place for reception, the deposition of stores, as well as for the manufacturing of some articles, will always continue necessary at Dublin: but yet I apprehend, when the canal becomes navigable for some miles from the Dublin termination westward, that the situation of the Dublin store-yard will be more proper at the canal near the city bason, than in St. James's street.—The very grounds and buildings intended for wharfs and warehouses may very properly be used as a store-yard, and store-houses, till they are wanted for their proper use, and, as said before may be added to, as occasion may require. The store-yard at Stackumney, being near the Gollerstown quarrys, and central to the works between Dublin and the Liffey, I apprehend to be very properly chosen.

To the 22d.

The materials that I have seen made use of by the company are in general of the best quality; only I have thought proper to recommend the use of pozzelana for mortar, in preference to terras; the reasons for which, as well as a full answer to the query, is found in article 20th. of my former report. I will only add, that Pozzelana mortar, does not spoil like terras by becoming dry, or wet and dry, nor does it grow in the joints.

To the 23d.

This is also fully answered in article 20th.

To the 24th.

This is fully answered in the 8th. article.

To

To the 25th.

In regard to the fituation of mills, bleach yards, &c. thefe are fecondary and local confiderations, about which my time and information are not fufficient; but as the company can hardly avoid making a road fufficient for a turnpike, upon at leaft one bank of the canal, nor can they avoid this being made a road of, it is lucky that their act enables them to erect bars, and take a toll.

To the 26th.

The water that will neceffarily be brought down the canal by the paffage of veffels will be far more than will fupply this contract; and even if the canal were hereafter continued down to the Liffey at Dublin, the partial navigation of veffels that will deliver and load cargoes at the prefent termination, will, in my opinion, be fully fuffi-cient; and for which purpofe it will always be the intereft of the company to continue their wharfs and warehoufes there. As the Morell water appears to be fine and clear, and is faid to be of the beft quality, it will be proper to let no more come down from the canal of partition than neceffary, and to prevent it from mixing with the Morell water, more than is abfolutely unavoidable; it will be proper to fix an overfall at the weft end of that ftretch of the canal, where the Morell is introduced, by which means all the water coming from the canal of partition that is not neceffary for the navigation of the veffels to Dublin, will be difcharged before it mixes with the Morell water.

To the 27th.

Having carefully perufed a paper containing the reafons for endeavouring to keep in the bog of Allen, as much as poffible, rather than avoid it, I muft obferve, that if thofe reafons were valid, it would appear that an undertaking to drain and improve the bog of Allen, or certain parts thereof, would be far more lucrative than that of making a navigation from Dublin to the Shannon; for I could chalk out a far lefs expenfive way of draining the bog, than by digging a large navigable canal through it.

As it is here ftated, an acre of land is to grow from the annual value of one penny to forty fhillings, without the ftate of one penny expenfe to be laid out upon it. It is faid, indeed, that clay and manure may be brought by water to cover that part of the bog next the canal, at a fmall expenfe, and in very large quantities; but then the quantities neceffary to cover an acre, together with the expenfes of water carriage,
shou1d

should have been stated.—It is very true, that carriage from great distances in general, is cheaper far by water than by land, but for small distances it is not. Suppose that lime-stone gravel from the deep cutting at the Downings is to be had for nothing, it is yet to raise and to wheel on board a vessel; the expense of wear and tear, or hire of the vessel is to pay, and the men's wages to attend it, and horses to draw it. It is then again to be put into wheelbarrows, and run out to the mean distance of 175 yards, which alone upon great quantities will be a very considerable expense, after which I suppose it is to be spread over, and regularly disposed of. Manure may indeed be brought hither from Dublin by water far cheaper than by land; yet I suppose it is at Dublin to be paid for, to be paid for loading from different parts of the city to the canal, to be put on board, and the freight to be paid for; and though the company will reckon nothing for tolls, yet the wear and tear of their lock gates, banks, and works, in common with other vessels, will in reality be so much out of their pockets. Again, the wheeling it 175 yards, and spreading together with the proper seeds, will be a further expense; all this put together will be the purchase of an acre of land: and though I doubt not but that the land may be thus made in favourable situations to advantage, supposing the canal actually made and navigable, yet, if in making the canal navigable very great (and at present uncertain and unknown expenses) attend it, by forcing a passage through the bog, in a direction ill adapted to the same, it will be very easy to bury all the profits of the land in the extra work, upon a very few miles of this kind. It is said that the digging through the bog is far cheaper than in the firm land, not above one-third: I shall not at present dispute the disproportion, and agree, that where the bog will stand there is no cheaper digging; but if I dig out a cube yard of soil from the firm land it is done, but under an idea of digging out a cube yard of bog, I have to dig out that and nine others, that will follow it, before I can make good the yard I want, I apprehend the balance will be on the other side of the book.

The query states, whether it will or will not be much for the advantage of the company, to carry the line as far through it as can be done with propriety? I apprehend it may; but the question turns on this, where it can be done with propriety, and from what is here stated, and from what has been already said in answer to query 8th, it is very plain that it will on no account be to the company's advantage to go through the bog, where it will require deep cutting; and to fix some idea to this term, I mean where it will exceed ten feet deep; yet this greatly differs according to the natural texture and firmness of the bog; my answer therefore is, that where the company have their choice, in going through the bog of Allen, or not, without deep cutting or high banking, I apprehend it may be for their advantage, and, confequently, for the national advantage; but I can by no means advife the company to attempt to carry the canal through every thing that may oppofe in the bog of Allen, merely for the fake of gaining and improving ground. For, at any rate, it does not feem that a navigation can be carried on from Dublin to the Shannon, without going through fo much of the bog of Allen, as will afford ample opportunities of trying thofe experiments.

ADDITIONAL CONSIDERATIONS.

Having now difpatched what occurs to me in anfwer to the feveral queries, in order as they ftand, I beg leave to add fuch further matters as have prefented themfelves to me fince my former report, which could not fo directly be given in anfwer to the queries.

In my former report, article 10th, I have ftrongly advifed the company firft to complete a navigation to the Barrow, before they begin to work from the Togher of Graig, towards the Shannon, for reafons given in the 11th, 12th, 13th, and 14th articles, which advice I now repeat; and, befides thofe reafons, I beg leave to add the following one, that I am now fully convinced, that this is the only executable and advifeable road to the Shannon, and by which means not only the junction of the Shannon and the Barrow with the Grand Canal, will be united in one intereft, but alfo produce the moft direct communication of the Barrow and the Shannon.

From what has been already ftated, in anfwer to the eighth, ninth, and tenth queries, it does not appear likely, that any executable and advifeable fcheme can be formed to obtain a direct paffage from the Togher of Graig to Philipftown, upon one level, nor indeed without extreme difficulty upon any number of levels. If, therefore, at any rate, we muft adopt at leaft two canals of partition, it will be more eligible to do it in fuch a way as to draw every poffible ufe from this conftruction. It has been already ftated, report, article 10th, that the junction of the Barrow with the canal, by means of the Blackwood river, will be perfectly eafy, and in point of probable profit advifeable prior to, and, independently of all the reft; fuppofe then, after the meeting of the canal with the Blackwood river, which I will call the fixed point, we carry the canal down the declining valley of the Blackwood by Rathengan, as has been already propofed; but, inftead of inclining it towards Monaftereven, it be inclined towards the joining of the Blackwood, with the united ftream made by the Cufh and Figuile rivers, which will be

upon

upon ground very little elevated above the Barrow, at Portarlington, it being over-flowed to a great diftance when I faw it; fo far then we may lay it down at all adventures expedient for joining the Barrow, and taking in a navigation from Portarlington, as well as from Monaftereven, the beft adapted poffible. This courfe, from the fixed point to junction with the Figuile, will be, I fuppofe, about nine miles, from thence to the Bar-row about two, but afcending from the faid point of junction of the Blackwood and Figuile, by the gentle inclined valley of the Figuile, we fhall, in a courfe of about eleven miles, come to Philipftown bridge, fo that this will be a ftretch of twenty miles from the fixed point to Philipftown bridge. According to the prefent line, through the middle of the bog, it is fifteen, fo that when you are arrived at the joining of the Black-wood and Figuile, you are then four miles nearer Philipftown by the courfe of the Figuile, than at the fixed point in croffing the bog according to the prefent line. Ac-cording to Mr. Trail's plan of the branch canal to the Barrow at Portarlington, which is the fhorteft communication of all, it is nearly feven miles; therefore, Philipftown bridge and the Barrow, taken conjunctly, in either cafe, are joined with the fixed point by twenty-two miles of canal. It is true that the courfe from the fixed point, by the Black-wood and Figuile to Philipftown, in a paffage to the Shannon, being twenty miles, is five miles more than according to the prefent line, which is only fifteen; but then, as I apprehend, no proportion of diftance fubfifts between a practicable and an im-practicable paffage. The ineligibility, not to fay abfolute impracticability, of any paffage from the fixed point to Philipftown acrofs the bog, has been fufficiently pointed out; what renders the now propofed line a practicable and an executable fcheme, is, that while you are following the valleys of gentle declining rivers, you are fure the general furface is inclining one way, and gradually either rifing or falling, according to the courfe you are going; fo that if in bog you can always choofe your ground, and by making locks at proper places, can always avoid deep cutting and banking, for at any rate you can take the trough of the river for your canal; and it happens very luckily here, that rivers' run upon fo flat a bed, that according to the levels taken there fhould not be above nine or ten feet difference between the levels of the united ftreams of the Blackwood and Figuile, and at the croffing of the branches of thefe rivers feparately in the bog; fo that I compute there will not be required above one lock more in completing the junc-tion of the Shannon and Barrow, taken together, than would be required according to Mr. Trail's fection.

In taking the diftances above mentioned, I have meafured them by the courfe of the rivers, not indeed according to their fmaller meanderings, but their general direction as their borders are moftly flat; but when this fcheme comes to be furveyed by the
level,

level, it is very probable that the courses may be shortened, without encountering any deep cuttings and high bankings; and particularly that the canal may be taken round, so as to effect a junction of the Blackwood and Figuile, by a nearer course than by their natural junction.

Respecting the passage of the Downings, I by no means look upon it as a very formidable obstacle in so great and expensive an undertaking. The soil is near upon the most favourable of all for such a purpose, and which having reduced to calculation, in my opinion, may be done for the sum of £13,500 extra to the expense of the same length of plain canal. In fact, I look upon the deep cutting at Knockballyboy to be the more formidable work of the two, on account of the uncertainty of the deep cutting in bog, no less than thirty-one feet deep; however, as the ascent is steep, and, consequently, the bog very much upon the decline, it is naturally firm and dry, and as the ease of tapping the under parts is proportionable, I hope, as all the difficulty after passing the Downings will be reduced to this one, that it will not be found insurmountable, yet, still if possible, it will be well to avoid this also; and though, from the views I have had, I have seen no probability of avoiding this passage, if the canal goes by or near Philipstown, as I have reported in answer to the latter part of the tenth query, yet, as at the time of making this view, I had no idea of taking Rathengan in the course from Dublin to the Shannon, I did not reconnoitre what might be done in effecting a direct junction between the rivers Barrow and the Shannon, but now remembering that the country was pretty flat in travelling from Banagher to Portarlington, and observing that the Barrow, (which runs upon a dead course for several miles above Portarlington), together with its branches, form near approaches to the head of the Maiden or Tullimore river, to Lough Pallas, whose waters fall into the Maiden, and to Lough Haunch, whose waters fall into the Lower Brusna, and by a pretty direct course by Birr to the Shannon, observing also that Mr. Trail, in his report, has mentioned an information of a passage by Lough Pallas; it is possible, that though such a passage may not be eligible from the bog of Allen, yet it may be so from the Barrow; I say, taking all these considerations together, it seems well worth the company's while to order the part of the country now pointed out to be reconnoitered; and if any passage seems more favourable than that at Knockballyboy, to have a particular survey, level, and section taken; and in case the company approve of this proposition, either by way of the Figuile and Knockballyboy, or any passage from the Barrow that may be more favourable, it will also be proper to have a survey, level, and section, taken of all the parts which have not been already reduced to paper, and then I have not the least doubt but that the whole work, the bog cutting at Knockballyboy, (or something that may occur of the same kind in making good

any

any other paffage at the weftern fummit), excepted, may be reduced to a fair and grounded eftimate, fo that the company may fee what they have in reality to do, in order to effect this great and very defirable work.

According to this conftruction, as well as that propofed by Mr. Trail, another collection of water will be neceffary to fupply the weftern canal of partition, but as the eaftern will have the united trade between Dublin and the Barrow to Lock as well as that of the Shannon, and the weftern will have that of the Shannon only, with poffibly a fmall trade between the Shannon and Barrow, it is poffible that much lefs water will be wanted at the weftern fummit than the eaftern, and to this end the head of the Figuile river above Philipftown will conduce, which, as I obferve, turns a mile near Philipftown; this, probably, with proper refervoirs, will be fufficient, yet it will be advifeable, (if this at laft is found the moft favourable paffage), to reconnoitre the country, to fee whether the head of the Tullimore river, the head of the Efher, or the water from Lough Pallas, or any other, cannot be conducted by aqueducts to the weftern canal of partition.

After the junction is made with the Figuile from the Blackwood, (fuppofing the Figuile the moft eligible paffage), and the navigation effected from thence to the Barrow, at fome convenient point between Portarlington and Monaftereven, the company will then be in poffeffion of the Barrow trade. For as there is already fubfifting a fort of navigation from Portarlington to the bridge of Athy, by openings made in the mill weirs for that purpofe, and though this kind of navigation is very imperfect, yet it may, with fome improvements, for a time be fufficient for the trade, or as the mode of making a navigation from Portarlington to the bridge of Athy is no way difficult or expenfive, the company will have it in their power, if they find it expedient, to perfect this branch at once, and at the fame time proceed from the aforefaid point of junction with the Figuile for the Shannon.

As, according to the now propofed plan, the canal is to be carried no further in a direct line into the bog of Allen, than the meeting of the Blackwood river, it feems that the method of joining the Boyne, by means of the Blackwater river, according to the fcheme propofed by Major Valency, will beft fuit the ground by which a direct paffage from the Barrow to the Boyne will be formed by means of the Blackwood and Blackwater rivers, united near the Togher of Graig in the bog of Allen

J. SMEATON.

Austhorpe, 3d April, 1775.

TYRONE

TYRONE CANAL.

Mr. SMEATON's opinion on Mr. JESSOP's REPORT on the Tyrone Canal.

To Mr. JESSOP,

Dear Sir,   Brough, 30th April, 1774.

MY journey into Yorkſhire gave me an opportunity of reading and conſidering your report, by way of repreſentation of the ſeveral matters relative to the works of the Tyrone canal ; and having been at home but one day, I now take the opportunity of a leiſure hour to give you my opinion thereupon, ſo far as I can do it without entering into calculation.

Had this canal been unbegun, the circumſtances attending it are ſuch, that I never could have recommended a canal of any kind, as I am ſenſible it never could be made in any degree to anſwer the expenſe : but if I underſtand you right, the bulk and ex-penſe of the work is in a great meaſure incurred, there being little to do but to complete the inclined planes between the different reaches of the canal. I think, therefore, that it would be a great pity after the public has been at ſo conſiderable an expenſe in con-ſtructing the apparatus, but that the experiment was made, and the validity of this mode of navigation put to a thorough proof, eſpecially as it may be a moot point, whether it may be cheaper to navigate upon this canal, ſuppoſing it ready made, and given for nothing, or to begin a new work, which, at any rate, will coſt a conſiderable ſum ; for by this means the public will at leaſt be put into poſſeſſion of this piece of knowledge, i. e. how far a ſcheme of this nature is likely to anſwer in other places.

I think the beſt chance of its ſucceeding will be by obliging the loaded boats going down to draw up the empty ones returning, which may be done by a ſimple pulley, with a brake or convoy upon it, to moderate the velocity of the deſcending boat, and that a great deal of friction may not be wanted upon the pulley, as you ſay the plane riſes one in five, I fancy they will act very ſufficiently, if they are made to ſlide upon the ſolid, rather than upon rollers ; but I fancy that light wooden rollers will be wanted at proper diſtances to keep the ropes from rubbing upon the gravel or ſolids between the ſlides ; and as I do not apprehend that above one boat can be let down at a time, for greater diſpatch I think it will be better to make the boats carry two tons rather than one, by making them of double length. I do not, however, hint theſe things by way of adopting the ſcheme, nor can I anſwer whether it will do or not ; but as mat-ters are carried on in ſo great a length, by way of ſuggeſting the likelieſt means of getting an effectual trial of Mr. Dukart's ſcheme.

If

If his fcheme does not on trial anfwer, I am of opinion, that a rail road of wood, in the manner of thofe at Newcaftle and Whitehaven, is much to be preferred in fuch a fituation to either a canal or a gravelly road, efpecially as you fay it can be done with a length of two miles, and all or moft of the way downhill: this being the cafe, the waggons will go down loaded by their own gravity, and the horfes will have nothing to do but to draw them empty back again, which, in a rife of 192 feet in two miles, will be eafy work.

Rail roads of this kind are in common fituations executed in the neighbourhood of Newcaftle, fo as to carry coals at the rate of 2d. per neat ton per mile, the original charge of making, maintenance, of wear and tear of the roads, waggons, horfes, and drivers included. But as the cheapnefs of things greatly depends upon a country's being ufed to the execution, which can never be the cafe in the firft inftances, if we fuppofe 6d. per ton per mile, of the Tyrone tons, they will be no more than 1s. per ton for the two miles, and in all probability will be cheaper than it can be done upon the canal after the whole of the original expenfe has been given.

You will be pleafed to communicate my fentiments to my Lord Archbifhop of Armagh, with my moft dutiful refpects, and am,

Sir,

Your moft humble fervant,

J. SMEATON.

P. S. I apprehend it will be neceffary, befides a fimple pulley for returning the rope, to have fomething of a tackle, affifted by a winch or windlafs, by way of bringing the boats upon the inclined plane, and till they are launched upon the down-going fide.

DRAINAGE

DRAINAGE of NORTH LEVEL FENS.

The REPORT of JOHN SMEATON, engineer, concerning the drainage of the north level of the fens, and the outfall of the Wisbeach river.

AS I perceive that much has been said and wrote upon the subject of the great level of the fens, and that a variety of opinions has been entertained concerning them, I find it necessary, in order to clear my way, to deduce what I may have to offer upon the subject of the north level, from its original, and though I set out with an hypothesis, which may be granted me or not, yet, as it is to be considered in no other light than a means of introducing certain matters of fact, which being in themselves incontestible, what is founded thereon will be equally true, whether the supposition by which they are introduced, be so or not.

I suppose then, that there has been a time when the whole of the great level of the fens, together with the adjacent low countries, has been one great open bay, where the tides have freely flowed and reflowed: I suppose also, that for a long course of years, far beyond what any present accounts of this matter reach to, a cause has subsisted, by which a quantity of sand and earthy matter has, by the action of the winds and tides, been gradually brought upon this part of the coast, and has particularly lodged itself in this great bay, as a receptacle where it could lie at rest: this being granted, it would follow that this bay would gradually become more and more shallow, till the matter deposited from time to time should become higher than the rise of the neap tides, in which case a country would be formed, and would grow over with grass, and form what we now call salt marshes: but as the waters of the upland country must continue to have a passage to sea, which had their outfall into this bay, they, conspiring with the flux and reflux of the tides to cover and uncover the surface of the land, must, while the very land is forming, make the channels necessary for the discharge of the land waters into prodigious large and deep gullies, whose bottoms we may suppose far deeper than the low-water mark at sea; and which, as soon as grass was formed upon the level surfaces near their borders, would put on the appearance of rivers. This being the case, the inhabitants would gradually observe, that by raising very inconsiderable banks against the rivers, to shut out the spring tides, the land might be gained from the sea; and, admitting the inhabitants of that age to have the same natural sagacity that we have, we cannot suppose they would

find

find it difficult to contrive expedients to let off the downfall and foakage waters into the oppofite rivers, when the tide had retreated, by means of fome kind of tunnels laid through the banks, equivalent to our fmall ftop fluices, which would be the whole of the art of drainage the country, in this condition, would want. By thefe means, a very great tract of country would be gained in a little time, which being immenfely fertile, inhabitants would of courfe flock hither, and multiply apace; villages and towns would be built (as in the moft eligible fituation in a flat country) on every little eminence or rife of the furface, originally produced by fome ftorm or tempeft, that had thrown up and amaffed a greater quantity of matter in thofe places than in others.

It is evident, that, after the tides were fhut out from flowing over the furface of the lands, the quantity of flux and reflux of fea-water at each tide being greatly leffened, and in confequence of its velocity, the power of fcouring would be thereby weakened; the channel of the rivers would therefore filt and contract from the very fame caufe by which the country itfelf was originally formed: but, as thofe rivers would, as already ftated, be very large and very deep before the imbankation, and as they would ftill contain a large quantity of falt-water flowing and reflowing, and thereby ftill exert a very confiderable power of fcouring; it would be a courfe of years before the rivers would become contracted in any confiderable degree: and this contraction, though we may fuppofe it apparent enough, in the compafs of one age, yet making its appearance at firft more in width than in depth; and as thereby a quantity of land would be further gained from the rivers, it would be far from being viewed in an unfavourable light, and the rivers being ftill fuppofed to be fo deep as to ebb every tide far below the furface of the land; as mankind could not then be fuppofed to be furnifhed with the experience of after ages, or to fee the refult of nature's tendency, the gradual changes that they might from time to time perceive, would not prevent them from fuppofing, that things would always continue in tne fame flourifhing ftate: and, in confequence, coftly churches, abbeys, and monafteries, would be built, and great ecclefiaftical eftablifhments made, in fo fertile a country.

So far we have proceeded by manifeft deductions from two affumed pofitions, viz. that the whole level was originally fea; and that nature has been gradually making an addition of matter to this coaft.

That the latter is and has been the cafe in our own times, and as far back as any accounts of hiftory reach, is very certain: we may therefore infer that the fame procefs has taken place, in times antecedent to the written accounts, and for a time backward, to which we need not prescribe limits: this being the case, the first assumption must be presumed, as the most natural state from which the effects of the second could proceed or be manifested.

The flourishing state of the fen country, as above described, is sufficiently testified in history; now, if it shall appear that the levels of the present surface of the fens are such, with respect to the present low-water mark at sea, that upon a supposition of large and deep rivers, the fens would naturally be in that flourishing state by the simple means of drainage before described, we may infer, that there actually were then those large and deep rivers: and if large and deep rivers, in a country where there are not powers sufficient to maintain them such, are the natural consequences of a country arising out of the sea, we may then fairly infer, that the country had its origin that way: and if, from the same causes still acting, this same flourishing country becomes a drowned and desolate country, as we are most sure it has been, we may conclude that the whole mutation has proceeded from one regular process of nature, gradually operating, and not from any unusual convulsions of her frame, nor from the superior skill or industry of former ages, nor from the particular neglects of those who succeeded them; and hence we may conclude, that we are not to expect to recover this drowned and desolate country, by partially restoring things to the same situation as they were in, when in that good state; for without we could restore the whole, and at the same time put a stop to nature's process, the same causes, still producing the same effects, would again reduce us to the same deplorable condition: we must therefore rather look out for such artificial remedies, as shall from time to time countervail nature's gradual process, so far as it is in our disfavour.

That the general surface of the fens is sufficiently high to drain, in the easy and natural way above described, upon the supposition of a large and deep river, is hence apparent. According to the levels and section of the river that have been taken by Mr. Elstobb, the bottom of the river Nene at Peterborough bridge, is now somewhat below the level of the sea apron of Gunthorpe sluice; but a little above the low-water mark at Sutton's washway house: it appears also that the low-water mark at the washway house, is near six feet above the low-water mark at the Eye, or road where the ships lie, and which is supposed to be the same as the low-water mark at sea: it appears also, that the surface of the lands in Wisbeach high fen, near Narr lake, which are supposed the lowest in the north level, are eleven feet and a half above low-water at sea, and the surface of the lands near Jacklin's house, opposite Thorney lordship, is

thirteen

thirteen feet above the level of the fea. Now, when the river Nene ran in a large and deep channel, and before the general outfall, called the Metuaris Æftuarium, was choaked with fands, as at prefent, we may fuppofe that the level of low-water at the river's end, was then the fame as the low water at fea; and as we alfo may very well fuppofe that the channel of the river might, in general, be ten feet deeper than low-water at fea, in this cafe it would ebb out fo much, even at a great diftance from the outfall, as to want but a very inconfiderable declivity per mile to fea at low water. The fall therefore from the loweft lands in Wifbeach high fen to the river's end, would be eleven feet and a half, the fame as it is now, to the upper beacon at the Eye; a difference in extent of ten miles according to the low-water channel; and fuppofing the water in the drains to have flood two feet below the general furface of the loweft lands, there would ftill have remained nine feet fall at a fpring tide low water; and fuppofing the ebb of a neap tide to be nearly as now, viz. four feet lefs, there would have yet been left five feet and a half fall from the furface of the drains to low-water. The furface of the lands mentioned, near Thorney lordfhip, being ftated to be thirteen feet above the level of low-water, would have a greater fall in all cafes than the other, by one foot and a half, but being farther diftant from the river's mouth, would have no preference in point of drainage; but having full eight feet for the declivity of the river's furface, even at neap tides, this is fufficient to carry it with a brifk current, in fuch a channel as we fuppofe, at three inches per mile, to a diftance of thirty-two miles to fea: and I doubt not, but that in thofe days, the furface of the water at low-water at Peterborough bridge, ran as low as the prefent bottom; and that the bottom then was confiderably deeper than at prefent, being now filled with land mud.

It is true, that the fuppofition of fo large a channel, would bring a much greater influx of tide into the country; for it appears that the high fpring tides flow at fea higher, by near five feet, and the ordinary fpring tides above one foot higher than the ordinary furface of the water at Peterborough bridge, as it was when the level was taken; yet, by the fame rule, that the water requires a declivity to feaward at low-water, on account of diftance, it would require a declivity to landward on tide of flood; that is, diftance requiring time, it would be confiderably ebbed at the river's mouth, before it would be high-water at a diftance up the river; we may therefore fuppofe, that according to the curved courfe in which it appears the river Nen anciently ran, the fpring tides would never rife fo high at Peterborough, by full fix feet, as the fea level, and the neap tides be but barely fenfible. That this would be the cafe, appears from prefent experience; for the fpring tides, though in a fhorter, but more obftructed courfe, feldom flow higher than Knarr lake, where the furface is lower by almoft one

foot

foot than at Peterborough bridge, and by two feet than the ordinary, and by five feet and a half than the extraordinary, spring tides.   Now, as the ordinary tides never could rise at Peterborough so high, by several feet, as the present ordinary surface of the water there, and as the land flood waters seldom rise in the open tides-way of a large river much higher than the tides themselves, it will follow, that the country about Peterborough might be an unimbanked country, and that very moderate banks would be a defence against every thing lower down, the neap tides scarcely rising higher than the surface of the lands.

According to this state of the rivers and tides, we may suppose those level countries to have continued very flourishing for many years; but as there does not appear to be a quantity of land water flowing through them, capable of maintaining the channels of the rivers so large and deep, as is necessary for their perfect drainage, upon the easy and simple principles already described, the sand gradually accumulating upon the coast, and thereby choaking up their mouths, and removing the point of low-water to a greater distance, would gradually check the reflow of the tides; the channel of the rivers would therefore by degrees contract, both in width and depth, till they would with difficulty convey the land floods with sufficient speed to sea; and in consequence, breaking their banks, and overflowing the country, first in a less, then in a greater degree, and more and more frequently, till the inhabitants becoming greatly annoyed thereby, and their crops destroyed, even before the drainage was totally spoiled, would, on finding lesser expedients insufficient, apply themselves to greater.   The great land floods would therefore naturally be regarded as the most formidable enemy; and the inhabitants being ignorant of the consequences, which the experience of after-ages has furnished the knowledge of, how to get rid of them would be the first and principal question: and if there appeared a practicability of turning them in other channels, where they might spread and become dissipated, this expedient would of course be eagerly embraced, in preference to that of raising the banks of the river, in order to confine them and force them to sea; by this means, having lost the operation of that great and powerful engine of nature, towards scouring and keeping open of rivers, viz. the great land floods, the river would then begin to silt at a great rate, and thereby, the bottom becoming so high as to spoil the drainage, ruin and destruction would come upon the level in a manner at once; and the inhabitants and land-owners being totally unprepared to unite in one general scheme for their relief, things must in a small course of years, come from bad to worse, so as to bring this fertile country into the deplorable condition it has been represented: and so it must by nature continue till fresh combinations

binations of men could be formed, with frefh fpirits, and frefh purfes, in order to fi t about thefe works which were likely to give a general relief.

Something feems to have been attempted very early, to remedy thofe evils, by cutting new leams and channels for the rivers, in order to give them a fhorter paffage to fea, and thereby to gain defcent for the drainage of the lands; but as they feem never to have fufficiently underftood, or attended to, the point of confining the land floods, and keeping them as much as poffible in a body, in order to fcour and keep open the outfall, inftead of diffipating them by different channels, all their fchemes became abortive by the outfalls growing fo much worfe, while the new leams were cutting, that by fuch time as the work was completed, or foon after, they found themfelves in as bad a condition as when they begun.

Having endeavoured to account for and eftablifh the reafon for the ancient flourifhing ftate of the fens, and the caufes of their fucceeding deftruction, let us now take a view of what has been principally done by the undertakers for regaining them.

The original of the modern projectors in draining came out of Holland; and their eftablifhed maxims feem to have been to imbank the rivers, fo as to prevent the land floods and high tides from overflowing the lands to be drained, but to leave the rivers open to the free action of the tides: then to conduct the downfall and foakage waters of the country to be drained by a fewer to fome place at or near the low-water mark at fea, that the waters of the country might run off at low water, by a natural defcent, and to place fluice doors at or near the mouth or outfall of the faid fewer, fo as to prevent the tides water from reverting: alfo, in cafes where it is found or thought expedient, to make new channels or courfes for the rivers, in order, by diminifhing their diftance, to increafe their proportional fall.

Upon this idea (which is very complete when it happens to take in all the circum-ftances) the modern works of the north level, under confideration, feem to have been carried on. With a view to fhorten the courfe of the river, I fuppofe originally Morton's leam, and afterwards the new leam, have been cut, and the matter thence arifing employed to form banks to prevent the land waters from overflowing the level. In order to receive and carry to fea the drainage waters from the fouth Eau, and other drains interfecting the country, the Shire drain has been cut, or enlarged, and adapted to that end: upon this fewer a fea fluice, now called Hill's fluice, was put down in the 1637, by King Charles I. then at the head of an undertaking for the drainage of thefe

levels;

levels; all which works feem tolerably well defigned to anfwer the end, at the time they were put in execution; and I doubt not would have anfwered the end very well, if nature would but have lain ftill and been quiet, and left things in the ftate in which thofe undertakers found them; but the misfortune is, nature has ftill alfo been bufily carrying on her work, of making addition to the fea-coaft; fo that Hill's fluice, which, at the time of its erection, muft be prefumed, was fufficiently near the tides way to have anfwered the end propofed, viz. of a fea fluice, by degrees got fo remote from the main channel, that the outfall being found too liable to filt and choak, by the fands depofited therein by the tides, another fluice was obliged to be built nearer the tides way, called Gunthorpe fluice, which alfo, in time, proving not fufficiently effectual, another fluice, bearing the fame name, was built not many years ago ftill nearer the fea; but this fluice, like the former, by the gradual progrefs and working of nature, has loft its due effect; for when I was there in September, 1767, I found, when the tide was out, and the land doors fhut, that there remained five feet four inches dead water, upon the fea apron; whereas, to produce its due effect, Gunthorpe fluice ought to run at two feet five inches water, upon its fea apron: it is faid, indeed, that before the laft fpring tides, it ran at three feet ten inches, but then the bottom was too high, by one foot five inches, to produce a fufficient drainage: and even the uncertainty is a great detriment; for if after the top waters are run off, a fpring tide comes, and lands up the outfall, the drainage may be loft for a feafon.

When the new fluice of Gunthorpe was built, the old one was demolifhed; but Hill's fluice is yet maintained in repair; the purport being to hold up a quantity of tides water, to be taken occafionally into the drain between the two fluices, as a refervoir; and therefrom at low water, to difcharge the fame, by means of the draw doors at Gunthorpe, to fcour out the outfall channel between that fluice and the main channel of the river. This difpofition has been well intended, and, fuppofing no other way practicable, would have been a very commendable expedient; but as it appears that there has been, in all modern times, a fufficient depth and certainty of channel, and ebb of tides, at Sutton's wafhway houfe, to drain the whole level, it is a great pity, that when Hill's fluice was found ineffectual, and a new fluice muft be erected, that it was not immediately carried down to the wafhway houfe: and it feems next to infatuation, that when the firft fluice built at Gunthorpe was rendered ineffectual, and to be taken down and removed, that it was not then carried to the lower place. However that might be, it is certain that the further increafe of matter thrown up, and left by the tides, produce ftill the fame effect of choaking the outfall channel; and that the operation of the refervoir has not been able to keep it open and free.

The

The reason of which, as I apprehend, has been, that finding the large tides taken into this reservoir, to have deposited so much silt as to endanger the choaking of it as a drain, it has occasioned that this matter has not been steadily, thoroughly, and sufficiently tried; for though by taking in the moderate tides only, to prevent sediment, and those but now and then, the effect has been found inadequate, and a quantity of silt gathered in the reservoir; yet, had the greatest tides been taken in, and that sufficiently often, so as to clear the outfall channel down to low-water mark at the washway house, the water held up in the drain beyond Hill's sluice, would then have so much fall in passing through the reservoir, as to produce an effect, in scouring it from the silt left by these great tides: whereas, I can readily conceive, that the taking in the middling tides only, and those in such degree as to fall far short of a thorough deepening of the out channel, would be attended with a bad effect, and thereby discourage the further prosecution of the experiment: but were the use of the reservoir prosecuted with vigour and judgment, and hedgehogs used, to assist the waters in scouring both the outfall channel and the reservoir, it seems to me that great assistance might be derived from these operations.

In case the use of the reservoir were to be set seriously about, it is quite necessary to place the land doors upon the land side of the sluice; for, being placed on the sea side of the sluice, along with the sea doors, they have not sufficient firmness to resist the pen of the great tides left in the reservoir, without danger of derangements: it appears also proper, that, to procure scours for the reservoir at times when the land waters of the drain are not of a sufficient height, it would be proper to put down a staunch, or slight stop sluice, somewhere about Tretham bridge, in order to take in a body of water, to be held up between the staunch and Hill's sluice, for scouring the reservoir when a sufficient fall is obtained by deepening the outfall channel to seaward: this body of water may be taken in from the small tides, when they are clear, and consequently little or no silt will be deposited from them: and those secondary scours should always be made as soon as may be, after the great tides have been used for scouring the outfall, that the silt deposited may not acquire a degreee of firmness by lying.

These methods, in whatever degree they might succeed, it is very evident, will require to be incessantly repeated; and that from the constant addition of matter to the coast, they will every year become less and less effectual; yet, considered as temporary expedients, to preserve the level in a tolerable state till something more effectual can be done, they may have their use, and be deserving of further trial: but as it appears

to

to me, that no effectual method can be taken of confining the channel of the river among the broad and loose sands, so as to keep close to the Lincolnshire coast (as I am informed it has sometimes done) and thereby to afford the Gunthorpe waters a short outfall channel; nor of preserving the outfall channel from silting, and its bottom frequently getting higher between the sluice and low-water mark, than is consistent with the good drainage of the level; it follows, that the only safe, certain, and lasting way of procuring a good outfall, will be at once to prolong the drain, and build a new sluice as near as possible to, but above the washway house. At this place alone, all accounts agree, that the main channel of the river has been constant, or with very little variation, for a long course of years. The channel here runs the nearest the shore, of consequence the outfall channel of the sluice will be short, and, if obstructed, will be the most readily cleared, whether recourse be had to nature or art. The ebb of the water here is also sufficient to make a complete drainage of the level, which appears as follows.

The waters of the drains having been rendered stagnant for some days, by shutting the land doors of Gunthorpe and Clow's cross sluices, and the weather being calm when I was there, I found that the top of the threshold of Clow's cross sluice lies two feet one inch higher than the sea apron of Gunthorpe sluice; and the sea apron of Gunthorpe sluice, according to Mr. Elstobb's levels, being nine inches higher than low-water mark at washway house, it follows that the threshold of Clow's cross sluice is two feet ten inches higher than low-water mark at Sutton's washway house.

It has been observed by Mr. Wing, of Thorney Abbey, whose business it has been to observe and to know the state of this level, that when there is four feet water upon Clow's cross sluice, then the general surface of the north level is in a state of drowning; and it has been remarked by the reverend Mr. Dickinson, of Wisbeach, who has been curious in observing the state and condition of the neighbouring levels, that when there is from two feet to two and a half feet water upon Clow's cross sluice, then the lowest lands in the north level, west of south Eau bank, which is Wisbeach high fen, begin to be under water. These observations nearly agree with Mr. Elstobb's levels, who makes the land there somewhat higher; but as the observations are of a nature from which the general surface can be judged of, better than by levels taken with instruments, I shall found myself thereupon.

Taking therefore the medium, viz. two feet three inches water upon Clow's cross sluice, to be even with the surface of the lowest lands in the north level, we shall have

the

the furface to be five feet one inch above low-water mark; and fuppofing the furface of the water in the drains to be reduced two feet below the general furface of the loweft lands, in order to make a complete drainage, we fhall have remaining three feet one inch of fall from Clow's crofs to the wafhway houfe; which being a diftance, by the courfe of Shire drain, of fomewhat more than twelve miles, will allow a fall of three inches per mile; a quantity not only fufficient to run off the water, but to vent the downfall waters with fo much fpeed (upon a fuppofition that the drains, fluices, and bridges are made anfwerable) that this level may once again emulate its ancient ftate of fertility.

This point being eftablifhed, I trouble not myfelf with computation, as to the lands betwixt Wifbeach fens and Peterborough, becaufe the general rife of the country that way, will countervail the greater diftance from the outfall: nor with the drains that are neceffary to be made or opened, in order to give paffage for the water through the level from weft to eaft; not doubting, but that if the waters are reduced at Clow's crofs, nearly even with the prefent threfhold of the fluice, the country wants neither fkill nor induftry to conduct them thither.

It has been commonly obferved, that Lutton's leam is the beft outfall of the whole country; and it has been made a queftion, whether it would not be beft of all to carry out the north level waters by way of Lutton's leam. It is certain, that the outfall channel of Lutton's leam falls into the main channel, near to the Eye, that is, at or near to low water at fea, and therefore has the greateft advantage of defcent: but it is equally certain, that the coaft is here advancing into the fea very faft; that Lutton's leam fluice is now nearly two miles from the main channel; and that, out of this diftance, the leam makes its way over broad fands, a mile and a quarter; in confequence whereof the outfall channel of that leam is fo far choaked, that though the apron of the fluice lies two feet two inches higher than that of Gunthorpe, yet, as it never runs lower than about fourteen inches water upon that apron, it will follow, that the outfall channel of that leam is not better than that of Gunthorpe, than by about one foot: it is true that an addition of water will mend it; but then we cannot tell how much. For thefe reafons I am of opinion, that an outfall at Lutton's leam, will be greatly inferior to an outfall at Sutton's wafhway houfe. It has alfo been queried, whether an outfall at Weftmeer creek might not be preferable to the wafhway houfe, as being nearer the low water at fea; but it appears from the levels, that the lowwater mark in the channel oppofite Weftmeer creek, is not above one foot lower than that at the wafhway houfe; and as the main channel is now departed from the fhore, to above a mile diftance, I do not think the channel, or gulley, that is commonly found near the fhore, fufficiently to be depended upon, for an outfall to the north level; and for thefe reafons think an outfall at Weftmeer creek far lefs eligible than at Sutton's wafhway houfe.

It may naturally be afked me, whether, from my own pofition of a gradual increafe of the fea fhore, the outfall at Sutton's wafhway houfe may not, in time, come to be choaked up; and confequently, whether it may not be preferable to go to Weftmeer creek, or Lutton's leam, as getting at a greater diftance from the enemy? I anfwer, that fo long as the upland waters maintain their courfe to fea this way, according to prefent appearances, there feems a probability, that an effectual outfall will be maintained, for a greater courfe of years, at Sutton's wafhway houfe, than at either of the places above mentioned; which I expect to make appear more demonftrably, when I have treated upon the river, which now comes in courfe to be confidered.

In drainage it is not only neceffary to procure a good outfall, but to defend the lands from being overflowed, by the foreign waters coming down from the upland country; in fhort, to maintain a good barrier. According to the Dutch maxim, imbank the rivers very well; but after the rivers have been imbanked, the reflow not being able to carry out the fands brought in by the tides, the river's bottom becomes raifed, till in many places it is higher than the furface of the land to be drained; in fhort, till it is fo choaked, that the great land floods, inftead of finding their way to fea with a good current, and thereby clearing out the fands brought in by the tides, when the land waters are weak, they overflow and break the banks, and drown the country, which again tends to increafe the evil, as thereby their ufe in fcouring is loft: things brought into this fituation, what is then to be done? The moft obvious and (as it would feem) natural means are, to dike out or otherwife deepen the river, agreeably to its former ftate, or to raife the banks in proportion, fo as to be able to confine the land floods, and force them to fea, which will of confequence deepen the river; and, at the fame time, to take all means to increafe the reflow, fo as to prevent, as much as poffible, the fands from returning: this looks very well upon paper; but as to raifing the banks, that, it feems, has already been done till, in many places, from their weight, and the foftnefs of the ground they ftand upon, they fubfide upon being raifed higher; and from the great weight of water conftantly upon them, and fapping their foundations, they become liable to breaches, when the extra-weight of flood waters comes againft them. It is true, that fpeaking theoretically, banks may be made of fo large a bafe, as to bear being

carried

carried up to the neceffary height and ftrength, upon any kind of fubftance that can be called ground; but in fact, where a want of matter is the natural defect of the country, the charge of amaffing together fo much matter as would be fully equal to this purpofe, would be fo enormous, that the propofition muft die in the idea: befides, even fuppofing fuch banks made, which would, in confequence, deepen the river, yet, as nature ftill keeps at work in lengthening the courfe of the land waters to fea, the declivity would, in time, become fo much leffened, that the fands would once again increafe upon the power of the reflow; and the banks that now are fufficient, would, in a courfe of years, become infufficient; fo that the banks would need raifing higher and higher without limitation.

In regard to the diking out the river, this, if done fo as to make a fmall channel, the fame caufes ftill producing the fame effects, will foon be landed up again; and to make a great channel, fuch as I have fuppofed in the flourifhing ftate of the fens, would not only be an enormous expenfe, but if made, as the ordinary quantity of land waters is not fufficient, and the land floods not fufficiently frequent to maintain fuch a channel, it muft by degrees fill up again, as it has done before: and though a large channel, from the fuperior action of the tides, muft fill flower than a fmall one, yet, if it could not maintain itfelf when the general outfall was lefs obftructed with fands, it cannot be expected to do fo when more obftructed: nor, indeed, can I apprehend it practicable to perform any material deepenings of the river in its prefent channel: for if the water were diverted from the river to another outfall, while the work was doing, in all probability, for want thereof, the outfall would be entirely landed up and loft.

I do not apprehend, that any great matters can be effected in deepening the channel by any other means than men's hands; for though all attempts to help the natural fcour by hedgehogs, &c. are very laudable, as they tend to prevent bad from growing worfe, yet, to perform any great matters this way is not only attended with great labour, but that labour muft be continually repeated, in order to maintain the improvement before gained.

When the rivers Oufe and Welland went out at this outfall, as well as the Nen, here was a confiderable power to contend with the fands; but the two firft fources being (I fuppofe) for ever loft, no improvement, in point of quantity, offers itfelf, fave the unneceffary diverfion of the waters at Stand-ground fluice. Where a power is in itfelf weak, but ufeful, every thing ought to be avoided which renders it weaker. Yet I
can

can by no means apprehend, that if a dam were put acrofs at Stand-ground, and no water to pafs at all that way, the difference would, as matters now ftand, become fenfible at the out-fall.

It feems, then, that things are preparing for the laft remedy, and I am glad to have it in my power to fay, that a remedy every way adequate and certain yet remains, equally beneficial to the drainage and to the navigation, and that is no other than to build an out-fall fluice upon the mouth of the river.

I know that fo much has been faid and wrote againft the putting fluices upon rivers, and ftopping the tides, that there may be thofe who will ftart back at the very found of the word fluice, as applicable to the out-fall of the river Nen, as now it may properly be called.

But be not alarmed; let us hear what may be faid in favour of a fluice. I am not, indeed, for putting fluices upon rivers which are in a condition to maintain themfelves open by nature; but, in the prefent cafe, nothing is more plain, than that the fands which are depofited in the river, from its out-fall to Knarr lake, and thereby obftruct the paffage of the land flood waters to fea, caufing them to break the banks, and over-flow the whole country, are originally brought from the fea; it is alfo plain, that if thofe tides, whereby thefe fands were admitted, had been fhut out, the fands would not have been there; it is equally certain, that if for the future the fands are fhut out, it will prevent their increafe; and if the flood waters fhall roll any particle of fand that is now lodged in the channel of the river, from its prefent place towards the out-fall, that there will be no power acting in a contrary direction to caufe it to return; it therefore follows, that the effect of the power of the land waters, whether greater or lefs, is to carry out the fands that now are lodged, and to deepen the channel of the river within the fluice.

Without the fluice, if the doors were to be always fhut, a quantity of fand and filt would be very quickly depofited; but at a diftance, where the river is of confiderable breath, and confequently the motion lefs rapid, the difference will be incon-fiderable: for, whatever fand may be depofited without the fluice, that would have been depofited in the river within the fluice; as that water which is checked by the fluice, and fo depofits its contents without, will check the influx of fo much water as would otherwife have come in, and depofited its contents in place of the former: it hence appears, that, except near the fluice doors, no more filt will be depofited than would have been in cafe the water had obtained its free paffage; nay, even lefs, for whatever

checks

checks the rapidity of the tide, prevents its bringing so much turbid matter along with it: it is true, that the power of the reflux is weakened also, even if we suppose the doors of the sluice to open, whenever the water without falls. below the level of the water within; because the former part of the ebb is weakened by the want of the water that the sluice has stopped out; therefore, balancing the less quantity brought in against the less quantity carried out, we may lay it down, that the channel without the sluice will remain nearly in the same state as if there were no sluice at all; that is, in case the sluice be suffered to run whenever the level of the water will admit it; and as there is always something more of run near the tail of the sluice, than at a distance, this will again counterbalance the greater deposition near the doors; but if, instead of permitting the sluice quietly to run whenever the water without ebbs below the water within, we suppose that the land waters are kept pent in by land doors, till low water, and then the quantity of water that has been collected, discharged in a body; or when the supply of land waters are short, to discharge what has been collected, in two, three, or four, or even ten or fourteen tides in one: we shall hereby obtain an artificial application of a natural power, which will turn the balance greatly in our favour, by deepening the out-fall channel without the sluice. It is clear from reason, that the operation of a sluice thus worked is to deepen the out-fall channel as well as the inner; but whether in a considerable degree might be doubted, if this were the first of the kind; but that the operation of sluices, properly managed, is very great, is proved by incontestible experience. The out-fall channel from the sea to Dunkirk is maintained wholly by sluices, the effect of which, in the time of Queen Anne, was so great as to make a passage by which large ships of war were enabled to enter that port, and the erection and destruction of such sluices as have been from time to time adapted to the purpose of scouring and deepening the out-fall channel of this port, has been a bone of contention between England and France ever since. The canal of Middleburgh in Zealand, by which East Indiamen are carried up with their cargoes to that city, is kept open wholly by sluices; and where they have but little help, in point of depth, from the tides. The port of Ostend is kept open, and the canal from thence to Bruges, capable of navigating the largest colliers from Newcastle, is maintained by the operation of sluices. And as this is a case in point, I will beg leave to mention, that the old sluice of Ostend having been blown up and destroyed some time about the year 1751, in or about the year 1752 deputies were sent to England to view and inform themselves, whether they might not save the great expense of re-building this sluice, by allowing the tides to ebb and flow up and down this canal, as is the case with almost all the rivers in England; but there being here little land water to assist the ebbs in scouring, experience quickly convinced them, that without re-building the sluice, they would soon have had both port and canal silted

up,

up, and the country drowned; accordingly, they proceeded to re-build and finifh the fame in the years 1753, 1754, and 1755, upon a moft magnificent plan, and at a proportionable expenfe. Indeed, there is fcarcely a port in Flanders or Holland, where fomething of this nature is not applied; but I mention the above, as having, with many others, been eye-witnefs to their effects, and as being the moft confiderable in fimilar cafes; but, without going abroad, the great fluice lately erected at Bofton, with the ftate and condition of the river before and fince, will, I believe, be a ftrong illuftration of this point.

Having mentioned Bofton fluice, it will naturally be inquired, that, fince I look upon an out-fall fluice as neceffary for the out-fall of the river Nen, why I do not, like Bofton, propofe to carry out the drainage by the river's fluice, and thereby fave the expenfe of erecting a new drainage fluice, and carrying down the fewer to the wafhway houfe. I anfwer, for thefe three reafons: in the firft place, I propofe the river's fluice to be built betwixt Wifbeach and the fea; and, therefore, to preferve the navigation even fuch as it is, on fhutting out the tides it will be neceffary to keep up the water higher within, than is confiftent with the good drainage of the north level; and, indeed, if exclufive of the navigation, the waters were run fo low as to drain the north level, they would be too weak to fcour the out-fall channel; that is, as much as to fay, the out-fall would then have the fame defect as that of Gunthorpe now has, and, in effect, the north level would not be drained. 2dly, That fuppofing by the fuperior action of the land waters the out-channel could be kept open to the neceffary depth to drain, yet the quantity of land flood waters coming down the river Nen is fo much greater than thofe of the river Witham, that the drainage would be much more fubject to be interrupted, by being over-rode by the river, and would be much lefs perfect on this account. 3dly, In cafe of drainage by the river's fluice, the river muft be diked out, or a new river cut from the fluice to Guyhern, of a very large and deep channel, to anfwer this intent; and as a confiderable fluice would be ftill needed upon the out-fall of the drains at Guyhern, in order to prevent the Nen's waters over-riding the drains in time of land floods; and, on the other hand, as the back drainage, by way of Shire drain, is already performed, and in a condition to perform its office, with a little help in the way of fcouring; that is, after the defect of the out-fall is remedied; it follows, that the expenfes of draining by the river would far exceed thofe neceffary to complete it in the other way, and at laft be lefs perfect, as by the back drainage the fewers can never be over-rode by the river's floods; and in cafe of any accident to the river's fluice, the drainage would not immediately be interrupted.

When

When a fluice is eftablifhed upon the river as aforefaid, it is very clear, that the accumulation of land waters will be highly neceffary, and of great ufe; for this reafon it will then be of great confequence, that no water is loft at Stand-ground fluice; and though I am not fo penurious of water as to defire to debar the veffels penning through that fluice, of the fupply of water neceffary for that ufe, which, as I am informed, fometimes amount to fixty in a day, yet I would by all means recommend that the fluice be not run, as it fometimes is, two hours in a day, in dry feafons, by which, according to my computation, is wafted as much water as would pen through it 168 veffels: this I call an unneceffary wafte of water; for though it may be neceffary to the navigation below that fluice, upon its prefent conftruction, yet the fhoals ought to be deepened, or the works otherwife altered, while they can work without flafhes.

The great and powerful agent upon which I principally depend, to cleanfe the river and open the out-fall, is the great land floods, which, being freed in a great meafure from the checks of the tides, will, by their action within and without, gradually wear down a deep channel to fea, and I doubt not will in time enable veffels drawing eight and nine feet water, to go up to Wifbeach at common fpring tides, and to give paffage to fuch as now navigate at fpring tides only, to navigate at all tides, and with a very little help to make a conftant navigation for lighters to Peterborough. Thus will the navigation be greatly improved; the level fecured from inundations, without raifing the north bank higher than it now is, and will be a means of maintaining the out-fall channel of the drainage fluice free and clear of fands for a long feries of years; and even of maintaining, in all probability, a good paffage out to fea, even after the coaft is extended much further out than its prefent limits, and thereby giving the whole every degree of permanency that the nature of the fituation will poffibly admit.

To this fcheme I can fee but two objections; firft, that to maintain the navigation as it now is, to Wifbeach, before the deepening takes places we muft conftantly hold up the water as high as the ordinary fpring tides rife at Wifbeach, that is, about four feet above its ordinary level there, and that in confequence of this, we may prevent the mills upon Walderfea fen, and Wifbeach hundred, from throwing their water into the river, or oblige them to throw it higher: I did not happen to think of examining this point when there, nor do Mr. Elftobb's papers fix it; nor yet do I know how far their right of fo doing is eftablifhed, fo as to put a negative upon any other works to be erected upon the river; but fuppofing it fo, the out fall and defcent in Shire drain will be fo much improved by the methods propofed, that the whole may be drained that way, the water of Walderfea fen being brought in a tunnel under the river.

The

The fecond objection is, the danger of the fluices choaking by ice, upon the break of a froft, when fucceeded by the great land floods that are generally confequent thereon; this is perhaps the only folid objection to fuch a work, but then it is an objection to every work of the fame kind; and as I do not know any other lafting way of keeping open this out-fall, I do not fee any way of totally avoiding this rifk; however, to make it the leaft poffible, I would propofe to make the fum of all the openings of feventy feet wide, equal to the width of Wifbeach bridge; but as I fuppofe that a far lefs quantity of opening would be fufficient for navigation, and the difcharge of the ordinary quantity of land waters, I would propofe to make one of the openings at leaft thirty feet of clear water-way; the doors of which not to be opened or fhut by the tides or land waters, but by proper tackles and machines for the purpofe, worked by men; and thefe doors not to be opened at all, but on occafion of great floods, when the others are found, or expected, to be infufficient; and though this may require vigilance and labour, yet, as the occafions of opening it will not be frequent, it may well be difpenfed with, to avoid a greater rifk. By this opening, will, at fuch times, be the principal run, which will therefore draw the largeft pieces of ice towards it, and which may be fo much the more eafily broken and managed, than in paffing fmaller openings, that I apprehend the rifk will not be great upon the whole, or any real objection to the undertaking.

After the firft erection of the river's fluice, fome little help may be expedient, to give the land floods more advantage in the beginning; and that is, to dike out a channel of the river from Guyhern, or a little below, to Knarr lake, and to take up the gravels that are lodged in the leam or river above; as alfo by the ufe of the hedge-hog to take proper times of loofening the fands between the horfe-fhoe and the fluice; but I apprehend that after the firft winter, nature will perform the reft, when properly directed by art, in the management of the fluices.

Having advanced thus far, the placing of the fluice and expenfe of the works propofed, will be next inquired after; but as the eligibility of the propofition is the firft thing to be determined, I muft referve a minute difcuffion of thefe matters to a future confideration; and, indeed, it will require a re-examination of the premifes, with this particular view, to be able to do it with precifion; however, to give the beft idea thereof that I now can, I would place the fluice fomewhere between Walton dam and the river's end, and lay its apron almoft as low as low water at the wafhway houfe.

It

It is said, that there was once a sluice built across the river at the Horseshoe, but as it failed in a few tides after it was opened, its good or bad effects could not be known from experience; and as its failure was undoubtedly owing to insufficiency in its construction, no argument against the proposition will fairly arise from thence.

A probable ESTIMATE of the proposed works.

	£
The great sluice, with its proper appendages, - - - - -	8,000
Expenses in first deepening the river, - - - - -	2,000
Expenses in scouring out Shire drain, putting it in order, and laying Clow's cross sluice three feet lower, - - - - - - -	2,500
To cutting a new sewer, supposed two miles in length, from Gunthorpe sluice to the washway house, partly through the salt-marshes, partly through the in-grounds, supposed at an average twenty feet bottom, sixty feet top, and ten feet deep, at four-pence per yard, including drainage, comes to - - - - - - -	2,614
To a new drainage sluice at washway house of three seven feet tuns, the apron to be laid two feet (if possible) below low-water mark, - - - -	2,500
The land for the aforesaid cut and cover will contain about forty acres; the value I am not a judge of, but if, for the sake of filling up a blank, we suppose it £30. per acre, this article will be - - - - - - - -	1,200
To contingencies upon the above articles, at ten per cent., - - -	1,881
	20,695

J. SMEATON.

Austhorpe, August 22, 1768.

P. S. In answer to the question concerning the enlargement of Gunthorpe sluice by way of relief, till something more effectual can be done, I am of opinion, that if the execution of the scheme of carrying the out-fall to the washway house is speedily resolved upon, as this is the first part of the work that should be put in execution, it is scarcely worth the while to alter Gunthorpe; but if that meets with difficulties and delay, then I am of opinion, that the floor of Gunthorpe, from the information I have had, is large enough and strong enough to carry three six feet water-ways; and I am also of opinion, that the affair of scouring by the reservoir should be again tried.

I am of opinion, that the embanking of falt marfhes is a means of more fpeedily gaining land, and removing the low-water mark to a greater diftance; but I am alfo of opinion, that if the falt marfhes were not to be embanked, the fhore would yet keep advancing into the fea; and therefore fome more effectual means than the non-embankation of falt marfhes muft be ufed to keep the out-fall open, and which, if put in practice, it is not much concern, whether the falt marfhes are embanked or not.

J. S.

HATFIELD

HATFIELD CHASE LEVEL.

The REPORT of John Smeaton, engineer, upon the means of improving the drainage of the level of Hatfield Chase.

THE perfecting of the drainage of Hatfield Chafe, whether confidered with relation to the proprietors of lands therein, or the public in general, is undoubtedly a great object; for if we confider the vaft quantity of good ground in a high ftate of cultivation, whofe crops are rendered uncertain by the want of a complete drainage, or the vaft extent of common, confifting in general of naturally good land, and which, if completely drained, might be taken in and inclofed with great advantage to the owners, together with the great quantities of meadow and pafture ground, that, by an uncertain drainage, yield but an uncertain profit, one is naturally led to conclude, that if nature admits of this drainage being made certain upon fo great an extent of furface, it will amply repay any moderate degree of expenfe.

According to my information, and indeed the thing itfelf fhews it, the evil attending the draining of thefe levels does not fo much confift in this, that the water cannot be got off the furface in dry feafons, as that the rains fend down fo great a quantity of upland rivers by the courfe of the river Torne, during the winter, which overflowing the banks thereof makes it way over, and upon the furface of the levels, fo that thofe parts of the levels that lie low relative to their diftance from the out-fall, cannot be got clear of water fufficiently early in the fpring to undergo the proper cultivation, or to be in a proper condition to bear a crop; but when this happens to be the cafe, that the latter end of the winter and the beginning of fpring proving dry, the lands are in proper condition for a crop; yet, if the fummer or autumn proves rainy, the Torne again, by overflowing its banks, and otherwife obftructing the courfe of drainage, lays many of the loweft parts of the level under water, fo that the crops thereon are either loft, or greatly damaged, which was indeed the condition of it at the time I viewed it, and made my obfervations, fo that, relative to the bufinefs I was upon, I could not have feen it at a better time.

Seeing the levels in the fituation I have defcribed, one is naturally led to examine, whether this imperfect ftate of the drainage be owing to an original imperfection in the conftruction of works of drainage, or have grown imperfect in length of time; and

on

on this head I am clearly of opinion, that in their firft execution they were greatly imperfect; for though there would appear a ftriking difference between a drowned country during the whole of the year, (fome fmall tracts of rifing ground here and there excepted), and its prefent ftate, whereof every part bears crops of fome kind in fome feafons, yet it is manifeft from certain figns, that the drainage could never be much better than it is, and feveral parts could never be fo good as they now are.

It fo happens, that between the part of the levels the moft oppreffed with water, and the river Trent, which is the natural out-fall, there is not only an interpofition of higher ground, but in this higher ground, at a certain depth, there are ftrata of rocky matter, interfperfed with plaifter or alabafter, which, though not very hard, is yet confiderably more hard to work upon than common earth or clay, and whether the expenfe of digging in thefe ftrata, in order to fink the drains and aqueducts in this matter to a greater depth, or the want of fufficient fore-knowledge and experience to fee the neceffity thereof, is at this length of time perhaps altogether uncertain; yet true it is, that from hence the greateft part of all the evils that have attended this level from the firft improvement to this time, have had their fource; for, in order to avoid the neceffity of cutting into thefe ftrata of rocky matter, which appear in the bed of the prefent artificial courfe of the river Torne near the Hurft, they have been under the neceffity of confining the waters of Torne in a courfe by artificial banks, through that very part of the level country that was moft oppreffed with water, fo that the furface of the Torne being at all times above foil, and in time of frefhes from downfall water in the upland country, confiderably above the level of the country to be drained, it is eafy to fee the confequences; for as the foil, during almoft the whole of the tract, extending for fix miles in length, is a black peat earth, thofe banks, (however they may have been raifed at firft), being very liable to be trod down and poached with cattle, the fact foon would be, and now is, that not being high enough to confine the waters of the Torne after any thing of confiderable downfall, the river then overflowing and breaking them down, covers the furface of the low grounds that moft want to be drained. Thofe low grounds that are near the courfe of the Torne, are fubject to be the moft immediately and in the greateft degree affected, that is, whofe waters are difcharged at Althorpe fluices, by which the Torne has its out-fall into Trent; for thefe waters, thus got over upon the furface of the loweft grounds, gradually pafs by the drains to the common out-fall, and in their way keeping the common drains gorged with water, they, in confequence, gorge thofe that lead from parts remote from the Torne, and all meeting together at the out-fall, very much retard and prevent the drainage of the whole level. Thefe, I beg leave to obferve, are very capital errors in the original conftruction, and though the enlarging

and

and deepening the north river, (in fome places between two and three feet), by cutting into thofe beds of rocky matter and plaifter which had never been ftirred before, is in itfelf a great improvement, which in the more dry feafons is, and will be very manifeft, as it was to me at my firft vifit to the levels; yet, after the Torne has topped its banks, and all the drainages meet therein, as was the cafe when I was laft there, the immediate benefit hence arifing was not fo great as might have been expected, and as they will be when the whole is made fuitable to that part already fo judicioufly done.

From what has been faid, it may be in general gathered, and in which I doubt not I fhall be corroborated by thofe that beft know the levels, that the main propofition for perfecting the drainage of Hatfield Chafe, will be that of conveying to the Trent the waters of the Torne, without overflowing its banks, or over-riding the internal drains at any time, except thofe of extreme wet feafons in the winter.

In order to effect this, nothing is more clear than that there muft be a fufficient capacity of aqueduct to convey, and of fluices fufficient to let out the water fo conveyed, in both which effential particulars the original fcheme has been fo very remarkably deficient, that in this, as in many other inftances, I have had the opportunity of feeing they have drained as if they were in fear of doing too much.

For remedy of the evils complained of as above mentioned, a new out-fall drain and fluice have been fuggefted and pointed out to me, to a part of the Trent, about feven miles below the prefent out-fall of Althorpe fluices, to a place a little above Waterton, to which place I am informed there is a fall at low water of the Trent's furface of above four feet, and which, having been repeatedly levelled by different perfons, I take for granted. Now, an addition of four feet, added to the prefent fall into the Trent, from the feveral points of the furface of the level to be drained, is, undoubtedly, a very valuable acquifition, and which obtained, and every advantage drawn from it that it is capable of, will certainly render the reft of the bufinefs more eafy, and more nearly reduced to plain failing, which every thing of this kind ought to be, as much as poffible. There is alfo another very defirable circumftance attending the place of this new out-fall, for, as I am informed, fo low down in the Trent as the point near Waterton, as above fpecified, a flood in the Trent is fcarcely fenfible at low water, fo that in the greateft downfalls the fluice there would run at every low water, whereas, according to my information, the flood water of Trent fometimes continues fo high at the loweft ebb of tide, that for feveral tides together it over-rides the drainage water, fo as not to fuffer the fluice doors at Althorpe to open, and the fluice-keeper fays he has known

them

them to have been kept fhut for five days together. Thefe are advantages of the new fituation, very material, and not eafy by other means to be as fully procured ; fo that though
according to my eftimation, it will coft the fum of £6165, in addition to what I would
recommend to be done, yet, if all parties are agreeable to this being done, and the additional money can be raifed, I think it will be well laid out, as being of permanent
utility to the whole country, which may have occafion to utter their waters thereby ; and
though to the execution of this I fee no natural difficulty, yet, as I underftand that the
whole courfe of the propofed out-fall drain lies through parifhes and lordfhips, wholly
unconnected with participants concerned in Hatfield Chafe, in cafe difficulties fhould arife
in reconciling thefe different interefts, it feems to me of confequence to that body of
gentlemen, and the country at prefent depending upon their undertakings, to fhew how
the level of Hatfield Chafe may be drained in a very competent manner, without
going out of their own boundary. And here I muft beg leave to remark, that having
been confulted in the year 1764, upon the means of improving the drainage of that part
of the level of Hatfield Chafe*, dependant on fnow fewer, which conducts the water
fouth of the ifle of Axholme to the Trent at Ferry fluice, in like manner as the north
and fouth rivers conduct the waters north of the faid ifle to Althorpe fluices, I fay, had
the gentlemen concerned been fo lucky as to have given full credit to the doctrines and
directions contained in my report thereupon, they would, before this time, have prac-
tically feen the advantages in fo ftrong a light, that, applying the fame reafoning to the
north fide drainage, as is there applied to the fouth, they would have feen their way
through this alfo. What I have to do, therefore, is to enlarge upon what is there contained, and adopt the fame reafoning more particularly to the circumftances of the
prefent fubject.

SCHEME OF IMPROVEMENT.

FACTS.

1ft. The threfhold of the big fluice at Althorpe, I found to lie fix inches below low
water of the Trent there, the evening of the 16th of September paft ; but the preceding
feafon having been fhowery, I conclude that, in dry feafons, the Trent ebbs down fo as to
be even with the threfhold of that fluice at low water.

2dly. I found the thicknefs of the water over Althorpe fluice threfhold, one hour before low water, the fame day, to be one foot, and in paffing the fluice there was one
foot fall from the furface of the north river, at the flow running water a little above the
fluice, to the furface of the Trent at that time. From whence I conclude the
fluice

* See Vol. I. page 130.

fluice was then vending at the rate of 6000 cube feet of water, amounting to 166 tons per minute.

3dly. The water's furface was then fallen two feet two inches below its mark before the doors opened, which mark was four feet $1\frac{1}{2}$ inch above the threfhold.

4thly. At the Torne end $2\frac{3}{4}$ miles above the fluice, the water did not fettle above nine inches in the whole tide.

5thly. In viewing the different parts of the level, I found the ground about the Tunnel Pit, and particularly that part north of the Torne and weft of the New Idle, to be the moft oppreffed with water, and, therefore, relative to its diftance from its out-fall, the loweft; here they were cutting corn up to the boot-tops in water. What therefore will effectually drain this farm, will furnifh the means of draining all the reft.

6thly. I found the furface of the river Torne at the Tunnel Pit bridge, two feet $6\frac{1}{2}$ inches above the furface of the ftagnant waters upon the faid farm, and as I judged it would take a reduction of eighteen inches to run the water well off from the general furface of the loweft parts, and one foot more to produce a competent drainage; hence then it appears there will be a fall of three feet eleven inches from the furface of the drains neceffary to drain this farm to the threfhold of Althorpe fluice.

7thly I found the furface of the river Torne, at the Tunnel Pit bridge, eight feet $11\frac{1}{2}$ inches above the level of the threfhold of Althorpe big fluice, to which I fhall conftantly refer.

8thly. The Torne was, during the five days the above levels and obfervations were taken, within a few inches of over-topping its banks, quite away from the eaft end of the new cut, a mile below Tunnel Pit, to the Crooked Dyke End, and continued at the fame height at Tunnel Pit bridge, with one-eighth of an inch.

Now, had the big fluice tunnel at Althorpe been one foot wider, that is, fifteen feet wide inftead of fourteen, and its threfhold laid two feet deeper then with the fame defcent from the drain into the Trent of one foot, this fluice would have run 21,000 cube feet of water, (amounting to 583 tons), per minute, and would have run 6000 cube feet, the quantity difcharged as above fpecified, with one inch of fall. Hence this circumftance alone would have reduced the water over the whole body eleven inches.

I would,

I would, therefore, propose to rebuild the leſſer ſluice, whoſe clear water way is ſcarce eleven feet, and to make it fifteen feet clear, and to lay its threſhold two feet lower than that of the preſent big ſluice. The ſouth river, which anſwers to this propoſed new ſluice, is now diking out to a twenty-feet bottom, and to be ſunk one foot below the preſent ſluice-floor head, which is nearly on a level with the big ſluice threſhold. This enlargement to go from the ſluice to Durtneſs bridges; but to make this drain, called the South River, ſuitable to the new ſluice, it ſhould yet be deepened one foot more, and widened ſo as to preſerve a twenty-feet bottom at that increaſed depth, and to be kept upon a dead level from the ſluice to Torne End. The further communication with the ſouth, as well as the north river, to be ſhut off by croſs banks, ſo that this part from the ſouth river, from Torne End to the ſluice, will be wholly appropriated to the diſcharge of the river Torne. From Torne End to the Great Elbow, at Hurſt, I propoſe to enlarge and deepen the preſent courſe of the Torne, anſwerably to the former, that is, to a twenty-feet bottom, with proper batters, carrying the bottom from the ſouth river no further dead level, but upon a plan gently aſcending, ſo as to riſe two feet at the Tunnel Pit; that is, the bottom of the river at Tunnel Pit, and the preſent threſhold of the big ſluice at Althorpe, (or low-water mark in Trent), will then be upon a level.

In this caſe the rock and plaiſter bottom, near Hurſt, will be to deepen in ſome places 4½ feet, and the water of Torne, (running the ſame quantity as I ſaw it when near bankful about Tunnel Pit), will have its ſurface reduced below the preſent bottom.

From the Great Elbow, at Hurſt, I propoſe to make a new cut through the common in a ſtraight line to the elbow below Roſs bridge, in length about two miles, which will cut off all the moſt diſadvantageous angles, ſhorten the whole length half a mile, and leave the preſent courſe of the Torne between theſe two points, to act as a drain for purpoſes that will be ſhewn hereafter. This new cut to be carried on with a twenty-feet bottom, and, after joining the elbow of the preſent courſe below Roſs bridge, to widen and deepen the preſent courſe to a twenty-feet bottom, inclined as before ſpecified to Tunnel Pit.

At Tunnel Pit, the new river being admitted three feet ſix inches deep of water, its ſurface will then be near 5½ feet below its ſurface when I levelled it the 18th, 19th, and 20th ult. It will be two feet eleven inches below the ſurface of the ſtagnant water upon the Tunnel Pit farm before mentioned, and one foot five inches below the loweſt

parts

parts of the general furface of the faid farm, that is, five inches lower than as above ftated, to produce a competent drainage there; that is to fay, coming to a fixed mark of which the country will have an idea; its furface will be within fix inches of the threfhold of the fecond ftop-gate, entering the tunnel under the bed of Torne, at Tunnel Pit. At the fame time there will be a fall of three feet from the new river's furface at Tunnel Pit to Trent's furface at low water, as I found it the 16th, and this, in a courfe of $8\frac{1}{4}$ miles, will allow four inches fall per mile, befides a fall into Trent, in paffing the fluice of three inches, and this is taking the water of Torne to be difcharged into Trent, (that is Torne bank-full), as I found them; whereas, in all dry feafons the furface of the Torne will be confiderably lower, as above fpecified; fo that we may fay in all ordinary feafons, all the grounds adjacent to Torne on both fides, will drain by ftop tunnels, or holds, as they are called, in this level into the courfe of the river Torne. However, as the river is liable not only to be fwelled confiderably by rains, but to be obftructed by the growth of weeds, which, as they are taken more or lefs care of, are liable to hold up the water more or lefs in the river; and though I think it very practicable to make a good drainage of all the lands on each fide the river into Torne, yet, as there are fure means of doing it otherwife, I would by no means propofe thofe lands to depend wholly upon the exact keeping of the river. I would propofe, therefore, that all the foil that will be dug out of the courfe of the river, which will be a very fufficient bulk, to be difpofed, bank-fafhion, on each fide the river, whereby the Torne, thus embanked, will be fuffi-cient to hold in the flood waters at all times from ever getting upon the furface of any of the lands weft of the high grounds at Hurft, fo that the flood-waters being conveyed away as they come, through the low country before defcribed, thofe low grounds will, at the decline of the upland floods, have nothing to vend but the downfall water upon their own furface, and thereby having little to drain themfelves, will not opprefs and gorge the common drains for a long while together, and thereby prevent, in like man-ner, the drainage of the more diftant, though higher parts of the country.

Having thus, in all common times, reduced the river Torne at leaft one foot five inches within foil, at the Tunnel Pit farm above mentioned, and ftill more fo with refpect to the grounds on the fouth fide of the river, which I efteem, at an average, to lie fix inches higher, I propofe to remove the tunnel entirely, and to make the New Idle communicate with the river Torne on both fides, by ftop-gates, fo that whenever the furface of the Torne is lower than the furface of the New Idle, the drainage will be into the river, and whenever the contrary, or fo on either fide, then the refpective gate to be fhut, and the drainage go on by other means that will be defcribed. But I judge thefe gates will be of further great ufe in very dry feafons, in letting frefh water out of the Torne into the drains for watering of the cattle, and thus the Torne, inſtead of being the bane and peſt of theſe levels, will become a valuable refreſhing ſtream.

That it may not be doubted whether the Tunnel Pit farm contains the loweſt grounds in the level, relative to drainage that lies near the river Torne, I beg leave to ſtate theſe further facts :—That upon the ſurface of the Torne I found a riſe from Tunnel Pit bridge to Gate High bridge, of no leſs than two feet ſeven inches, in a courſe of about four miles. The ſurface of the water at Gate High bridge being no leſs than eleven feet 6½ inches above Althorpe threſhold, and a further riſe to the Crooked Dyke End, (where the Torne enters the level of Hatfield Chaſe), of 4½ inches. So that from the entry of Torne into the level of the Crooked Dyke End, to its departure into the Trent at Althorpe, being a courſe of 12¾ miles, there is a fall of no leſs than eleven feet ten inches, a very ſufficient allowance certainly to make a very complete drainage. I muſt, however, obſerve, for the ſake of reconciling my levels with others that may have been taken, that the ſurface of Torne at Gate High bridge was higher by 2½ feet than when taken by your ſurveyor Mr. Scott, in a dry ſeaſon. This premiſed,

I obſerved, that the grounds that were the moſt oppreſſed with water in the neighbourhood of Gate High bridge were thoſe lying to the north of Torne, and eaſt of Gate Wood road, between the Torne and the drain called Ring Dyke, and as thoſe grounds did not appear more overflowed, and about the ſame depth below the ſurface of the Torne, it will follow, that they lie higher than thoſe of Tunnel Pit farm by the aforeſaid riſe of two feet ſeven inches, ſo that when the Torne's ſurface is reduced at Gate High bridge, they will even drain into Torne there ; but much more ſafely and effectually by the courſe of the Ring Dyke, which goes from thence through the Tunnel Pit farm, and falls into the New Idle there, which drain will then want nothing but a more perfect ſcouring and cleanſing to do its intended office, in draining thoſe grounds, and thoſe ſtill weſt, which lie proportionably higher. And N. B. The lands on the ſouth of Torne lie univerſally higher than thoſe on the north ; they, therefore, will in like manner, by a proper cleanſing and ſcouring, convey all their waters by the preſent courſes to the New Idle ſouth of Tunnel Pit, that the others are deſcribed to do reſpecting the grounds on the north. Conformably to what has been already ſtated, it will be abſolutely neceſſary to prevent the Torne overtopping its banks in this diſtrict, to reduce it alſo within ſoil, and at the ſame time procure matter to form ſufficient banks to confine it in extremes of land floods. For this purpoſe, I propoſe to carry the Torne in its preſent courſe nearly ſo, that its bottom ſhall riſe from Tunnel Pit to Gate High bridge two feet ſix inches, and contract from a twenty feet bottom to eighteen, that is

to

to fay, to continue an eighteen feet bottom to Wroot Brick bridge; a nineteen feet bottom from thence to Fulfick Nook, and an eighteen feet from thence to Gate High bridge, which will be nearly fix feet lower than I found it; a pit of two feet, which will be fix inches per mile nearly, the fame as at prefent, there will then be three feet depth of water at Gate High bridge, which will be nearly fix feet lower than I found it, and will be at leaft two feet within foil of the loweft adjacent grounds, and even below the bottom of the prefent river. From what has been faid, I truft it will appear that the difcharge of the waters of the Torne has been very well provided for in all common times, fo well, that the only diftrefs that can happen will be when the river Trent is in fuch a ftate as that the doors at Althorpe continue fhut for five days, as has been reported; but it muft be obferved, that as at prefent the banks of the north river are incapable of penning in the water to above five feet above the threfhold of Althorpe fluice (the north and fouth river having an open communication between them); it will follow, that wherever the Trent in time of flood continues more than five feet above the threfhold, though it was but an inch during the whole five days, yet the doors will never open, as the flood waters of Torne will have an opportunity of expanding themfelves over the level; whereas, it is poffible, that could the Torne have been confined, fo as to have rifen fix feet above the threfhold, then the doors might not have miffed opening above two days of the aforefaid five: but to give this matter a much better chance, and, indeed, a very good one in fuch cafes, with the matter that comes out of the fouth river from Althorpe to Torne End, I propofe to make good the banks of that river, fo as to be of nine feet above Althorpe threfhold, which is full four feet higher than the banks are now capable of penning, and which is as high as the high water of neap tides in Trent, at Althorpe; fo that unlefs the river Trent in time of floods fhould continue at low water as high as the ordinary heights of the tide at high water neap tides, the gates will run once in every tide, and, as I apprehend, will as rarely be ftopped by a fingle tide when the fouth river banks are fully charged by the land waters, as they now are for five days together; and when the little fluice is re-built as above mentioned, it will difcharge fuch an amazing quantity of water, upon a declivity of three or four inches, that it will feldom happen when the fluice runs at all, but that it will difcharge as much water in an hour as the Torne accumulates in a tide*, and after the floods begin to abate the water will be run off fo quick, that as much reduction will be made in ten days as now will take two months.

* If the prefent fluice runs 6,000 cube feet per minute, it will run 360,000 cube feet per hour, and 2,160,000 cube feet in six hours, fuppofed the average time of running for a tide; the new propofed fluice would on a defcent or difference of four inches from the drain to the Trent furface, and furface in the cafe mentioned, difcharge no lefs than 2,640,000 cube feet per hour.

What

What principally remains is to fhew how the internal drainage is to be provided for, when over-rode by the Torne; and the principal part of this difficulty confifts in running off the waters collected in the New Idle at the Tunnel Pit, both on the north and fouth of the Torne, when the ftop-gates there are obliged to be fhut. I fhall begin with the fouth fide : and for this purpofe,

It is to be obferved, that the prefent courfe of the Torne, from the elbow next below Rofs bridge, to where the new river will join it at the great elbow weft of Hurft, will be deferted; this, I would propofe to dike out and fcour, which it will want in places chiefly from Shore Nook bridge to the faid elbow below Rofs bridge and from thence to continue a new drain nearly parallel to the courfe of the Torne, to join the Gadinftake drain and New Idle near the prefent ftop-gates of the Tunnel. This will be capable of running off the downfall water nearly to the level of the Torne's reduced furface at the Hurft elbow, and where I would put a fmall ftop fluice to prevent the Torne's reverting in floody feafons, at which time all the internal drainages will be ftopped, but which, (by the aforefaid provifion) having nothing to run off but their own downfall waters, will foon be done; that is, as foon as the flood water is run off in Torne, and that will be as faft as in Trent, upon which laft, in fact, the whole will then depend.

The moft overflowed part of the level fouth of Torne, and the moft difficult to be drained, is generally reckoned to be the Mizen Deeps and the grounds about the Bull Haffocks. The Mizen Deeps is a remarkable flat track of ground, very far from being the loweft in the level, but being remote from its out-fall, and near that part of the level bordering upon the New Idle, where the water runs indifferently towards the Tunnel, and fo to Althorpe, or towards Snow fewer, and fo to Ferry fluice, according as it makes the beft defcent; and as the waters from the New Idle fouth of Torne are as I now find only fuffered to run at the Tunnel two days per week, and the works formerly propofed for Snow fewer are now only executing, but have never been executed with the full effect; this track has hitherto remained the opprobrium of this level's drainage. Now, from my obfervations, it appears that the general furface of the track of ground called Mizen Deeps lies at leaft one foot higher than the general furface of the low parts of the Tunnel Pit farm, and two feet five inches above the propofed reduced furface of Torne at Tunnel Pit as above fpecified, and two feet eleven inches above the threfhold of the ftop-gate of the prefent tunnel; and further, when the deferted courfe of Torne is fcoured out, and a fuitable drain brought up to the New Idle at Tunnel Pit fouth fide, as already mentioned, the furface of the New Idle will be capable of being reduced by

this

this drain to a level with the threshold of the tunnel's stop-gate, so that there being a descent of two feet eleven inches from the surface of the Mizen Deeps, along the course of the New Idle, in the space of about two miles (which is already a good and sufficient drain) there can be nothing more wanting effectually to drain the Mizen Deeps, and if those, all the grounds less oppressed of course.

I must further observe, that on examining Ferry sluice, I found it substantially rebuilt according to my former report; it was discharging a very good body of water with only a single inch of descent in passing through the sluice, and but five inches fall from the drain to the Trent at low-water, the other four inches being occasioned by the outfall passage between the doors and the low-water mark of Trent, not being proportionably deep and clear of silt. There were also some inches deep of warp upon the floor, and still more in the drain above the sluice, occasioned chiefly by the sluice doors being out of repair and leaky, which would not be the case were every thing made suitable, in order, and the proper means used; from the sluice, for a mile upwards, the Snow sewer drain has been deepened according to the proposed dimensions in my former report, by sinking into the strata of rocky matter, and alabaster, as described near the Hurst, that had never before been stirred; yet above this deepening, the water was so held up by want of capacity in the drain, that at two miles from the sluice, the surface of the water of the drain was full two feet six inches above the surface of the water of the said drain, at the entry of the river; and had the drain been completed in like manner, to the place where Monkholm drain takes off (above which Snow sewer, in its present state, seems very good) this alone would have reduced the water over the whole depending drainage, at least two feet in its then floody state, and in dry seasons considerably more. From the above facts and observations from the Snow sewer to its head, where it joins the New Idle south of Torne, I conclude, that the surface of the grounds of the Mizen Deeps, lies full three feet above the low-water of the Trent at Ferry sluice, so that the course not being (by estimation) much above four miles, it is not to be doubted, but that when Snow sewer is completed, and the leading drains put in order, but that this track of ground will admit of a comtent drainage also.

The country south of Torne being thus adequately provided for, we shall now go to the north side; and here, after a total separation of the Torne from the drainage as before proposed, there is little remaining to be done; the north river is already deepened, and rendered of sufficient capacity, being a thirty-two feet bottom, and carried to Durtness bridges, at least six inches below the threshold of the big sluice

at

at Althorpe; the Old Dun has been diked out and deepened, with fufficient and proper capacities as far as the double bridge, within a mile and a half of Torne : the Anchor drain has alfo been diked out and deepened, from Durtnefs bridges to the elbow at John Coulmon's fhuttle, and from thence acrofs into the drain that leads the water by the fide of the Brier hills, from Woodhoufe common, which laft, on my laft view, appeared to be the moft oppreffed with water in all the diftrict of the levels, yet, from my obfervation, its mean furface lies at leaft five feet above Althorpe threfhold; fo that during my ftay there was nothing to hinder all the grounds dependent on thefe capital drains from being in a complete ftate of drainage, but the continual pouring in of the river Torne at Torne End, into the north river, which bringing down water nearly as faft as the fluice * could vend it, kept the north river, though otherwife a very capacious drain, fo conftantly filled up and gorged, that the water at Durtnefs bridges was from four feet to four feet one inch and a half, and never lower than three feet nine inches above Althorpe threfhold, the whole time I was there, and yet the furface of Brier hills was nearly clear, and Woodhoufe common with a few inches ; now, if the Torne had been entirely feparated from the north drainages by the north river, and carried into Trent by the fouth river with a feparate fluice, I look upon it as a certainty, that in the ftate of feafon I found it, the big fluice at Althorpe, that anfwers to the north river, would have run off all the other drainages ; fo that the water at Durtnefs bridges fhould never have been more than two feet above the Althorpe threfhold, and there would have been a better fall at a medium by two feet, from the furface of all the grounds that drain hither, and confequently a very complete drainage ever to Brier hills and Woodhoufe common.

Furthermore there is dependent on this drainage, a long tract of low grounds that runs up toward Armthorpe, called the Uggins Carrs, which, though oppreffed with water on account of their being more remote from their outfalls, yet, as this tract of country appears to rife even more than in proportion to the diftance, what will effectually drain Woodhoufe common, will furnifh the means, by fufficient leading drains, to drain thefe alfo ; but as I think the drain along fide of Brier hills not fufficiently capacious to receive the drainage of fo extenfive a track, it appears proper, that the crofs drain leading from this into the Old Dun, at the place called No Men's Friend, fhould be deepened, and alfo that the continuation of the faid drain fhould be enlarged and deepened from where it turns off into the Anchor drain, to its other outfall into the New Idle, between Sand toft and Durtnefs bridges.

* N. B. The fouth river being at this time diking out, very little water was difcharged by the lefser sluice at Althorpe, which anfwers to the fouth river.

It

It yet remains to obferve, that the moft material of all the works that want to be executed, is for the drainage of thefe parts of the levels depending upon the New Idle, north of Torne; and that is, the diking out, deepening, and enlarging of the New Idle itfelf, between the Tunnel Pit and Durtnefs bridges. This drain, like feveral others, paffing from a low through a high ground, as about Sand toft, is much ftraitened in dimenfions both of width and depth, paffing this high ground. This drain fhould be taken up at the depth of the prefent bottom of the north river, and carried on fo as to rife but fix inches at the Tunnel Pit; that is, fo as to be there upon a level with Althorpe threfhold, the fame as the bottom of the river Torne : but fup-pofing the water at Durtnefs bridges kept within two feet above Althorpe threfhold, and allowing the drainage water in this drain a fall of three inches per mile, this, in three and a half miles, will be ten and a half inches; that is, the furface of the water in the New Idle at Tunnel Pit, north, will be two feet ten and a half inches above Althorpe threfhold; that is to fay, eight and a half inches below the furface of the river Torne, in the ftate we have fuppofed it; two feet one inch and a half below the general furface of the loweft parts of the Tunnel Pit farm, and two and a half inches below the threfhold of the ftop-gate of the tunnel; and which furface, in dry feafons, or rather in the begin-ning of a dry feafon after a wet one, may be further reduced if found neceffary. It appears to me, that the New Idle fhould be made an eighteen feet bottom to the New drain, and a fifteen feet to the Tunnel Pit.

I mention only fuch matters as happen to occur with refpect to the particulars of the improvement of the internal drainage, becaufe it would require much more time, and more obfervations, to fay exactly what fhould be done, or what dimenfions given in the more minute parts; my purpofe at prefent is to point out the general outlines of a fcheme, whereby the whole of thefe extenfive levels might be put in a complete and fecure ftate of drainage, leaving the detail to be more minutely fpecified hereafter, in cafe the general plan be approved to be carried into execution: becaufe, if not carried into execution, the entering into thefe particulars would only be fo much lofs of time; and I truft I have been fufficiently explicit, that the validity of the fcheme itfelf may be judged of: for the fame reafon alfo I avoid entering into the detail of what appears to me might be done in the way of improvement of the works, in cafe the bufinefs fhould not be judged proper to be entered into upon fo large a fcale as I have endeavoured to explain; the prefent propofition, as I have underftood it, is to give a fcheme that if poffible may be fufficient effectually to drain and fecure the whole level; and in doing this, as boundaries and properties are become in a great meafure fettled by the prefent courfe of the drains, I have endeavoured, as much as poffible, to

keep

keep the fame courfes, without confidering how far the works might have been laid out to more advantage, fuppofing nothing had been done.

I fhall now conclude by obferving, that from the capacity of the new-propofed fluice at Althorpe, that in all moderately dry feafons, it will run off the Torne, and all the fouth-fide drainages, confiderably beneath thofe of the north river by the prefent big fluice ; I would, therefore, as an occafional help to the north drainages propofe to make a communication between the north and the fouth river, a little above the fluice houfe at Althorpe, with a ftop-fluice thereon, with pointed doors towards the fouth river ; and, furthermore, to have a ftaunch erected upon the fouth river of equal width with the propofed fluice above the communication, capable of penning occafionally the waters of Torne to the height of four feet above the threfhold of the prefent big fluice, which ftaunch will have a threefold ufe. 1ft. It will, by fhutting down when the out-fall fluice doors are fhut, prevent the warp that will always, more or lefs, get through the doors from getting further up the river. 2d. By fhutting down occafionally to ftop the Torne, in order to give the north river's waters a better opportunity of efcaping by this deep fluice through the communication. And 3dly, by penning in the Torne at convenient feafons, and letting it go at low-water, to produce a fcour, in order to drive out the warp and filt brought in by the tide, and depofited between the fluice and the ftaunch.

J. SMEATON.

Aufthorpe, 7th September, 1776.

ESTIMATE for the execution of the new works propofed for the improvement of the drainage of the level of Hatfield Chafe, exclufive of a new out-fall, by JOHN SMEATON, engineer.

	£	s.	d.
To enlarging and deepening the south river to one foot deeper than the prefent undertaking, that is, fo as to be two feet below the level of the threshold of the big sluice at Althorpe, to be twenty feet bottom, and carried upon a dead level to Torne End, this is fuppofed in addition to what is now doing, being in length 2,207 chains, containing 27,630 cube yards, which, including drainage and banking, at four pence, - - -	460	10	0
To digging out by the courfe of the river Torne to a twenty feet bottom, inclined fo as to rife from the former two feet at Tunnel Pit, the length from the Torne end to the great elbow weft of Hurst, being 101 chains, this fuppofed the same as if dug out of the solid, on account of the managing the Torne's water, will amount to 97,062 yards, which, on account of the depth and rocky bottom, including drainage, is reckoned at four-pence per yard, - - - -	1,617	14	0
Carried forward	2,078	4	0

	£ . s. d.
Brought over	2,078 4 0

To digging a new course for the river Torne, from the elbow at Hurst, to the elbow next below Ross bridge, to a twenty feet bottom, with proper batters conformably to the depth before specified, being in length 16,655 chains, or 2 miles 655 chains, will contain 150,942 yards, which, including drainage, at three-pence half-penny, — 2,201 3 6

To digging out the river from the elbow before Ross bridge to the east end of the New Cut, from the former length, being the further length of ninety seven chains, or one mile seventeen chains, keeping a twenty feet bottom, conditioned as before, will contain 87,336 yards, at three-pence half-penny, - - - 1,273 13 0

To digging out the river Torne, and embanking the same according to its present course, from the east end of the New Cut to Tunnel Pit bridge, being in length 105 chains, that is, 1¼ mile and five chains, to a twenty feet bottom, conditioned as before, which, on account of trouble in diverting the Torne's water, is supposed the same as dug out of the solid, amounting to 48,474 yards, which, including drainage of the water, at three-pence half-penny, - - - 706 18 3

To digging out the river Torne, and embanking the same according to its present course, from Tunnel Pit to Gate High bridge, so that the bottom may rise in that length two feet six inches and diminish in breadth from twenty feet at Tunnel Pit to eighteen feet at Gate High bridge, being for the same reasons estimated as dug out of the solid, which, being 335 chains, or four miles fifteen chains, will contain 114,708 yards, at three pence half penny, - - - - 1,672 16 6

To digging out and embanking the river Torne, according to its present course, from Gate High bridge to Crooked Dyke End, so as to conform to the new river at Gate High bridge, and to the old one at Crooked Dyke End, containing 6,844 yards, at three-pence half penny, - - - - - 99 6 2

Total of digging of the Torne, - —	8,032 1 5

To cutting a new drain for the south drainage from the New Idle, near the stop gate at Tunnel Pit, to the present course of the Torne, at the second elbow below Ross bridge, being in length 218 chains, and on a fifteen feet bottom, containing 53,289 yards, at two-pence half penny, - - - - 551 1 10¼

To diking out the river Torne where wanted, from where the drain falls into it, to the Hurst elbow, being supposed in length 110 chains, that will want at a medium two feet deep, upon a fifteen feet bottom, will amount to 9,142 yards, at two-pence, - - - - - 76 3 8

New drain making, -	631 5 6½
To digging of the Torne,	8,032 1 5
Total of spade work,	8,663 6 11½

CONSTRUCTIONS.

	£	s.	d.
To building a new sluice at Althorpe south river out fall, of fifteen feet clear water-way, and the threshold to be laid two feet below that of the big sluice on the north river, including the present sluice, taking the same up, the dam to Trent, &c.	1,500	0	0
The communication sluice and cut,	500	0	0
The staunch on the south river, of fifteen feet water-way, and threshold as deep as the out-fall sluice,	250	0	0
To six carriage bridges of £120 each over the Torne,	720	0	0
To six private bridges of £60. each over ditto,	360	0	0
To a road bridge over the new drain,	80	0	0
To a private bridge over ditto,	40	0	0
To a stop tunnel to the old course of Torne at Hurst elbow,	40	0	0
To two stop gates at Tunnel Pit, £60. each,	120	0	0
Constructions,	3,610	0	0
Spade work,	8,666	6	$11\frac{1}{2}$
Neat estimate,	12,273	6	$11\frac{1}{2}$
To allowance of ten per cent. for contingencies,	1,227	11	1
	13,500	18	$0\frac{1}{2}$

N. B. There is nothing allowed in the above estimate for land converted from the commons to a new river or drain, or taken from the inclosure in widening the present courses from the river or drains, nor for surveyors' salaries, nor any other kind of expenses, except what attend the articles set down, nor can the article of constructions, amounting to £3,610, be properly estimated till the proper plans for the execution of the respective articles are fixed, which cannot at present be done, which would take up a good deal of time, and if not executed, would be so much time and labour lost; nor is the deepening of the New Idle from Durtness bridges to Tunnel Pit, or the scouring or deepening any of the internal leading drains, considered in the above estimate, those being supposed works to be done by the present course of business.

ESTIMATE of the extra expense of the proposed new out-fall.

Now supposing a new sluice to be built at the proposed new out-fall near Waterton, at the same expense as the proposed out-fall and communication sluice at Althorpe, then

all

all the other articles will be equally neceſſary to both ſchemes, and the difference will lie in the cut, and as the cut will paſs through other lordſhips, the value of the land muſt alſo be conſidered, which, if not paid for, other conceſſions muſt be made to the amount.

	£	s.	d.
Now, suppoſing this new out-fall cut to be a twenty feet bottom, and to riſe two feet in its whole length from low-water mark of the Trent towards the north river, which is the leaſt dimenſions it can be ſuppoſed to have, the length being ſix miles and three chains, from Mr. Scott's meaſures, and levelling notes, I find it will contain 276,716 cube yards, which, at three-pence half-penny, is -	4,085	8	10
And there will be cut twenty-ſeven acres of land, and near twice as much covered; but if the land cut is valued at £20. per acre, this will amount to -	540	0	0
Suppoſe four road bridges, at £120. each, - - - -	480	0	0
And three communication bridges, at £60., - - - -	180	0	0
And eight ſtop tunnels for the four interſecting drains, at £40. each, -	320	0	0
Neat eſtimate, - - -	5,605	8	10
To ten per cent. contingencies on the above,	560	10	10½
Additional expenſe of the new out-fall, -	6,165	19	8¼

J. SMEATON.

Austhorpe, 7th September, 1776.

DRAIN

DRAIN FROM KNIGHTSBRIDGE TO CHELSEA.

The REPORT of JOHN SMEATON, engineer, upon the expediency of carrying the outlet of the drain or sewer from the new street carrying on by Mr. Holland from Knightsbridge to Chelsea, into the brook below the turnpike road bridge, at the Cheshire Cheese, Chelsea.

HAVING carefully viewed and levelled what relates to the premises in question, it appears to me, that the surface of the water of the new sewer, where it is interfected by the drain running across the street that brings the surface and drainage water from a considerable track of diftant grounds, is nearly upon a level with the bottom of the vaults of the adjacent new erected house, which is three feet four inches below the ground floor, or lowest floor of the said house; but as I found a point of the surface of the meadow opposite the said house, and between the street and the brook, to be elevated two feet eight inches above the surface of the drain, which point of the meadow is said to be covered several inches deep of water in time of great land floods coming down the said brook, it will follow that the ground floor of the said house is but little elevated above the surface of the flood water of the brook; while the area between the house and the vaults, as well as the vaults themselves, will be full three feet below the flood mark without; and hence it appears necessary to seek for some outfall more eligible and convenient than any that can be procured into the opposite part of the brook.

The fall that can be obtained by carrying the tail of the fewer down to an elbow, being the loweft part of the brook that lies open to the road about sixty or seventy yards below the bridge, is, according to my level taken, four feet nine inches, and which, being several yards lower down the brook than the point to which Mr. Holland levelled, agrees sufficiently near therewith to prove both, who made it four feet six inches, but being shewn the marks by the inhabitants there of the highest tides, and of the highest rise of the water in the time of the greatest land floods respectively; it appears that the highest tides rise at this place four feet six inches, and the land floods five feet, so that the highest tides will be within three inches of the bottom of the vaults, and the land floods three inches above them. The question that remains therefore is, whether, if the tail of the sewer be carried into the brook above the said brook at

the

the Chefhire Cheefe, which being carried through grounds unembarraffed with houfes, ftreets, or roads, will be more eafily done, may not be fufficient for the drainage of thefe new buildings ?

The declivity of the furface of the water between the top of the conduit above the bridge, and the loweft point at the faid elbow to which the level was taken below the bridge, was nine inches, but in time of great land floods from the obftruction and want of capacity, not only under the faid bridge, but in the paffages immediately leading to it, and from it, I can readily conceive that this fall at the furface of the water may at thofe times be doubled, and become eighteen inches ; and this being the cafe, if the outfall of the fewer was above the bridge, the water would, in confequence, be liable to be pent therein eighteen inches higher than before, that is, to the depth of twenty-one inches above the bottom of the vaults, and as this would put upwards of a foot depth of water into the areas between the houfe and the vaults, this would not only render the vaults ufelefs, but cut off the communication between them and the refpective houfes, which would continue for fuch time as the land floods remained at the height above mentioned ; and though land floods, at that height, do not happen very frequently, yet, as fuch have happened, and may happen, in an undertaking of fuch extent and confequence, where even a mifapplied apprehenfion or alarm of fome of the firft inhabitants might bring a difcredit upon the whole, and where it can be but barely made fafe when the tail of the fewer is carried to the moft advantageous outfall, I cannot hefitate to recommend that it be fo carried accordingly, that is, to the elbow of the brook, about fixty or feventy yards below the bridge, at the Chefhire Cheefe, being the loweft point where it lies open to the road.

I would furthermore recommend, to prevent all annoyance by land floods, that provifion be made for hanging a door at the tail of the fewers, and that the flood-waters brought down from the country by the interfecting drain firft mentioned be not admitted into the fewer now recommended for building ; but as by the gaining this outfall the fewer of the building might be laid fo deep as to go under the former, I would recommend the faid interfecting drain to be tunnelled over it, and conveyed upon its own foot to its former or prefent outfall, becaufe, if the country waters were to be admitted from the interfecting drain into the propofed fewer, it might in fudden downfalls fo gorge the fewer with water, when obftructed by the brook's water at the tail, as to produce the fame ill effect as would probably refult from carrying the fewer into the brook above the faid bridge as before ftated.

I obferve

I obferve that the courfe of the brook below the faid point recommended for the outfall of the fewer, is obftructed and narrow in feveral places, between that and the outfall of the brook at Ranelagh Water Gate Stairs, I would therefore advife, for further fecurity, that the proper means be taken that the courfe of the faid brook be kept properly cleanfed and fcoured.

J. SMEATON.

London, 8th April, 1778.

CARLISLE

CARLISLE QUERIES.

Carlisle Queries, with the Answers.

Query 1ft. Whether the bank made by Mr. Milbourn, in 1770, and continued ever fince, has not, in a great meafure, been the caufe of the breach in Mr. Milbourn's ground in 1771; alfo, by throwing more water into the channel of the river, has not greatly contributed to the wafting of Mr. Milbourn's ground fince that time?

Anfwer. It is impoffible for me to fay pofitively, that Mr. Milbourn's bank of 1770 was the fole caufe of breaking his ground at the fouth-weft end of the corporation wier, becaufe other caufes might concur, but the fituation of the bank was fo directly adapted to produce this effect, that unlefs fome other fufficient caufe had appeared, I fhould not have hefitated to have attributed the whole of the damage thereto.

A bank, extending more than half way acrofs a valley compofed of fine earth and loofe gravel, the whole of which was, in great floods, overflowed with water, in the flanking pofition in which it ftood and ftands, muft directly tend to wafte the ground in the way it appears to have done, from the endeavour of the water again to fpread itfelf after leaving the bank, and with an addition of rapidity arifing from the contraction of its paffage.

Query 2d. Whether the fouth-weft end of the corporation's wier is not confiderably higher than the fouth-eaft end, confequently, will tend to throw the water more to the eaft fide of the river?

Anfwer. The fouth-eaft end, and indeed the greateft part of the length of the weir, is feveral feet lower than the fouth-weft end; indeed, in its prefent ftate, it is fo low as hardly to be confidered as a wier or obftruction at all, which circumftance contributes very much to keep the current to the eaft fide, but its floping pofition upftream, like the former wier (fo far as its effects can amount to) will co-operate with the effect of Mr. Milbourn's prefent bank, tending to fpread the water againft the weft border of the river.

Query

Query 3d. Whether the corporation's prefent wier be not in as good a fituation as the old one, and will do as little damage to Mr. Milbourn's grounds, if Mr. Milbourn neglect to put in defences to his ground ?

Anfwer If the wier were again rebuilt in its old pofition, it would be attended with every ill effect of the prefent one, as relative to Mr. Milbourn's grounds, with this very bad one in addition, that to raife the water into the corporation's mill leet (or dam, as here called) it muft be raifed juft fo much higher than the bed of the river at that place, as the bed of the river there is lower than the bed of the river in its pre-fent fituation ; it would therefore act more forcibly on Mr. Milbourn's grounds, and create a greater neceffity of making defences, unlefs, by way of compenfation, Mr. Milbourn removed his bank, and reftored every thing to the fituation of the year 1770 before his bank was raifed.

Query 4th. What quantity of water will be fufficient to fupply the corporation's mill in its prefent fituation and ftructure ?

Anfwer. The Barrow, in her prefent fituation and ftructure, I find is effectually worked with a quantity of water amounting to 600 cube feet per minute.

Query 5th. Whether the wafte in the Brumel and Barton s bleach field would be of any material fervice to the corporation's mills ?

Anfwer. The wafte at the Bleach field or Stampery amounts to three and a half cube feet per minute, which is lefs than $\frac{1}{170}$ part of the above quantity, whofe effect cannot be perceived in the working of the mill.

Query 6th. Whether there is not more than fufficient at Milbourn's fluice to furnifh his mills with water, on their prefent conftruction ?

Anfwer. In the ftate in which I found the river, it afforded 2560 cube feet per mi-nute, down to the corporation's, and Mr. Milbourn's dams, befides a quantity that overflowed the corporation's wier, and went down the channel of the river ; but from the beft information I can collect from the minutes of thofe who took meafures thereof during the drought of laft fummer, it afforded full 1300 cube feet per minute ; if therefore 650 cube feet were conftantly taken for the fupply of the corporation's mills, as leakages and evaporations will unavoidably happen, and they appear clearly to have

the

the firſt right of being ſerved, there will ſtill remain 650 feet for Denton mill in the drieſt ſeaſons, which, as it appears, will render it equally capable of working one wheel abreaſt.

Query 7th. Whether you think it would be prudent in the corporation to purchaſe thoſe fields of Mrs. Lind's, and take in the water at Rocky Bank?

Anſwer. If the corporation could purchaſe Mrs. Lind's lands at a fair valuation, they lie ſo that ſometime they may be of uſe; but as I think the corporation cannot ſafely waive their right of attachment of a wier upon Mr. Milbourn, the diſagreeable circumſtances reſulting from that, as alſo from the partition of the water, would equally remain.

Query 8th. Whether the tucks were not neceſſary to preſerve the mill race, and not to injure Mr. Milbourn's grounds?

Anſwer. A defence is neceſſary where the corporation have made the tucks, and as they are in the mode of the country, I apprehend they are not under obligation to find a better, but in my own practice I recommend defences made parallel to the ſtream; I do not ſee they have any tendency to injure Mr. Milbourn's grounds in the poſition they ſtand.

Query 9th. Whether the abutment made at the high ſluice is not abſolutely neceſſary to prevent the water race being choaked with gravel?

Anſwer. I apprehend a work of this kind to be abſolutely neceſſary.

Query 10th. Whether can the corporation's wier be more properly placed to do leſs damage to Mr. Milbourn's ground?

Anſwer. I do not ſee that it can, without the riſk of producing freſh damages, fully adequate to what are likely in its preſent ſituation.

Query 11th. Whether the bank, in Mrs. Lind's ground, be not lower than Mr. Milbourn's bank, and neceſſary to prevent the ground from being overflowed, and the mill race deſtroyed in time of floods?

Anfwer. This bank appears to be a full foot lower than the oppofite one of Mr. Milbourn's; in the prefent fituation of Mr. Milbourn's I apprehend it to be expedient as a fecurity, though perhaps not politively neceffary; but in my opinion it would be much better for the prevention of damages to all the lands, if all the banks were away, and the river fuffered to take its free courfe over the whole furface of the valley, as it did before the year 1770.

J. SMEATON.

Carlisle, 1ft November, 1781.

THANKS

THANKS EMBANKMENT.

The REPORT of JOHN SMEATON, engineer, upon the practicability and expense of embanking a track of mud lying in the bay before the House of the Honorable Captain Graves, at Thanks, upon the river Tamar in Cornwall.

THIS track of foil, confifting of twenty-five acres, according to the admeafurement, lying directly in front of the manfion houfe, and betwixt that and the beautiful river Tamar, though it affords an addition of profpect when it is covered with water at high-water fpring tides, yet being imperfectly covered at high-water neap tides, and at all times deferted and left bare for the greateft part of the interval between tide and tide, it muft upon the whole be confidered as an object very defirable to be converted into a different appearance. I therefore apprehend the value thereof as a piece of rich meadow is not the whole of the confideration in the prefent cafe; if it were, I could not advife, along with the certain expenfe that muft attend the undertaking, to run the rifk of its fuccefs.

A ftrong embankation againft a large tide river, to hold out againft the greateft extremes of winds and tides without any other foundation than a bed of mud, fo foft that it is impracticable for a man to make his way over it, and in all probability the deeper the fofter, may be reckoned among the extreme cafes of practicable embankations; however, I look upon it in the prefent cafe, there is fo great a probability of fuccefs, that if Captain Graves fhall think proper to hazard the expenfe, I fhall not be unwilling to rifk my reputation as an artift upon it; the prefent and previous queftion therefore is, what the probable coft may be; and having confidered and made my own fuggeftions as to the mode of execution, and regulated the meafures by the plan and fection put into my hands, and having carefully computed every article, the fum total in materials and workmanfhip, exclufive of temporary damage or fpoil of ground for the obtaining matter from each fide, and exclufive of any charges that may arife by fupervifal thereof, will amount to £1193, which call £1200. If, therefore, the purchafe of this piece of ground as meadow is worth £1200, I muft advife Captain Graves, for the fake of the ornament, to ftand the rifk, and whenever the execution fhall be refolved on, I fhall be ready to prepare the proper plans and fections, and lay down the mode of proceeding that I would advife to be purfued.

J. SMEATON.

Austhorpe, 15th May, 1778.

COQUETT

COQUETT DAM.

(See plate VI.)

EXPLANATION of the design for a dam for the iron works upon the river Coquett, with directions for putting the same in execution.

Fig. 1ſt. is a general plan of the whole dam, with its land walls and conduits.

A B C D E F G—Shew half of the body of the dam, and the ſouth land wall, as it will appear when completed wherein.

A B—Shews the ſloping caping.

C—The rough ſetting, or continuance of the ſlope.

D—The rubble ſloping apron.

E F—The caping of the dam's end land wall.

G—Rough ſetting within the caping, which, together with the caping of the return walls, is to be ſloped upwards towards the land, in order to throw the ſtream as much as poſſible from the land.

H—The ſtart and collar of the draw-ſhuttle, for cloſing or opening the conduit of the ſouth-end wall at pleaſure ; the dotted lines ſhew the direction of the conduit.

I K—Shew half the body of the dam, as it will appear when got a courſe above the ground courſe, or ſet-off, wherein

L L—Shew the diſpoſition of the Tyes Bond-ſtones, or Headers, after every three ſtretches ; and N. B. The headers are diſpoſed, not over one another, but in the intermediate ſpaces, ſo as every third courſe to come ſomewhat about the ſame perpendicular again. They are to be ſnapped off at the inner end, to prevent the water in caſe of flood, while going on, from taking ſo much hold of them as to turn them up.

a a a, &c.—Shew trenails in the aiſler of the down-ſtream courſes, which will be more particularly explained, and

b b b b,

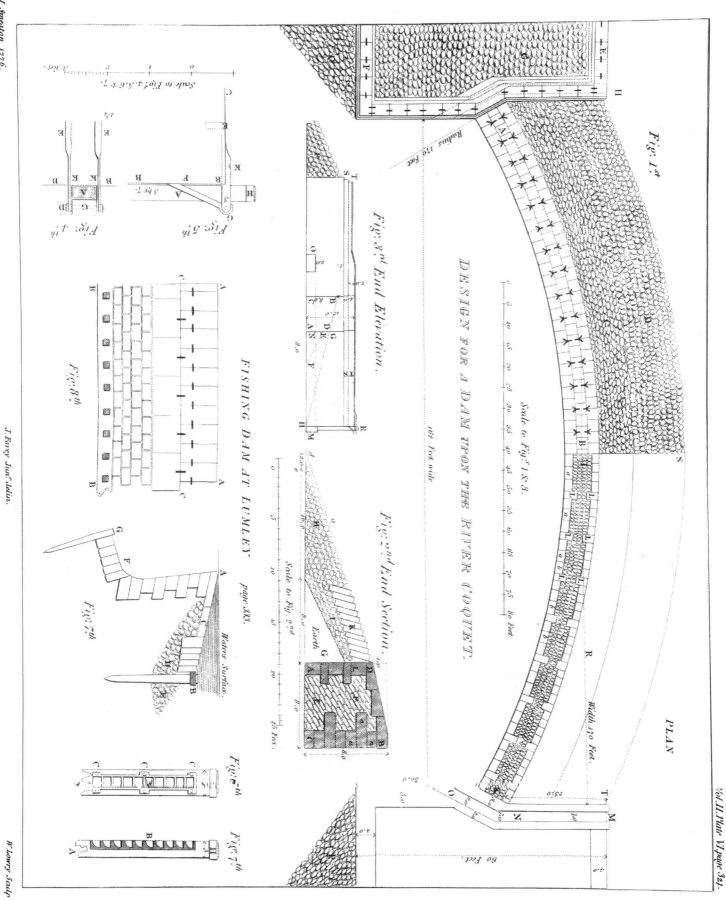

Fig: 1.ˢᵗ

DESIGN FOR A DAM UPON THE RIVER COQUET.

Scale to Fig.ˢ 1 & 3.

162 Feet wide

PLAN

Radius 110 Feet.

Width 110 Feet.

60 Feet

Fig: 3.ʳᵈ End Elevation.

Fig: 2.ⁿᵈ End Section.

Scale to Fig: 2.ⁿᵈ

Earth

FISHING DAM AT LUMLEY. page XXX.

Waters Surface.

Fig: 4.ᵗʰ

Fig: 5.ᵗʰ

Scale to Fig.ˢ 4, 5, 6 & 7.

Fig: 6.ᵗʰ

Fig: 7.ᵗʰ

Fig: 7.ᵗʰ

Fig: 8.ᵗʰ

J. Farey Jun.ʳ delin.

W. Lowry Sculp.

b b b b, &c.—Are wedges in the inner joint of each courfe, likewife to be more par-ticularly explained. The fpace between the two aifler faces to be backed with rubble, difpofed as will be alfo further more particularly explained.

M—Shews two projecting cheeks of ftone, containing the grooves for the fhuttles, and N. B. All the parts of ftone, againft which the fhuttle will rub or flide, fhould be made dry, and tallow burnt into the pores of the ftone with a hot iron, which will preferve not only the ftone but the wood from wear, and to make the tallow lead in the better, and ftrike deeper, it will be proper firft to oil over the furface with linfeed oil.

M N O—Shew the conduit before it was covered.

R—A flope of earth to prevent the water from penetrating into the ground joints of the body of the dam, and

S T—Mark out the extremity of the rubble upper apron.

P Q—The corners, filled with flopes of rough rubble to prevent the eddy water from returning and affecting the land.

Fig. 2d. is a fection of the body of the dam, to a larger fcale, wherein

A B—Is the folid body of the dam, to be firmly built with mafonry.

B C—A fection of the down ftream face.

A L D—A fection of the up-ftream face, and D B that of the caping.

E F—The rubble backing laid in floping courfes, as well as the nature of the mate-rials will admit, to what is reprefented, and to be worked in very folid, the utility whereof will be further explained.

a a a—The trenails in the down-ftream courfes of aifler to be further explained.

G—The flope of earth.

H I—The

H I—The rubble floping apron to protect the flope of earth from being wafhed away by the water, and to conduct heavy floating bodies from injuring the body of the dam or caping.

K—Rough fetting, the better to refift the action of floating bodies.

D L—Upright cramps to every other piece of caping.

  N. B. The dotted lines, a b, a c, denote the mafs of rubble ufed for a catch dam, and a d the flope of earth thereto.

Fig. 3d fhews the upright face of one of the dam's end walls, or land-walls, being to the fame fcale as fig. 1ft, wherein

A D B C—Shew the fection of the body of the dam.

E F—The flope of earth.

G H—The rubble flope.

M—The projecting cheeks for the fhuttle to flide in.

O—The tail of the conduit, appearing broader on account of its oblique fection by the face of the wall.

M N O—The conduit, defined by the dotted line.

M R—The fhuttle ftarts and cheeks of the collar.

P—The filling up of the angles with flopes of rubble.

S T—Set off on the face of the wall of fix inches each, the backfide is alfo fet off as reprefented at E F, in fig. 1ft,

Fig. 4th is the plan, and fig. 5th the upright of the top of the ftart of the fhuttle, wherein the faid letters refer to both.

A—Is the ftart of the fhuttle.

B B C—The

B B C—The face of the wall.

D E—The cheeks of the collar cramped into the wall.

D F—The ſtruts for keeping it firm and ſteady.

G—The roller, whoſe axis being releaſed by taking out the bolt, the ſtart is at liberty to be taken out.

H—A hoop to keep the head of the ſtart from ſplitting.

K—Is an indenture into which a ſhort round being laid will ſerve as a priſe to the lever.

Fig. 6. is the upright face, and fig. 7. is a profile ſection of the ſtart and caſt iron rack, to which apply an iron gave lock or lever to draw the ſhuttle, wherein both refer to the ſame letters.

B—Denotes the teeth of the rack, to ſupport which, they are to be caſt ſolid with teeth, and let into

A—The ſtart, and are fixed and prevented from drawing by the ears.

C C C—

H—The hoop at top.

DIRECTIONS FOR THE EXECUTION.

There is not a more difficult or hazardous piece of work within the compaſs of civil engineery than the eſtabliſhment of a high dam upon a rapid river that is liable to great and ſudden floods, and ſuch I eſteem the river Coquett, and ſuch the dam here propoſed to be erected; and when it is conſidered, that the performance of every part of the intended works depends upon the firmneſs and well-eſtabliſhment of the dam, and further conſidered what loſs, diſappointment, trouble, and vexation will attend a failure thereof, eſpecially in the winter ſeaſon, when ſuch a misfortune is more likely to happen than at any other time, it will readily be granted that too much care and cir-
cumſpection

cumfpection cannot be ufed in putting the defign here propofed into execution, and which, if duly and carefully attended to, will, in my eftimation, not only be proof againft all that can come againft it, but may be executed without any material derangement from floods, while it is going on, and efpecially if begun and carried on at a proper time of the year, with a competent number of workmen.

The time of the year I would recommend to begin is as foon in April as the weather appears to be tolerably mild and fettled, which generally happens about the middle of that month; and as the diftance between the land ftool walls on the upftream fide of the dam is propofed to be 170 feet, equal to the width of the river at the water line, when raifed to the height of the dam, the foundation pits will be required to be funk in the flopes of the banks, fo that they will be defenfible from the river in all middling fpeats, and this part of the work may be going properly on before the furface of the river is reduced to its low dry weather ftate. I would therefore advife, that the land ftool walls, with their refpective conduits, (which will be fully fufficient for conveying the current of the river in its ordinary dry weather ftate) be firft not only eftablifhed, but completely finifhed before the work of the body of the dam is begun; and this direction is the more neceffary to be obferved, becaufe I have known in feveral inftances, for want of attending to this, that the river has made a new courfe between the land or dam's end walls, and the main land; fo that the walls fhould not only be raifed to their height, and caped, but rough fetting, fuch as is reprefented at the fouth land ftool at G. fhould be completed.

It is to be noted, that the conduits being fuppofed two feet fquare in the clear opening, are to be built with well-jointed aifler; and in cafe the rock fhould prove too tender, jointy, or uneven, which will probably be the cafe, it will be proper to inlay well-hewn floor ftones of fix or eight inches thick, and at leaft three inches broader on each fide than the conduit, fo that the fide walls may ftep or tread thereon, and the covers muft alfo be of fingle ftones, to reach acrofs, fimilar to the floor ftones, and the whole jointed with good water cement of pozzelana mortar, fuch as hereafter will be defcribed. The dam's end walls being thus fubftantially conftructed, and made ready to receive the river's current, the water is to be turned through the conduits as follows. Let a part of the mafs of rubble ftone be depofited as reprefented at H, fig. 2., fo as to form a kind of bank $2\frac{1}{2}$ feet high, and floping each way, and arching conformable to the bafe of the dam, before which depofit a quantity of earth, which, by treading or working, will ftop the water and form a catch dam fo as to turn the current of the river through the conduits. And as this catch dam, as to

the

the rubble part of it, will, in fact, be so much done towards the general construction, it will be proper at first to give it a good base, and slope downstream, for it will thus be enabled to resist a speat without material derangement, which is always to be expected and provided against during the going on of the work : by this means, and the aid of a few sodds, and earth to prevent the water reverting, after it has passed the conduits, the whole area, whereon the dams are to be built, will be laid dry.

The outline of the downstream side of the dam is then to be traced out by a radius of 170 feet, (which may be done by forming together a number of small slips of deal, like pantile laths, having men to support them at proper distances from the ground), and the upper side by a radius of 178 feet. The stony bed of the river for the breadth of the aisler respectively forming the upstream and downstream face of the wall, is then to be reduced into level stretches, not all reduced to the same level for the whole length of the dam, or breadth of the river ; but as far as each can be conveniently pursued upon one level, rising or falling by a step from one level stretch to another : and the thickness or height of the aislers are to be so adapted, that they will all come to one level at the top of the first, or if need be, the second course, and then upon the downstream face you set off a couple of inches ; but the upstream face wall is to be built perpendicular, without any set-off.

The whole ground bed of the aisler upon the rock must be laid with the best cement, and the front of every course upward, for at least six inches inward from the face of the wall must be jointed with the best mortar also ; but still better, if the aisler were wholly bedded therein from top to bottom.

The whole row of aisler in the ground course of the downstream side should be set and closed in before any part of the first row of aisler of the upstream face wall is set, and as the first row of aisler on the downstream side is carried on, it must be backed up upon a sloping surface, a little short of the upstream course up to the full height of the downstream course ; and the whole of that course being thus finished, and backed up, then begin to bring on the second course downstream, and the first course upstream, completing the backing upon a slope as before from the top of the first course upstream, to the top of the second course downstream ; observing always, that these slope surfaces be as well closed in, and laid as snugly and smooth as possible, with this intent, that when a fresh or speat comes down while the body of the dam is building, that the water may glide over it with the least impediment possible, by which means it will be subject to the least derangement ; and to the intent that the water may affect the whole as equally as possible. A new course is never to be begun till the course preceding is completed, observing always, to keep the upstream face a course lower than the downstream; and every Saturday evening, or upon the apparent approach of rain in an evening, not to leave, if it be possible, a course incomplete in that situation.

As a flood going over the dam in an incomplete state is most likely to take down the aisler of the downstream face, by way of security and prevention, I propose that every other stone should be pinned down by an oaken or fir trenail of one inch and a quarter diameter, reaching through the upper, and at least six inches into the course below, and a fir wedge being moderately driven into the back side upper corner in each joint of the aisler, this will steady every stone between two of its neighbours that are trenailed down, and thereby the whole will become fixed, as it were, in one arched mass, and fast to all under courses. This trenailing may appear a good deal of trouble, as it is only meant as a temporary security, while the dam is building, yet as the hindrance will be nothing, provided additional people are appointed to this service, and the cost trifling in proportion to the damage that might otherwise be done by a single flood coming down during the course of the building, I cannot but strongly advise it; and as by this means, together with the former directions, the work will be carried on with perfect security, all that hurry and bustle will thereby be avoided, that alone renders the work considerably more expensive. The practice should be this: the holes in the course setting should be bored through before the stones are sett, with jumpers gaged to bore a hole $1\frac{3}{8}$ inch diameter, and after the stones are set, another person with a jumper gaged to $1\frac{1}{4}$ inch, continues the whole six inches deep into the stone below; which done, (a number of trenails being ready prepared of a sufficient length, and truly planed to an octagon formed from a square gaged to full $1\frac{1}{4}$ inch), and the leading end of the trenail being a little snaped off, to give it a clear entrance, it will easily drive through the upper stone having a hole of $1\frac{3}{8}$ inch, and the angles of the octagon left on will, by driving, compress and fill the under hole pretty tightly, which the swelling of the wood by wet will soon render still tighter; but as notwithstanding all care the jumpers may not all bore perfectly alike, nor the trenails planed up be exactly to the same gage, the under or entering end of every trenail must be split with a saw for about $1\frac{1}{2}$ inch, and an oak wedge of about $2\frac{1}{2}$ inches long, $1\frac{1}{8}$ inch broad, and about a quarter an inch thick at the head introduced into the slit, and then when the end of the wedge touches the bottom of the hole, the trenail, if not already jambed, will be effectually so, by the penetration of the wedge into the trenail; this done, the top of the trenail being cut off flush with the

upper

upper fide of the ftone, and then wedged fomewhat kindly, the whole is made level, and the trenail prevented from drawing either above or below. I hardly need fay, that the trenails fhould be made of clean ftuff, free from knots, otherwife they will break in driving; but it may be neceffary to fay, that they fhould be planed up upon a parallel, and not made with the entering end fmaller than the other; for every thing expected from this is done by the two wedges, and all danger prevented of their jambing improperly; it is alfo neceffary to fay, that they ought not to be driven by an axe, or any fmall round faced hammer, but by a flat faced iron maul, fuch as is ufed by fhip-carpenters for driving their trenails. The whole that I have hitherto mentioned is eafily executable by common hands: but the point that will require the moft attention will be in the perfon that fits the ftones to the courfes, to mark out the holes of fuch as he propofes to be bored through, that when they come into their places, the holes of the upper may mifs the joint of the under; but if this is thought too great a tafk, if every bored ftone is about four inches out of the middle of its length, it will always mifs a joint, by changing the other fide up when found neceffary. The upftream aifler courfe will need no trenails, being abutted upon the rubble, and defended by lying fo much lower.

As the ground courfe upftream is carried on, it fhould be immediately gently rammed behind with earth, and the floping apron of earth gradually brought on as the courfes are advanced; and the rubble backing fhould alfo advance proportionably with the main body between the catch dam and this flope of earth.

J. SMEATON.

Aufthorpe, 13th September, 1776.

Face mortar for downftream aifler.

To two bufhels of flaked lime in flour fifted,

One bufhel of Civita Vecchia pozzelana.

One bufhel of good fharp fand, beat the whole well together till tough like a pafte.

Face mortar for upftream aifler and backing the joints of downftream.

To two bufhels of common mortar prepared with its ordinary quantity of fand meafured when wetted up, add

Half a bufhel of quick lime in flour fifted, and

Half a bufhel of Civita Vecchia pozzelana fifted, the whole well beaten till tough, like a pafte.

Mortar

Mortar for the backing, or inside rubble work.

To three bushels of common mortar measured, when wetted up, add

One bushel of quick lime in flour, when sifted, and thrown in by degrees, with the neceſſary addition of water, upon making it up for uſe, beaten well.

The conduits to be jointed with downſtream face mortar.

The land walls below the level of the dam's caping upſtream, of upſtream face mortar.

The land walls to be one foot above dam's height for ten feet below the dam's crown, and three feet high at the tail of downſtream mortar, all above of upſtream mortar, the backing mortar as above.

J S.

FISHING DAM, AT LUMLEY CASTLE.

See pl. VI. figs. 7, 8.

ON viewing and examining the dam at Lumley, I find its original conftruction fo very bold, and, as I fhould efteem it, unfafe, that it furprifes me how it has ftood fo long. The breach made laft winter is now fo far ftopped as to pen the water to its ufual height. By way of further fecurity, I would advife to fill the pool-hole below with rough quarry ftone thrown in, and alfo to form a fufficient flope on the upper fide in the fame manner, and then to make good the wall and penning at the foot, in the way that it has hitherto been; for, as it would coft a very confiderable fum of money to build a new dam there, upon folid principles, I do not fee that any thing can be done better with it, than to repair it as it ftands; and as it has ftood a great many years, notwithftanding the natural weaknefs of its conftruction, it is very poffible, with proper repairs, it may continue fo to do. The greateft deficiency that is experienced, arifes from the caping being taken off by the ice; as the top is compofed of ftones of very middling fize, they are to be kept down by laying a boarding of brufh-wood upon them, which brufh-wood is kept down by laying ftones thereupon; when this fails, the wall of the dam is liable to be taken down. The moft likely way to preferve the dam will be, therefore, to cap it in a more effectual manner, and I would recommend the method defcribed in the fketch; fig. 7, pl. VI. being a fection, and fig. 8. a plan of a portion of the dam, wherein A A A is the front or delivery nofe of the dam, B B B a half tree fixed with piles, at about ten feet up-ftream of the front, and to be fixed as low as it can conveniently be got in a dry feafon in the fummer, fuppofe $2\frac{1}{2}$ or three feet below the full dam's furface; inftead of the brufh, I propofe long ftones of eight or nine inches thick, A C laid floping towards the half tree, the longer thefe ftones are the better, but they fhould not be of lefs length than three feet; thefe to be cramped, every other ftone, to long deep fetters, and joggled every ftone to its neighbour, as fhewn, fig. 8. The reft of the fpace between the deep courfe of large fetters and the half tree, to be penned in with rough pen ftones, and the whole to be fupported by rough ftones as fhewn at D, and the flope E in the fection, thefe to be thrown in after the piles of the half-tree are drove; F fhews the prefent penning or apron fupported by the piles G.

J. SMEATON.

Austhorpe, 15th May, 1777.

N. B. A part of their caping may be done annually, and the reft fupported with brufh-wood till the whole can be executed.

KINNAIRD

KINNAIRD ENGINE.

The REPORT of JOHN SMEATON, engineer, upon the powers and improvements of the engine of Kinnaird, &c.

HAVING received the following propofitions or meffage, from the Carron Company, I proceeded to an examination of the premifes. The meffage is as follows:

After Mr. Smeaton has viewed the prefent engine, with the fituation of the houfe, and confidered any other facts he may think neceffary to be informed of, the company wifh to be advifed which of the following modes they fhould purfue, in order to be fully able to command the water, and fulfil their contract and after-agreement, with Mr. Bruce.

1ft. Would Mr. Smeaton advife the increafing the powers of the prefent engine, and to fink it to the Cox-road coal, and run another back mine?

2dly. Would he advife the prefent engine to ftand where it is, and erect a fmall additional engine upon No. 10. pit? and if that could be done, and leave room in the pit, (which is nine feet diameter), to draw the coal, and fave the expenfe of finking another pit?

3dly. Or would Mr. Smeaton advife the erecting a new engine upon No. 10. pit, or elfewhere, of fufficient powers to draw all the water from the depth of fifty fathoms?

N. B. The bore of the prefent lift is $14\frac{1}{4}$ inches, and it was imagined it would require another bore of eight or nine inches to command the water?

Having, in confequence of the above, examined the premifes, and made my own obfervations, and having alfo read a printed tract, entitled, " Memorial for the Carron Company, 23d Jan. 1777," and confidered the plan thereto annexed, my opinion upon the different articles is as follows:

Anfwer to the 1ft.—In perufing the printed paper, I find there has been much altercation, whether the engine that has been erected at Kinnaird, either in its former ftate,

with

with a cylinder of $52\frac{1}{2}$ inches, or as at prefent, with a cylinder of fixty-two inches, was or was not of fufficient power to draw water from the depths required; now, as it may give fome light into the whole of the matter before me, to clear this point, I will endeavour to do it before I proceed further.

I do not find that, in all the various reafonings about the power of this engine, regard has been had to any other circumftances than the diameter of the cylinder, and the diameter and perpendicular height of the pumps thereto annexed, fo as to calculate what neat burthen is laid upon each fquare inch of the cylinder or pifton's area, without paying any regard to the velocity of the engine's motion under fuch burthen, that is to fay, to the number of ftrokes made per minute, and length of the ftroke, without which it is impoffible to calculate the quantity of water drawn to a given height, and without which all reafoning about the effect of the power of an engine, is like attempting to afcertain the capacity or content of a folid, by having only two dimenfions.

In the courfe of my obfervations upon fire engines, through a confiderable feries of years, I have found engines calculated to carry a load, varying from under 5 lbs. to upwards of 10 lbs. to the fquare inch, thofe carrying a light burthen are expected to go with greater velocity than thofe carrying a heavy one; fo that if an engine, carrying 5 lbs to the inch, goes with double the velocity, or, as I call it, makes twice the journey per minute, to what is made by an engine whofe cylinder is of equal area that carries 10 lbs. the effects of the power or bufinefs done will be equal, that is, the water actually raifed from an equal depth will be equal. In the fire-engine, however, as in other machines, there is a maximum that without new principles of power cannot be exceeded; bad proportions of the parts, and bad workmanfhip, may make an engine fall fhort in any degree of what it fhould do; but which cannot be exceeded by the moft accomplifhed artifts.

Experience has, however, in fome degree, directed difcerning artifts towards a medium, as to the burthen an engine fhould carry upon the fquare inch. The original patentees, from fome of their firft performances, laid it down as a rule to load the pifton, fo as but little to exceed 8 lbs. to the inch; but, on more experience, they diminifhed that load, and amongft the beft articles of late years the practice has been to give them at or about 7 lbs. to the inch. Any of thefe will do, if the parts are properly proportioned, but, from a long courfe of very laborious experiments, I have fixed my fcale near upon, but fomewhat under 8 lbs. to the inch, including the raifing the

injection

injection water, which is a circumstance never brought into the question in the several computations mentioned concerning Kinnaird engine.

A pump $4\frac{1}{2}$ feet below the pavement of the Cox road coal appears to be 104 yards below the delivery drift of the engine pit, which, to allow for one or more cisterns, call 106 yards; now, according to my scale, an engine of sixty-two inches cylinder, will work a pump 106 yards in height, of near upon $14\frac{1}{2}$ inches, the present pump being but $14\frac{1}{4}$ inches, will, therefore, be under the proper load. An engine also so loaded will go $10\frac{1}{4}$ strokes per minute of eight feet two inches each, that is, it will make a journey of $83\frac{2}{3}$ feet per minute; whereas the present engine, in the condition I found it the 2d September last, was going, $11\frac{1}{4}$ strokes per minute, of six feet each, and then taking down its water, there being near twelve fathom water in the pit, occasioned by a preceding stoppage to put in a new set of larger pumps for the ground column: now, though it made $11\frac{1}{4}$ strokes per minute, yet they being only of six feet each, its journey was no more than $67\frac{1}{2}$ feet per minute, which is less than $83\frac{2}{3}$ by near upon one-fourth part of the present performance; and as the upper tier of pumps was no more than fourteen inches, we may safely lay it down, that the effects of the present engine may be improved without augmenting the powers, in the sense I apprehend is meant by the question (that is, by putting in a larger cylinder and pumps in the proportion of four to five), that is, by one full fourth of the present performance, and yet take the water from the pavement of the Cox road coal. This supposes also, that all the water is let down to the bottom of the Cox road coal, and then drawn up again, whereas the water being detained at its present random of the engine pump foot, it is probable, that a small bore would draw the water from the Cox road coal; and also in running a back mine underneath the former to the verge of the boundary of the estate, which is all I apprehend can be expected from the company in literal performance of their after-contract, as they cannot be required to find the main coal within the boundary of the estate, unless it were there; and if the company have nothing further in view than the literal accomplishment of their contract, this seems not only to be the best, but the only way by which it can be done.

In comparing the performance of an engine of sixty-two inches cylinder, such as I have stated it when improved, with what it performs at present, or probably ever has done, it may possibly be inferred, that the engine at Kinnaird has always been defective in construction, size, or strength, to what it ought to have been in conformity to the lease; but here I must remark, that if it was not defective in any of those respects,

according

according to what was deemed the best mode of practice at the time it was erected, it could not be incumbent on the company to perform what was then not known, as no one could, or can foresee what improvements in these machines could, or actually have been made; and as I date the improvements mentioned when brought into the field to so late a date as the beginning of the year 1774, this is long posterior to the erection of this engine.

Kinnaird engine appears from the first intended to work a fourteen inch pump, which is a larger size than has been commonly calculated upon, to wier a coal whose quantity of water was unknown: twelve inches in this case has generally been deemed sufficient.

A cylinder of $52\frac{1}{2}$ inches to work a pump of eighty yards in height, which appears to be sufficient for winning the main coal at the place where the engine now stands, and of fourteen inches bore, would lay no more, including injection water, than $7\frac{7}{10}$ lbs. per square inch upon the pistons, which is very nearly what I have since proved by experiments to be the best; and if a $52\frac{1}{2}$ inch cylinder were fitted to this work, as we may well suppose in the interval between Martinmas, 1760, when the lease commenced, and the 31st of December, twelve months following, when the visitors ordered the company to go down to the Cox road coal, and there run a level mine to cut the main coal to the dip; I say, if in the space of thirteen months, a $52\frac{1}{2}$ inch cylinder were fitted to this work, it was very natural, and I look upon it at the time adviseable, to try how far they could go with this cylinder, since no one can say with certainty what is in the bowels of the earth, either of strata or water, till they are pierced; and the rather, because if they then had in prospect, what I am informed was really the case, that they should take in a considerable feeder at the middle of the pit to the cistern of the upper column of pumps, then they had a probable chance to go down with a pump of less size to the pavement of the Cox road coal; and a less pump was actually put in, and there continued till the new pump was put in (as I am informed) the last summer. Now, if the upper column of pumps were fourteen inches, and a lower column of thirteen inches had been found sufficient to have drawn the water from the Cox road coal, these would have laid no more burthen upon the piston than $9\frac{1}{3}$ lbs. upon the inch, which is yet considerably under the limits of the burthen, which before that time I had seen in use, and at that time, so far as I know, no man had proved what was best. It was, however, then commonly known, that the engine under this load would go slower than if the burthen were lighter; but for ought that then appeared, or was known, it might be expected to draw more water than the same cylinder would have done, if fitted with twelve-inch pumps, which, from the Cox road coal, including injection, would have laid no more upon it than $7\frac{7}{10}$ to the inch and which, as already observed, speculatively confidered, might be deemed at this time an advifeable power to attempt the winning of a colliery whofe quantity of water is not known ; and had the ftrata laid as fuppofed in the after-agreement of the 31ft December, 1761, the work done accordingly, and the water grown eafier, by continuance of working, as frequently proves to be the cafe, then there had been no reafon of complaint of the infufficiency of the engine.

It happened that in the year 1769, curiofity and obfervation (being in thefe parts) led me to take a view of Kinnaird engine, which was then working with the fifty-two and a half inch cylinder; the minutes of this view are now before me, and it appears to me that this engine was doing as much work in proportion to its fize of cylinder, as the generality of engines at that time did, of which I examined a good many, being then preparing for my own experiments.

This engine, however, as I underftand, by unloofing more water, was afterwards overpowered, and the Carron company did then what is ufually done in the cafe, put in a larger cylinder, increafing it from fifty-two and a half to fixty-two; the pump being too great a load for the fmaller cylinder, remaining as before; the effect of which alteration would doubtlefs be by diminifhing the load upon each inch of the pifton's area, to increafe the velocity of the motion, and thereby with the fame pumps to draw more water; and which water having fince again increafed, the fixty-two inch cylinder is now in want of a further increafe, or of improvements that may be tantamount. It, however, may now be made a queftion, whether the building was originally made fo as to be likely to be ftrong enough for fuch an increafe of power, becaufe, if it was not, the Carron company muft be allowed to be blameable otherwife, in building an engine incapable of a greater power than that originally defigned, when it frequently happens that an increafe of power in thefe cafes is wanted.

In this refpect, as I find the beam wall above five feet thick, and the other walls proportionable, I conclude it built with intention, if occafion required, to receive a greater cylinder than the original one of fifty-two and a half inches; its dimenfions are fully fufficient for a cylinder of the fize put in, and of fuppofition of the foundation being good, I fhould not fcruple to put in a cylinder of feventy inches; a confiderable fettlement has however happened in the beam wall directly under the working beam, fince the putting up of the fixty-two inch cylinder; the queftion is therefore, whether this fett has happened through want of dimenfions, or infufficiency of the work, or from fome other caufe?

As

As I am informed, some feet under the beam wall there is a stratum of sand which, though sufficiently compact to bear weight, will not bear the least oozing of water, if it can get loose, as is commonly the case: this stratum I understand was sunk through in sinking the little staple pit for the injection pump; now, in wet seasons, a very small drainage of water from this stratum of sand into the staple pit will bring particles of sand along with it, and by continuance of the same for years, though almost imperceptible in a small space, is very capable of producing the effect now seen, and though it is probable that this would have appeared before this time, though the fifty-two and a half cylinder had remained, yet the greater the agitation caused by the larger cylinder, and perhaps wet seasons co-operating therewith, may of late have brought on this appearance in a greater degree; and it is an argument of the solidity wherewith the walls of this building have been raised, that the settlement is perpendicular, being in the middle under the beam, and near the staple pit, while the side walls at a greater distance, and less pressed, stand where they were built.

The properest way to put a stop to this evil, which, from the nature of it, above described, must be growing, will be to put into the staple a cradle or tub, close boarded, so as nearly to shut up the water, but effectually the particles of sand issuing from this stratum, and if need be to continue the boarding or sheeting on the side next the engine of the main pit. This effectually done, the settlement will go no further, but if it be not, recourse must be had to buttresses to discharge the weight and action of the beam sideways, which buttresses had best be founded near the surface, and if need be, supported by piles driven thereunder; it is impracticable to rebuilt the beam wall in a going engine, otherwise, if time would permit, an arch might be cast over the space liable to settle.

Furthermore, as it appears that the plan of the engine is such as to be capable of having as many boilers, and of as large size, as in event might be wanted; in this respect also it is adapted to an enlargement of the powers, and though it is by no means such a construction as I should now recommend, yet at the time of the erection, things appear to have been done and disposed in such a method, as was most generally approved; in short, though the Carron company do not seem to have been sparing in the execution of such things as from time to time have appeared for the best, yet they seem to have been particularly unfortunate in receding from the first proposed situation of the engine; for had they persevered in getting through the first difficulties of that situation, every thing after would have become easy, and many heavy expenses, and many disappointments

ments avoided. Having now difpatched the immediate fubject of the firft queftion, as well as what naturally arofe out of it, I proceed to confider the fubfequent.

In anfwer to the 2d and 3d queftions, which I fhall take together, becaufe they both of them feem to imply the giving up any idea of a fpecific literal performance of the after-agreement of the 29th of December, 1761, in refpect to the finking of the Cox road coal, &c. and to put the matter from what has turned, upon what is beft now to be done, in order to win out of the Kinnaird eftate as much coal as can be got out of the main coal, and fuch of the upper feams as it may feem eligible to get.

It is now ftated to me, that the pavement of the coal, at the pit No. 10, lies ten feet below the foot of the prefent engine pump; it appears therefore neceffary in order to get the coal there, to erect an engine of fome kind at the pit No. 10, and the queftion is, whether a fmall engine, with pumps of eight or nine inches, to draw fuch quantities of water as cannot be commanded, in point of level, by the prefent engine, or at once to erect an engine of fuch conftruction, power, and fize, as fhall command the water of the whole field at that place? and I cannot hefitate in faying, that I look upon it to be the ultimate intereft of the company, under the circumftance of waiving all other agreements, to build a new engine at pit No. 10, that fhall at once be likely to command the water of the whole field.

To begin a work with fuch a power as is generally found to be competent, though from unforefeen accidents it afterwards proves to be otherwife, this muft be looked upon among the number of uncertainties that attend thefe affairs; but now that the field is in a great meafure explored, and the quantity of water inveftigated, to erect any thing that has not the probability of being fully competent, would be unpardonable. It is ftated that a nine-inch bore may be wanted in aid of the prefent engine: to work this with proper effect at No. 10 pit, will require a thirty-feven or thirty-eight inch cylinder to be erected new, with a houfe, and all its furniture; but to do this work, and all that I found doing at the prefent engine will be done by a fixty-five inch cylinder, properly conftructed at No. 10 pit; yet to provide for all contingencies that can reafonably be expected, I would advife the company to put on an engine of the firft rate, viz. an engine of feventy-two inches cylinder, which will work pumps of feventeen inches bore, and draw one-fifth more water than the laft-mentioned engine, and in cafe there fhould be no increafe of water beyond the prefent, will do its bufinefs in lefs than fifteen hours per day: the great advantage of having time to fpare is well known.

Much

Much in cafes of this kind depends upon the eftimates of expenfes, but to make an accurate eftimate of two fire engines would of itfelf take much time, and when done, the balance of the account would be involved in the uncertainty that would attend the altering and putting the prefent engine into a good and durable way of performing its duty. I muft therefore, at prefent, content myfelf with fuch opinion as to expenfes, as arifes from a general view of the fubject, and in this way I cannot think that the company can poffibly, in the firft inftance, fave £500 by erecting a fmall engine upon pit No. 10, rather than the large one recommended, to fet againft which difference, they will firft have the materials of the prefent engine. 2dly. They will fave confiderably in engine keepers, for the prefent engine will require the fame number as at prefent, with an additional number for the new engine, whereas, an engine upon my plan of feventy-two inches cylinder, will confume no more coals than the prefent engine requires per hour, when there is ftandage for the water, it will be worked by fingle fhifts. 3d. The whole of the coals confumed by a leffer engine will, for the fame reafon, be faved. 4th. Which is moft material, as the prefent bufinefs will be done under fifteen hours per day, there will be more allowance for additional water, and confequently more certainty in carrying the works regularly on.

I have only now to add, that a fett of feventeen-inch pumps, in three lifts, may ftand in a pit of nine feet diameter, and after brattifhing them off entirely, leave five feet of the pit's diameter clear, for drawing coals.

J. SMEATON.

Austhorpe, 9th October, 1779.

YORK WATER WORKS ENGINE.

DIRECTIONS for working the engine at the York Water-works.

Management of the fleck.

OPEN the fire door when the steam is at the strongest, and with proper tofs of the shovel, spread thinly over the thin places, miffing the black ones of the former feed, and keep doing this conftantly, repairing the thin places, and keeping as good a body of fire as you can, which will be done by opening the fire door, and repairing more often than if you were working with coals; when the ciftern is full, take the opportunity of cleaning your grate, which muft never be fuffered to grow foul.

Management of coals.

Break every coal that is bigger than a goofe's egg, and the oftener you fire and the thinner the better.

Management of the cataract.

When the engine throws more water than the fervice takes, put on the cataract which is adjudged to eleven ftrokes per minute, if this does not keep up with the fervice, about fix inches before the ciftern is empty, throw off the cataract, and felf work the engine till the ciftern is full, then apply the cataract, and fo on.

N. B. When the cataract is at work, cafe the fire proportionally, and alfo take care to leffen the boiler's feed, lefs water being confumed in fteam.

Management of the air cock.

After the engine is got to work out of hand, open the cock till it begins to flirt a little water at each ftroke, if it comes over hard, give it more air; in fhort, give it all it will take without fpoiling the coming in of the engine.

J. S.

To

To the Gentlemen Proprietors of the York Water-Works.

GENTLEMEN,

FROM the experiments I tried, on Wednefday the 24th inftant, compared with an experiment made before the proprietors, the 25th Nov. 1779, the following turn out to be facts, by deduction.

1779, Nov. 25.	The engine raifed	244 hhds. per hour	160½ per mett of coals
1785, Aug. 24.	Per cataract	328	240½ of fleck
		328	298½ of coals
	Self-worked	442½	357½

DEDUCTIONS.

1ft. From the above it appears, that the improvement of the engine, from 1779 to the prefent time, is in proportion of 160½ to 357½.

2dly. That the engine, when felf-worked, makes a better produce than when worked by the cataract in the proportion of 357½ to 298½.

3dly. It appears from the account of fleck ufed from the 21ft Feb. to the 23d Aug. that there were confumed feventy-three waggons, that is, 2044 metts of fleck, and the number of working days in the above time being 158, that will amount to near thirteen metts per day.

If the engine is fuppofed to work at an average feven hours per day, and an hour be reckoned for lighting the fire, this will be eight hours coals, and allowing 1½ mett of fleck an hour, the confumption of fleck will be per day twelve metts.

4thly. According to the price of fleck, when depofited in the yard at 11s. 8d. per waggon, a day's work in fleck will amount to 4s. 6½d.

But, according to the price of coals, when depofited in the yard at 14s. 2d. the day's work of coals will amount to only 4s. 5½d. and if we could, in all cafes, take off the water as faft as the engine can draw it, when felf-worked, a day's work will be done with coals for 3s. 8½d.

5thly.

5thly. In 1779 they worked about $8\frac{1}{2}$ hours, and were an hour and a half raising the steam, and as they raised 244 hogsheads per hour, in $8\frac{1}{2}$ hours they raised 2074 hogsheads per day's service.

In 1785 we work seven hours, and raise the steam in three-quarters of an hour, but as we now raise 328 hogsheads at common rate of working with cataract, in seven hours we raise 2296 hogsheads per day's service.

J. SMEATON.

York, 29th Aug. 1785.

P. S. The cataract makes more of the coals than when self-worked, and the overplus thrown over into the Ouse, in the proportion of twenty-seven to thirty, or of nine to ten.

Date.	N.	Strokes.	Length of stroke.	Time.	Weight of coals.	Time of consumption.	Reduced strokes per minute.	Metts used per hour.	Hhds. raised in an hour.	Hhds. raised per mett.	
1779. Nov. 25.	1	12	5 3	1 0	24 0	1 19	$10\frac{5}{10}$	$1\frac{52}{100}$	244	$160\frac{6}{10}$	coals, self-worked
1785. Aug. 24.	2	10	6 0	1 3	11 $4\frac{1}{2}$	0 44	$9\frac{5}{10}$	$1\frac{28}{100}$	$328\frac{3}{15}$	$240\frac{3}{10}$	sleck, cataract
Do.	3	10	6 0	1 3	12 $5\frac{3}{4}$	0 $56\frac{1}{2}$	$9\frac{5}{10}$	$1\frac{10}{100}$	$328\frac{3}{10}$	$298\frac{7}{10}$	coals, cataract
Do.	4	13	6 0	1 2	11 6	0 $46\frac{1}{4}$	$12\frac{6}{10}$	$1\frac{24}{100}$	$442\frac{5}{10}$	$357\frac{9}{10}$	coals, self-worked

The above corrections were made in consequence of an error found out in the calculation, and were corrected at York accordingly, the 24th Sept. 1784.

If we compare the produce of the cataract with the produce self-worked, according to coals, say $442.5 : 328.3 :: 1.24$ to 9.22, but per No. 3. the quantity used was 1.1, therefore too much in proportion, but $1.24 \times 9.22 = 1.4428$, whose root 1.07, so that the saving by the cataract is nearly but rather short of a mean proportional between the reduction of coals, according to the quantity raised and self-working, if wasting the overplus.

N. B. An experiment was tried the 24th September, 1785, of the difference between working by the cataract and self-working and stopping; the trial of self-working and

stopping

stopping lasted three hours, in which there were three stoppages, which together made 47½ minutes, that is at the average sixteen minutes; the time going was therefore 142 minutes, or 47½ minutes each, say forty-eight, the time of going was therefore to the time of stopping as three to one, that is, they stopped one-quarter of the whole time. In this proportion, and the time so small of stoppage, the produce of self-working was better than that of the cataract in the proportion of twelve to thirteen; or one-twelfth part of the cataract's produce; but could the stoppages have been longer, the advantage would have been greater, as the damped fires grow foul, and extra coal is consumed by raising the fire.

LUMLEY COLLIERY ENGINE.

DIMENSIONS proper for a Fire Engine for raising water at the New Winning of great and little Lumley Colliery, belonging to the Right Honorable the Earl of Scarborough, to answer effectually the quantity required by Mr. Edward Smith, in his memorandum of the 13th May, 1777.

A thirty-two inch cylinder will, at fix fathom three feet, work a pump of twenty inches bucket, and deliver 1150 hogfheads per hour.

The fame cylinder will, at twenty-feven fathoms, work a pump of ten inches bucket, and draw 288 hogfheads per hour.

A cylinder thirty-two inches diameter fhould be tight feet long.

The diameter of the fteam-pipe 5½ inches.

The boiler eleven feet diameter in the bilge, and nine at the bottom, which is to rife nine inches.

N. B. If inftead of a twenty-inch pump for draining the upper main coal wafte, four pumps of ten inches are made ufe of, they will do the fame thing; and all the pump pieces can be afterwards applied in forming the pump to go down to the low main coal, the clack pieces and working barrels excepted.

J. SMEATON.

Lumley Castle, 15th May, 1777.

P. S. Having viewed the ground where a waggon bridge is intended to be erefted, I would advife to build crofs-walls, at the diftance of about twenty feet from each other, up to a certain level, and then to lay a couple of Riga or Memel whole balks of twelve inches acrofs upon the walls, and upon the balks to lay joifts, and platform with waggon rails in the ufual manner as upon geers. The walls may be each about twelve feet long, three feet thicknefs at bafe, and diminifhed to two feet at top.

J. SMEATON.

CHASE

CHASE WATER ENGINE.

CALCULATION of the effects to be expected from Chase Water Engine, and comparison with the former two engines upon the same mine.

See plates 6, 7, 8.

THIS engine is expected to go, with its full load of fifty-one fathom of main pumps, full nine strokes per minute, and being gauged, as per design, to $9\frac{1}{2}$ feet strokes, its common working stroke will be full nine feet.

The working barrels being full $16\frac{3}{4}$ inches; this, from the above data, will turn out per hour 880 hogsheads of water, wine measure, and I expect it to work at this rate with thirteen bushels of coals, London measure, per hour, supposing that all the steam necessary for working thereof be raised by boilers of a similar construction to what is proposed in the design to be placed under the cylinder; but as under this boiler alone, not above six bushels of coals per hour can be consumed with advantage, what is equivalent to the rest must be supplied from the proposed smelting furnaces; it is possible, however, by means of the Cornish cataract, to work it at the rate of four or $4\frac{1}{2}$ strokes per minute, with its full load, and, possibly, this boiler may furnish steam to work the upper column of pumps at its full rate of nine strokes per minute, till the second or middle column be prepared and fixed, without any help from the out-boilers.

Respecting the former engines upon this mine, it has been stated to me that

The larger cylinder of 66 inches drew an $18\frac{1}{2}$ inch bucket 24 fathoms.
And that the lesser of 64 ditto $17\frac{1}{2}$ ditto   26 ditto.

Total lift   50 fathoms.

That eight strokes per minute, of six feet each, kept the water down in summer, and ten in winter.

That engines of this size generally consume six London chaldrons per day, or 216 bushels, and that the two engines above mentioned consumed eleven chaldrons in twenty-four hours.

Now,

Now, as thefe two engines drew from one to the other, and took in no additional water by the way, the effect in reality will depend upon the lesser bore, and it will be found, by a calculation similar to the former, that a $17\frac{1}{2}$ inch bore, working a six-feet stroke and ten strokes per minute, will deliver 710 hogsheads, (wine-measure), per hour, which is lefs by 170 hogsheads than the calculated product of the new propofed engine, and is near upon one-quarter of the compound effect of the two former engines lefs than now propofed, fo that what was done in twenty-four hours by the two former engines, will now be done in about nineteen hours twenty-two minutes; and even if the effect was to be computed upon the larger barrel, without paying any regard to the lefs, (which ought not to be admitted), its effect will be only 798 hogsheads, which is lefs by eighty-two hogsheads per hour than the propofed new engine, being near upon one-ninth of the former product lefs than the prefent, which would reduce the time from twenty-four hours to twenty-one hours forty-eight minutes.

The coals confumed before being eleven chaldrons in twenty-four hours, this will be at the rate of $16\frac{1}{2}$ bufhels per hour; but as what was done in twenty-four hours will now be done in nineteen hours twenty-two minutes, the confumption per day will be only 252 bufhels, equal to feven chaldrons, fo that the natural motion of the engine being reduced by the cataract to do the fame work, that is, to make delivery of the fame quantity of water per hour, viz. 710 hogsheads regularly through the twenty-four hours, the coals ought to be reduced in the fame proportion; that is dividing 252 bufhels by twenty-four hours, and then the hour's confumption will be reduced to $10\frac{1}{2}$ bufhels, at the average of twenty-four hours.

Still, however, this is fuppofing the new engine always to be raifing as much water throughout the year, as has been found neceffary in winter, but the former confumption of coals being regulated upon the mean of fummer and winter, and the winter's water being to the fummer's as ten to eight, as found by former experience, the mean will be nine, and the work of the new engine being lefs in the fame proportion, that is as ten to nine, the coals will be again reduced from 252 bufhels per twenty-four hours to 227, equal to fix chaldrons eleven bufhels, that is $9\frac{1}{2}$ bufhels per hour upon the average of the year.

The performance therefore of the new engine, in point of effect, will be greater than the two former actually were in the proportion of four to five nearly, but the effect produced by the fame quantity of coals will be as $16\frac{1}{2}$ to $9\frac{1}{2}$, that is, nearly as feven to four.

The

The above computation of the quantity of coals for the new engine, is on fuppofition of coals of the beft quality of Newcaftle; and, having procured a fample of coals from Wales, from William's colliery, which, as I have been informed, are the beft engine coals ufed in Cornwall, I have found them very little inferior to the beft Newcaftle. Whether the coals ufed at Chafe Water, for the two engines, were of inferior quality, I cannot fay, but have no reafon to fuppofe it, as the wrong general proportions ftated, together with the errors commonly committed in the conftruction of thefe machines, are fufficient to account to me for all the difference.

It will, therefore, upon the whole, appear, that after the mine has been once cleared of water, fuppofing the quantity to be lifted, to be the fame as before, that whatever expenfe of coals is faved upon $9\frac{1}{2}$ bufhels per hour, upon the average of the year, will be in virtue of the new applications of the heat of the fmelting furnaces.

J. SMEATON.

Austhorpe, 14th Feb. 1775.

P. S. On perufing my former difpatches, I find fome fmall differences in the computations, but thofe have been chiefly occafioned by the different ftates of the former machinery and perpendicular height to be lifted.

J. S.

N. B. This engine, though not the largeft that has been built, will be of confiderably greater power than any I have feen, and when worked at its full extent, will work with a power of 150 horfes acting together, to keep which power throughout the twenty-four hours would require at leaft 450 horfes to be maintained. I have feen an engine of feventy-five inches cylinder, but that which worked with the greateft effect was feventy-two, and its effect equal to that of 108 horfes continually acting, or of 324 horfes to be maintained, which is lefs than the above in the proportion of feven to five nearly.

ADDITIONAL.

ADDITIONAL REMARKS upon particular parts of the engine for Chase Water.

PLATE VI. contains an upright section of the whole engine, which in general explains itself to those who are acquainted with the general principles of the fire-engine; what follows is descriptive of those parts, which require particular management in the execution.

Plate VII. fig 1, is a plan of the lower side or working surface of the regulator plate, which is situated in the top or crown of the boiler. Fig. 2, is a plan of the metalic plate, which works against it; and fig. 3, a section shewing the two put together. Figs. 4, 5, 6, 7, and 8, refer to the working geer, as will be more fully explained.

Plate VIII, fig. 1 and 2, shew the method of supplying the boilers of a fire-engine with fresh water. Fig. 3, is an elevation of the piston and chains for Chase Water fire-engine. Fig. 4, a plan of the piston. Fig. 5, a section of the cylinder and piston. Figs. 6 and 7, a plan and section of the injection cap; and figs. 8, 9, and 10, explain the manner of uniting the pump rods to the chains.

Fig. 4, of pl. VIII. shews a plan of the under surface of the piston's bottom, which, as there shewn, must be a planking of wood, and by the bolts and rings A A and B B, must be fastened to and under the cast-iron piston plate C D, fig. 3, in the manner shewn by the section fig. 5.

This planking to be of elm or beach, about $2\frac{1}{4}$ thick when worked, the edges of the two planks forming the cross, to be grooved with a plow plane, about three quarters of an inch wide, and as much in depth, and the corner pieces being tongued, so as to fit thereto, this, with a few rivets to hold the cross planks together where they are to be halved into each other at their intersection, being hooped with a good iron hoop half an inch thick and $2\frac{1}{4}$ broad, will bring the whole tight together; the ring out and out to be a quarter an inch less than the cylinder. The flat iron rings shewn in fig. 4, upon the under surface of the piston, are to be let in flush with the surface of the wood, and the bolt heads to be chamfered or counter-sunk, so as to be flush with the ring. The

planking

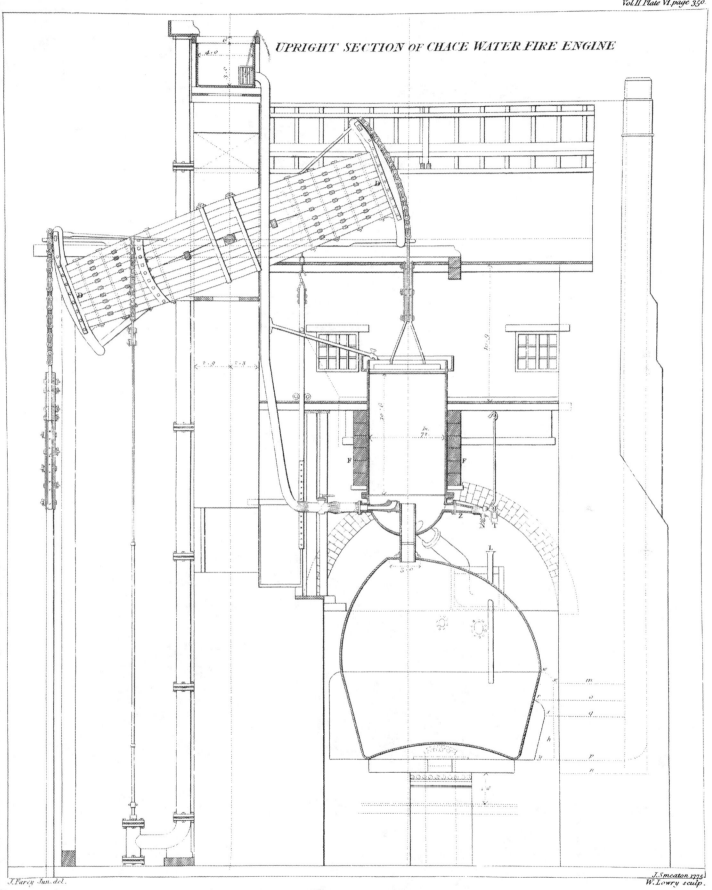

UPRIGHT SECTION OF CHACE WATER FIRE ENGINE

J.Farey Jun.del.

J.Smeaton 1775
W.Lowry sculp.

Published as the Act directs, 1812, by Longman, Hurst, Rees, Orme and Brown, Paternoster Row, London.

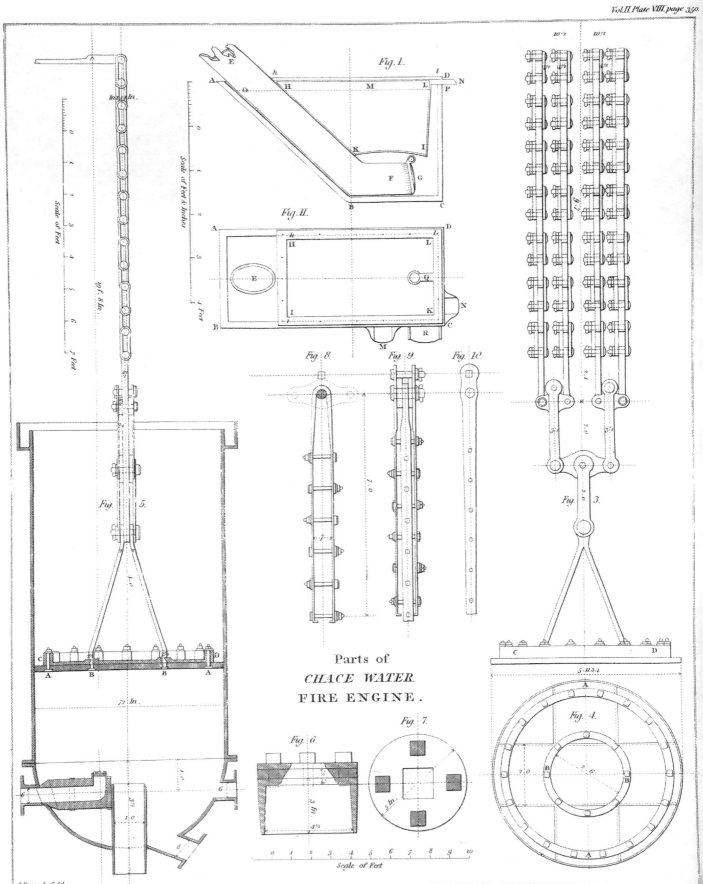

Parts of
CHACE WATER
FIRE ENGINE.

Fig. 1.

Fig. II.

Fig. 5.

Fig. 8.

Fig. 9.

Fig. 10.

Fig. 3.

Fig. 4.

Fig. 6.

Fig. 7.

Scale of Feet

Scale of Feet

Scale of Feet & Inches

J. Farey Jun.r del.

W. Lowry sculp.

Published as the Act directs 1812, by Longman, Hurst, Rees, Orme and Brown, Paternoster Row, London.

planking to be fcrewed on with a double thicknefs of flannel and tar betwixt that and the iron pifton plate, and in cafe of irregularities, the hollows to be filled up with addiitonal thickneffes of flannel, &c. fo as to exclude the air between the plate and the wood, the bolts to be alfo carefully fecured fo as to make a water tight joint from above. The planking being thus prepared, a fheathing of deal boards, fhot clear of fap, and three quarters of an inch thick, muft be nailed on to the planking, with tar and hair, or a fingle thicknefs of flannel, fo as to exclude the air between the planking and fheathing, and at laft the fheathing to be dreffed off, fo as to be perfectly flat and fmooth. N. B. If any parts of the pifton fhanks happen to be prominent, they may be let into the planking, and the air extruded from the cavity by fome proper matter applied, as tar and hair, white lead and oil, &c.

The fpace that may, and, indeed, ought to be left, in fetting the boiler and cylinder between the top of the regulator's fteam pipe, and the bottom of the cylinder's fteam pipe, may be fitted up with a ring of iron or lead, fo as to make a fmooth infide, and fupport the wrapping which was bound round it to make the joint.

The regulator is fuppofed to be fixed upon a copper plate adapted thereto, and to the top of the boiler; this very little to exceed one-eighth of an inch in thicknefs; the part covering the opening of three feet at top of the boiler to be fet (not concave or convex), but as little flat as poffible, by which means it will fpring up and down, and give way to the working of the engine, without ftraining the joints.

The upright feeding pipe L, in the general upright, pl. VI. is here defigned to anfwer, not only the purpofe of a lower gage cock, but of a fafety pipe, which will be done by boring two or three gimblet holes through it at the depth proper for a lower gage cock, and three inches lower than thefe holes, to bore a hole of half an inch diameter, and three inches below that to terminate the pipe altogether: by this means, when the water is got too low, the fmall holes will give notice by producing a rackling noife; when the fteam blows by the half inch hole, the noife will produce a greater alarm, and when fo low as at intervals to get to the bottom of the pipe, the water and fteam will iffue in fuch a manner as to awake the dulleft apprehenfions.

In the injection cap, figs. 6 and 7, pl. VIII., is fhewn a fquare hole of $1\frac{1}{2}$ inch fquare, which is fuppofed to be fufficient when the engine has got its full lift; but it is hardly neceffary to fay, that when firft fet a-going, it fhould be much lefs, by being plugged up, or or originally pierced to half an inch, and enlarged, as the weight of the column makes it neceffary. To avoid the trouble of drawing the pifton, to perform this enlargement gradually, often occafions much mifchief at ftarting; it is, therefore, of confequence to have the tackle for drawing the pifton, fuch as to do this bufinefs handily and readily; for this purpofe, let an iron crofs of a competent ftrength be provided, the whole length of one bar to be about two feet, and a fhorter welded acrofs it of about one foot, or fifteen inches; the two longer ends to be terminated with an eye or loop; laftly, provide a rope of competent ftrength and length, with a hook at each end, fuppofe in the whole about two yards long, then the fhorter two arms of the crofs being introduced between the branches that are united at top, the larger arms being in a parallel fituation to the cylinder beams, and the two hooks of the rope ap-plied to the two eyes of the crofs, hook the lower tackle block to the middle of the rope, and applying the purchafe, the chains will be eafed of the weight of the pifton, without altering their fituation, by which means the great bolt will be eafily drawn out, the pifton raifed in its proper fituation, and the bufinefs being done within, may be reftored to its place again, with little lofs of time or difficulty.

As the pifton has frequent occafion of being refted upon the cap at the top of the cylinder, and as the weight is apt to bilge it when made of lead, it would be proper to order it of caft iron; and if the pifton head water is fupplied by introducing a pipe through the fide of the cap, making the cock a ftop cock, then the top of the cap will be clear to reft the pifton upon without difturbing that pipe and cock, and fome precaution in thefe refpects is the more neceffary, on account of the wooden bottom.

Refpecting the great working beam, fufficiently fhewn at D D, pl. VI. The great plate for the gudgeon being five inches in thicknefs, the two pieces on each fide the middle are made of whole balks, of twelve inches fquare when dreffed, thefe are notched in three quarters of an inch on each fide to keep the gudgeon plate fteady; they are then fcarfed and fprung together as per defign, and moulded off upon the outfide towards the extremity by a fair curve, according to the defign.

It is beft to fpring thefe four pieces together before the reft, and leaving them a little longer than the others, confine them together by hoops at the ends, and the curve thereon being made fair, the plank pieces are to be brought on upon them by ftrong notched pieces of wood, which the carpenters call clamms, feveral on each fide, between

the

the places where the bolts are to be; thefe, by wedges, will bring all together, then the bolts and plates are to be fitted, and all brought tight; laftly, the plates for the drift keys are to be mortifed out, or, to fave trouble in boring and mortifing, a little attention to meafures will enable the ingenious carpenter to nick the pieces with a faw, and fplit out the cores, before they are brought together, and being made a little too fhort at firft, either in this cafe, or in that of actually mortifing, they are to be very carefully dreffed with a heading chifel; after this, the keys are to be adapted with a plane, one by one, eafing the parts that bear; the keys are to be made of heart of oak, perfectly dry and hard; thofe keys are to have no drift fideways, that is, to be quite parallel with refpect to their thicknefs, but to take their drift altogether endways of the mortifes, being fix inches at one fide, and five the other. And N. B. If any key fits well, but drives too far, or, by tightening occafionally after the engine is fet to work, gets driven too far, it is better to put it in with a ftrip of pafteboard, or other matter of an equal thicknefs, than make a new one, as hereby the parts are fure to fit. It is neceffary to fay, that the dreffing the heads of the mortifes as near as may be to a ftraight line is the principal nicety of workmanfhip that is required.

Refpecting the cylinder beams fhewn at F F, pl. 6, upon which the cylinder refts, and which are kept down by being entered into the fide walls of the houfe, it is to be obferved, that the key's being fixed in the fame manner and form, and for the fame ufe as the great beam; the fame care muft be obferved as to the matter and fitting thereof.

As the bolts of the fpear plates are the moft liable to fail, in figs. 8, 9, and 10, pl. 8. is fhewn a method of making them of a very increafed ftrength, without weakening the wood more than ufual, by making them fquare $1\frac{1}{2}$ inch by one inch, the breadth-ways upward, and by bringing the whole fubftance through the plate upon the nut fide as well as the head fide; but if this increafe of projection is objected to, it may be fufficiently reduced by ufing drift keys inftead of fcrews and nuts, which will at the fame time render the work more fimple.

The fame method as is fhewn in cramping the ends of the plates for the top of the main fpear will be proper alfo for the top of the fpear plates in ordinary, the cramp or cock part being not more than about five-eighths of an inch. The fpear is fuppofed to be fingle till you go down near the upper pump head, then it is propofed to be divided into three flatways by the ufual method: but I have to remark, that the uppermoſt pump ſhould anſwer the middle ſpear, by which means, the weight of dead ſpear on each ſide the middle will be the moſt equal poſſible, and on that account leſs liable to ſwag in working. And N. B. As the pumps being in different heights, will not interfere, the outſides may be brought nearer the middle, if found more convenient.

The upper end of the ſpear is ſhewn, fig. 8. pl. 8. to be ſeven inches ſquare, and ſo to continue to the diviſion into three; but I apprehend for the greater ſecurity in jointing, it may be more eligible to make it of eight by ſix, or rather ten by five, and at half way between the top and bottom to cut away half its breadth, reducing its breadth by a ſhoulder to five inches ſquare. Suppoſe the length of ſpear wood fifty feet, then the ſhoulder will be at twenty-five feet from the top; here bring on another piece of five inches ſquare and fifty feet long, which will exceed the firſt piece by half its length; ſecure the pump joint with three ſpear plates one on each ſide, and one oppoſite the ſide where the wood is whole, croſs bolting all together in the method already ſhewn; and to the ſhoulder formed between the two pieces laſt joined apply a third, exceeding the former by half its length, ſecuring the butt joint as before, and ſo going on till you have got your length. By this means, one half of the wood will every where be whole, and the other half ſecured by ſpear plates in the beſt way; and ſtill the better to keep the whole together, and from drawing, in caſe of any failure in either half, between two of the laſt bolts next the ends of the ſpear plates (that is, between the two five inch pieces), apply a key of dry oak, of one inch thick, and five inches high, as is ſhewn for the cylinder beams, and working beam; and alſo two more at equal diſtances between each ſett of ſpear plates, putting a clink bolt through both pieces, one on each ſide the key. The other ſpears after the diviſion I ſuppoſe to be made and ſecured in the uſual way.

As the wear of the leather of the bucket, and conſequently much trouble, depends upon a proper manner of ſizing and making the buckets' hoops, I ſhall here add ſome notes thereon.

The bucket hoops on the outſide ſhould be a cylinder, that is, the ſame diameter above and below their tapering or conical inſide, muſt therefore be formed by the different thickneſs of the metal, their external diameter ſhould not be more than one quarter of an inch leſs than their reſpective barrels, that is, they ſhould be as big as poſſible, ſo as to allow for the unequal thickneſs of the leather without jambing the

hoop,

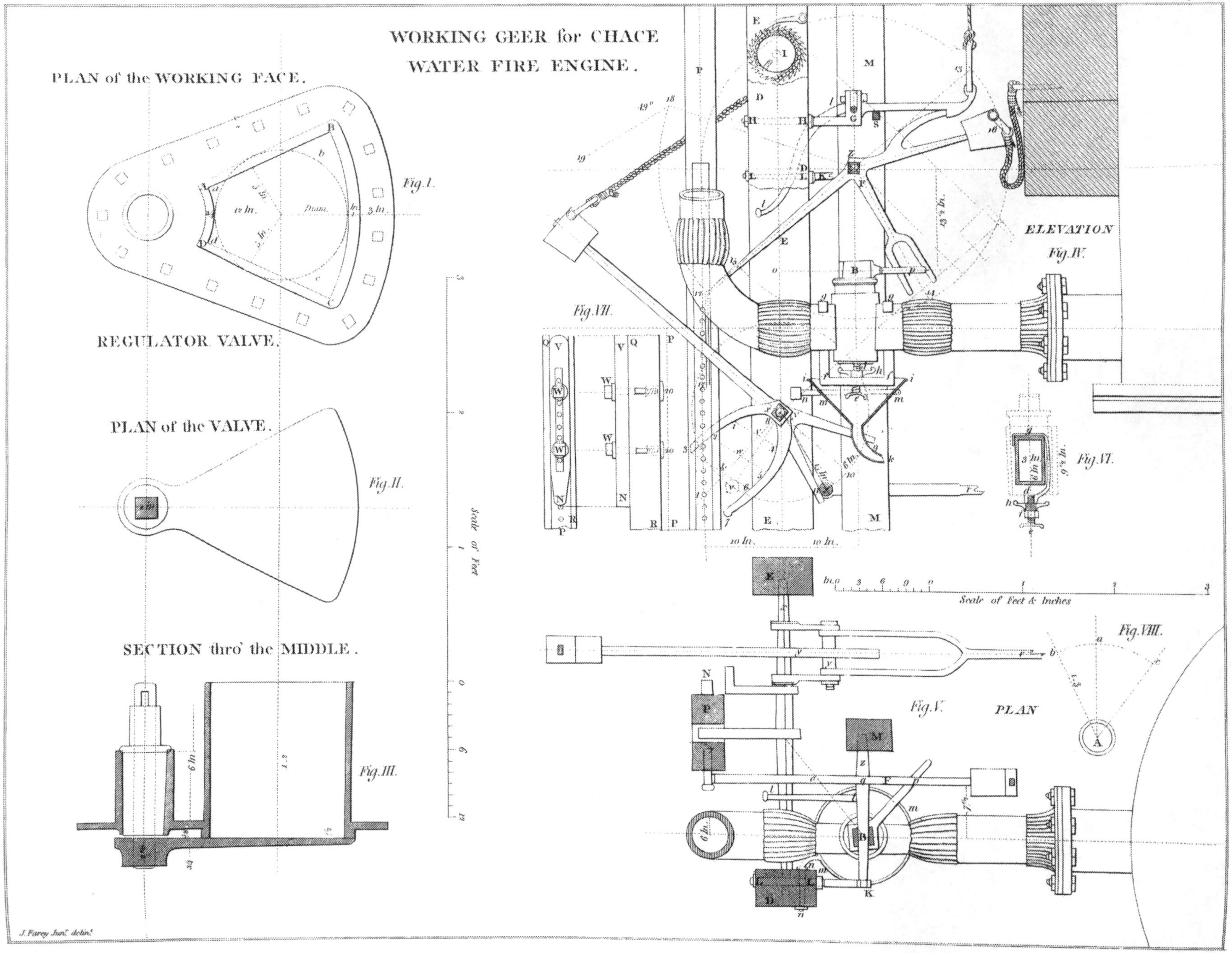

WORKING GEER for CHACE WATER FIRE ENGINE.

ELEVATION
Fig.IV.

Fig.VI.

Fig.V.

PLAN
Fig.V.

Fig.VIII.

A

Scale of Feet & Inches

Fig.III.

PLAN of the WORKING FACE.

Fig.1.

REGULATOR VALVE.

PLAN of the VALVE.

Fig.II.

SECTION thro' the MIDDLE.

Fig.III.

Scale of Feet

3   2   1   0   6   12

J. Farey June. delin.t   Published as the Act directs 1812 by Longman, Hurst, Rees, Orme & Brown, Paternoster Row, London.   W. Lowry sculp.

hoop, the hoop to be made as broad as can be allowed, but the leather need not ftand above an inch above the hoop; this being carefully attended to, the barrels will be as little as poffible fubject to chamber, either by the hoop or the leather.

And N. B. Once for all, that where I have given no particular directions, as I fuppofe myfelf addreffing my defigns to ingenious engineers, acquainted with the ufual practice, I fuppofe the thing to be done in the ufual way, or fubjected to convenience and their difcretion.

J. SMEATON.

Austhorpe, 13th February, 1775.

EXPLANATION of the defign for the working geer for Chafe Water Fire Engine, Figs. 4, 5, 6, 7, and 8, of pl. VII.

IT is unneceffary to enter into a minute defcription of the operation of thofe parts which are made as common, and are obvious to all practitioners in this kind of work, it is fufficient to explain thofe that are more uncommon and particular; and it is fufficient to fay in general, that the parts being exactly meafured off from the fcale, and enlarged according to the defign (paying due regard to what follows the pofition of the great parts firft fixed), all the pieces will come together, and do their refpective offices without material alteration, and that though many other and different proportions will anfwer the end, yet thefe will be found to do their refpective offices completely, and work kindly.

The cylinder, the boiler, with the regulator and the beam, with the plug frame, being fixed as per general defign, pl. VI. and directions; the angle muft be examined that the regulator and injection cock muft turn in, in order that they may completely open and fhut.

The geometrical figure in the plan, fig. 8. pl. VII. reprefents the different pofitions of the regulator's fpanner, in which it is open and fhut, wherein A reprefents the focket, in which the pin of the regulator turns, which (to bring the drawing into a lefs compafs) is fuppofed to be brought forward from its true pofition towards the working geer; the line A a is fuppofed to be fquare to the pofition of the great beam, and parallel to

the

the cylinder beams, the line A b is fuppofed the fpanner's pofition when the regulator is fhut, and A c that of its being open; the arc b a c will therefore be the arc of a circle comprehending its fhut and open pofition, which is here fuppofed fifty degrees, whereof the angle b A a is twenty, and a A c equal to thirty degrees. The-radius of the circle, or length of the fpanner A b, reckoned from the centre of the focket to the point in which the fliding rod joins, is propofed to be fifteen inches, and the ftirrup that fufpends the other end of the fliding rod x y, fig. 4, is to be of twelve inches length; the middle line, therefore, of the regulator paffing through the centre of the focket, and the centre of the fteam pipe, will be parallel to the line A b, and not to the cylinder beams, and the line A a. N. B. The angle b A a is made lefs than the angle a A c, becaufe the fliding valve is hardeft to move in its fhut pofition; in all others, it being but little confined, except by its own friction.

Fig. 6, is a fection of the fquare water-way of the injection cock, the dotted lines fhew the relative fize of the cock pin and focket, the mean diameter of the cock pin being five inches will allow to be pierced three inches wide, and $5\frac{1}{2}$ inches high, and a pin of this fize being of a confiderable weight, to eafe the friction it will be proper to fupport a part of the weight upon the point of a fcrew fupported by a kind of fpringing frame, or double hook hanging upon the fquare branches of the cock on each fide of the pin, which is reprefented in figs. 4 and 6, wherein d e is the fcrew pin f g, f g, is the fpring frame, and h is a counter nut to keep the fcrew from flacking by the circular motion of the pin; this frame hangs loofe, without any confinement, but is kept to a place by the point of a fcrew, and as the hooks are made flight and thin, to allow them to fpring, any part of the weight of the pin may be fupported till it is found to open with the neceffary freedom; it alfo prevents unneceffary wear, both of which are material to the performance of the engine.

All injection cocks fhould have their pins made to take out, in order to be frequently greafed with hogs-lard, or other foft greafe, with a valve at the top of the injection pipe, or fome other equivalent contrivance to take off the water while this is doing; and as experience fhews, that by the inceffant motion of the engine, injection cocks will not remain long tight, a contrivance to the amount of what is fhewn in fig. 4. will be neceffary to catch the water, wherein i i k is the fection of a tunnel of ftrong tin, or thin copper, fupported by an iron ring, whofe fection appears at m m, with a branch n, (fee alfo fig. 5.) by which it is faftened to D, one of the pillars of the working geer; the water iffuing at k may be received into a fmall wooden fpout, and conveyed to the wafte water.

An

An injection cock made in the proportions above specified, or almost any common cock, will completely open and shut in the angle shewn in the plan fig. 5, wherein B o is the position of the cock's spanner open, and B p that of it shut; and the angle o B p, the angle in which it turns, so as effectually to open and shut, being nearly eighty degrees.

The height of the regulator's spanner being given, as y y r in fig. 4, the height of the tumbler's centre x will be had, by setting up ten inches, as per design; and the height of the injection cock's spanner B being given, the height of the centre z of the piece marked and called the F, will be had by setting up B z, equal to $13\frac{1}{2}$ inches; the distance of the two points o p, in which the acting radius of the fork, z o, z p, meets the spanner, being equal to the distance of the same two points so marked in the plan, fig. 5. The distance B q of the plane of the F from the centre line of the injection cock, is $7\frac{1}{2}$ inches, but by bending the fork a little towards, or from the cock, it will cause the spanner of the cock to turn in such a greater or less angle as shall be sufficient.

In fig. 4, D D, shew a part of the fore pillar (marked in the plan fig. 5. D), the rest being supposed broken off, to shew the better what falls behind it; the rest of the upright piece E E E being the far pillar, marked in the plan fig. 5. E, the better to support the F, and keep it steady in its working; it is fixed upon an axis of a competent length, and its catch above is also fixed upon a like axis; the fore end of which is supported by an arm from the fore pillar D, marked G H H; the arm supporting the near end of the F axis, is perfectly similar, but here shewn broke off, as K L L, the more distinctly to shew the shape of the F; the far ends of both axes are supported upon a back pillar M M, as they could not go across to the far pillar E, on account of interfering with the tumbler, or y, so marked and called. The pin or stud at S, driven into the pillar M, is to support the catch from falling too far after the F is discharged, and the tail t t, attached to the axis of the catch, serves as a handle to strike off the injection in hand working.

In fig. 5, the catch, with its axis and arm lying directly above those of the F, is supposed removed, in order to shew those of the F, but the place of the catch tail is shewn at 1 l, it being not otherwise determined.

The only parts that require an exact form, are the F, and the two arms working in the plug frame, which may be moulded by being drawn upon a board or table as follows:

For

For the arms of the tumbler, firſt form a ſquare equal to, and repreſenting a ſection of its axis, draw a line, repreſenting x t, right acroſs the ſquare, according to the meaſures ſet off the points 8 v w and t, and at theſe points croſs the line x t, with the perpendiculars i v 4, 2 w 5, 3 and 6, t 7, ſetting off the points 1 2 3, on one ſide, according to the diſtances v 1 w 2 and 3, meaſured from the ſcale, and in like manner, the points 4 5 6 7, from the diſtances v 4, w 5, and 6 t 7, on the other; this done, draw a fair curve through the points 3 2 1 8 on one ſide, and a fair curve through 7 6 5 4 8, on the other; and laſtly, fixing on a ſquare piece of wood or iron, upon the ſquare, repreſenting the axis, the board is prepared; then the arms being ready formed in iron, and the ſquares made to fit the axis, apply the ſquare hole in the arm upon the prominent ſquare on the board, and having heated the iron as much as neceſſary, bend each arm till its working face anſwers to its curve reſpectively.

The legs of the y may be ſet to a proper opening, by drawing a circle as y y, 10 9, of a radius, equal to the length of the ſtirrup, viz. twelve inches, and from a middle line x 10, ſetting off five inches each way, from ten towards nine, and from ten towards eleven, and applying the y in like manner, upon a central ſquare, open the legs till the inſides touch the two points eleven and nine reſpectively.

The figure of the F will be conſtructed by drawing a right line, repreſenting 12 z 13, and making the length 12 z, equal to two feet two inches, then ſtriking the arc of a circle 13 14 15, draw the line z 14, ſo that the angle 15 z 14 may be, as here, of ſeventy-five degrees, and this line z 14, will give the middle line of the fork; and ſetting off the arc 14 16, equal to the arch 14 15, the line z 16 will give a proper poſition for the middle line of the faller. The angle z 12 17, or which is the ſame, z 18 and 19, is of 131 degrees, or forty-nine degrees, from the ſtraight line; laſtly, marking in the proper thickneſſes, and branch to lay hold of the catch, this will be a mould for making the F.

The poſition and ſhapes of the other parts will be readily determined from the eye, or by meaſurement from the deſign; but it muſt be carefully obſerved, that wherever the meaſures marked upon the plan differ from thoſe reſulting from the ſcale, the figured meaſures are to be adhered to; and N. B. the ſame letters denote the ſame parts in all the figures.

Fig. 7, ſhews the manner in which a ſlider may be fixed upon the plug frame, inſtead of the pin marked in the plan at N, and in fig. 7, is ſhewn by the rounded end
of

of the flider at N.   This figure fhews a part of the plug frame, with the flider annexed, in two pofitions, wherein the dotted line P P, reprefenting the middle line of the plug frame; the pofition No. 1, fhews the left hand half of it, as feen in front, the fpectator's back being towards the pit; and No. 2, fhews the plug frame as feen fideways, but on the oppofite fide to what is fhewn in fig. 4.   In both pofitions of fig. 7, Q R, fhew a part of the plug frame ; N V the flider; and W 20 W 20, fhew the fcrews for re-taining the flider in its place ; by this means, the ftroke may be adjufted, when the beam comes in, by fmaller differences than what the diftance of the holes would occafion, as is adjufted going out, by more or fewer faddle pieces of leather.

At the letter I. fig. 4, is reprefented a roller that reaches from the pillar D to the pillar E, furnifhed with a ratchet and catch, to adjuft the check rope to a proper length.   And N. B. a foft matted rope foaked in tar anfwers well for this purpofe, having but little elafticity, and the fame for the faller of the F.

A fliding rail for the regulator's fpanner to reft upon is not here reprefented, as being a thing of courfe.

N. B. If the weight of the tumbler falling towards the cylinder, fhould not fuffi-ciently fhut the regulator, the long fhank of the y may be bent over a little from a right line towards the cylinder, which will in effect counteract the weight of the arms, acting in a contrary direction.

J. SMEATON.

Austhorpe, 27th February, 1775.

CRONSTADT

CRONSTADT ENGINE.

DIRECTIONS to be observed in adjusting the Engine for Cronstadt to its work*

1ft. THE whole of the engine being put together and in place, but the buckets without leathers, and the pifton without wad or ramming, lay 32 cwt. upon the pifton, and try by levers or fmall tackle, whether the beam, with all its accompaniments, are in balance, that is, whether the refiftance to motion be equal both ways; if it be, it is well, if not, put as many of the balancing weights on one or the other of the great beam as fhall bring the whole to a balance, or indifference to motion either way, there fix the balancing weights and take away the 32 cwt. from the pifton, and the engine when at work will be in proper trim. N. B. It is expected that with full fteam the engine will go quicker out than in, and if fo the balance muft not be altered on that account.

2dly. Take the mean diameter of the valve of the puppet clack or fafety valve in inches and decimal parts, and multiply it by itfelf, and the integral parts of the product will be the number of pounds, and the decimals the parts of a pound, that is to be the whole weight of the puppet clack with its load; thus, if the puppet clack is $5\frac{1}{2}$ inches, it will require its weight to be $30\frac{1}{4}$ lbs. including the weight of the moving part of the valve itfelf.

3dly. A fmall cock, fuch as are ufed for wine or fpirit cafks, muft be fixed upon fome convenient part of the upper fide of the floping fink or eduction pipe, not only to give air to fill the cylinder when the work is over, but to give a fmall quantity of air while it is working. On fetting agoing this cock muft be fhut, and the fnift cock open; when the engine is got fteadily in motion, diminifh the fnift by clofing its cock till there is not more appearance of fteam in the whole of one ftroke, than would amount to the apparent bulk of a hogfhead; when this is done, gradually give vent by the air cock till the beam will but barely come down to the fprings at the cylinder end when the fteam is full, and when weak to ftrike fhort by about fix inches.

* This engine being constructed in the same manner as that of Chase Water, it would be unnecessary to give the design, it will be sufficient to state the following dimensions, in which alone it differs from the former.

Diameter of the cylinder	5 feet 6 inches
Length of the stroke	8 6

4thly.

4thly. Choose a coal that affords a bright clear flame, such as the splint coal of Scotland. Small coal, approaching to dust, or round, in large pieces, are equally to be avoided; coal, from the size of a sparrow's to that of a hen's egg, answer best in proportion to their weight; feed the fire a little at a time and often, spreading the fuel equally over the grate; it is no matter how few red coals compose the fire.

Keep up the steam so as always to shew a little waste when strong, and as soon as ever the engine is found to fall off from its motion, repeat the feed.

N. B. This engine is expected to go at a mean rate of $9\frac{3}{4}$ strokes in a minute, of $8\frac{1}{2}$ feet each, with $7\frac{1}{2}$ chaldron of coals in twenty-four hours, London measure, of the best quality of Newcastle or of splint coals, which I account for working a fire-engine as equal thereto. This will produce 27,300 tons of water in twenty-four hours to the height of fifty-three feet, and is equal to the labour of 400 horses. But, if the engine goes $9\frac{1}{2}$ strokes in a minute, of eight feet each, this will more than satisfy my proposition to Sir Charles Knowles; that is, to raise 24,300 tons per day to the height of fifty-three feet, with $7\frac{1}{2}$ chaldron, London measure, of the best Newcastle coals, in twenty-four hours. This is an effect considerably greater than that of the largest engine yet built in Britain.

J. SMEATON.

Austhorpe, 1st March, 1775.

GATESHEAD PARK ENGINE.

DESCRIPTION of the method of supplying the boilers of a Fire Engine with soft water.

See plate, 8, figs. 1, 2. (page 350.)

IT has been the general practice, and found very advantageous, wherever the situation will admit of it, to supply the injection water of fire engines with the water of a burn or rivulet of soft water, or from a soft-water spring, in consequence whereof the boilers of such engines become in course supplied with soft water, which is the only water that can be depended upon not to fur the boilers, the very great advantage whereof is well known and experienced; but where the situation does not admit of a supply from a rivulet or soft-water spring, as is generally the case, the quantity necessary for the supply of the injection of a fire engine is so considerable, that still fewer situations admit of making an artificial reservoir or reservoirs of sufficient capacity to supply the injection water in the interval between rains and great showers; so that for want of any tolerable certainty in this matter, recourse is usually had to the raising of water from the main-pump heads of the mine, which, in general, is not only of a hard nature, but frequently so corrosive and adhesive to the boilers, as to do them more prejudice in three months, than as many years would do them if worked with soft water. It, therefore, appeared to me as one of the greatest improvements that could be made to the fire engine, to be enabled to supply the boilers with soft or rain water in all situations; that is, particularly in those situations where, as above mentioned, a supply of soft water has not as yet been attempted; and this idea led me, in the course of experiments I went through some years since upon the fire-engine, particularly to investigate the quantity of water that was necessary for injection, in proportion to the quantity that was necessary to feed the boiler, and I had the satisfaction to find that the latter was in most cases less than one-twelfth part of the former. It also occurred to me, that since the boiler will not feed unless the hot-well water is let into it by the feeding cock, if the small quantity necessary for the feeding can be separately supplied from a reservoir of rain water, and the chill taken off, so the smaller quantity may be warmed in a degree by the great quantity of hot-water produced from the injection; then the engine may be worked by injection water of whatever quality, provided it be cold, and the boiler may be fed by such only as is of a good quality, without any communication except that of heat.

With

With this view I placed in the hot-well of my experimental engine, a pan made of tin, as large as could be placed therein, leaving a vacancy between the fides of the hot-well and the fides of the pan, for the injection water to afcend, and to pafs off at the hot-well fpout as ufual. This pan was then continually fupplied with cold water from a neighbouring well, in fuch quantity only as continually running off into the boiler kept up the feed therein ; I did not find any fenfible difference in its product of work in proportion to the coals ufed, than it did in its ordinary way of being fed upon the hot-well.

The following method, therefore, founded upon the above experiments, is what has fince occurred, as being the beft and eafieft way of applying thofe principles, and which, on fuppofition of a refervoir large enough to hold water betwixt rain and rain, cannot fail of fuccefs.

Fig. 1. is fuppofed to be a fection of the hot-well, wherein

A B C D—Shew the hot-well, fuppofed to be cut through the middle.

E F G—Part of the fink pipe, with the horfe-foot valve.

H I K L—Shew a fection of the copper pan made to fill as much of the vacant fpace of the hot-well as poffible, it may touch the fink-pipe, and but juft clear the joint of the horfe-foot valve, the bottom being a little rounding, and about an inch higher at the point I than the point K, in order to induce a part of the injection water to afcend towards the part H, which, without that advantage, it would not be apt to do, the conveyance-fpout being near the part L, but the three fides of the pan that apply themfelves to the three fides of the hot-well, are to be about $1\frac{1}{2}$ inch diftant therefrom at the top, and about two inches diftant at the bottom of the pan; and, according to thefe directions, the abfolute dimenfions of the pan is to be taken from that of the hot-well.

Now, the feeding water being brought and let fall into the pan by a lead pipe from the feeding refervoir, it will be continually receiving heat from its immerfion in the hot-well water, up to the dotted line O P, and muft be continually paffing off by two fpouts, one marked M, the other N, from whence the water is to be conveyed to the feeding pipes of the two boilers.

Fig. 2. is fuppofed to be a plan of the hot-well and pan therein, wherein A B C D reprefent the top of the hot-well, E is a fection of the fink pipe, and R the conveyance fpout for the hot-well water.

H I K L—

H I K L—Shew the top of the pan, which, being flanched an inch, or all the four fides, is to be rivetted to an iron frame h i k l, which is to lay upon the top of the hot-well, and intended to ftrengthen the pan, being of flight copper, fo that, when any thing is to do at the horfe foot valve, the frame and pan will lift out altogether.

M—Is the feeding fpout for the centre boiler, and if the engine has two boilers, N will be the feeding fpout for the off boiler.

Q—A branch upon the fquare frame, with a hole in it to receive the tunnel pipe, which will be further explained in the fequel.

Now, to give the water a degree of warmth before it enters the pan, it is brought from a fmall refervoir through many yards of lead pipe, immerfed in a lander trough, conveying the hot-well water (after having left the hot-well) throughout its whole length.

The lander trough is compofed of three deals, and fuppofed to be eight inches wide, and nine deep, infide meafure, made to convey the water of the hot-well by a fmall defcent from the hot-well (fuppofe about one foot in forty yards) into a little pond or receptacle, from whence the water runs away by a ditch.

In this lander trough is laid a lead pipe of two inches bore, made of fheet lead, turned and burnt, and extending through its whole length, which, being inferted into the wooden trees, bringing water from a fmall refervoir, containing foft water, will convey it into the feeding pan, and the leaden pipe being all the way immerfed in the current of hot water through the lander, will by that means get prepared by having the chill taken off before it enters the pan

After being further warmed in the pan, it runs off by the two fpouts M N, whereof the water from M falls into the wooden fpout, which leads into a tunnel upon a pipe of about 1½ inch bore, which is turned and joined to the upright feeding pipe of the boiler. In like manner, the fpout N is provided with a wooden fpout to convey the feeding water for the off boiler into a tunnel and pipe of the fame kind.

The floping pipe of wood for conveying the hot-well water from the hot-well down to the lander, for the fake of its being out of the way, is propofed to be carried under the timber floor that makes the boiler head floor over the ftoke hole, and then paffing by the lander.

At

At the distance of three or four feet, little ribs or bars are to be laid acrofs the bottom of the lander to fupport the lead pipe from the bottom, that the hot water may get round it, and alfo to impede the water from running in fo fwift a current as not to drown and cover the lead pipe.

Now, the end of the lead pipe at the place where the lander trough terminates, being feveral feet below the level of the top of the hot well, it is neceffary that it be joined to the wooden trees coming from the little refervoir, whofe bottom, or rather the entry of the wooden pipes, fhall be one foot higher than the top of the lead pipe at the cock delivering into the pan.

The little refervoir is fuppofed to be made artificially, or dug upon fuch ground that the mouth of the pipe at the refervoir fhall be about three feet below the furface of the refervoir when full, in confequence hereof the water will defcend through the wooden pipes, and afcend through the lead pipes, and pafs by the cock into the tunnel Q, fig. 2, where, defcending by the copper tunnel pipe thereto annexed of about two inches diameter, it will gradually afcend towards the furface of the pan, and it paffes off by the two fpouts to the boilers, as before defcribed.

Befides the two inch cock at the pan, there is propofed a $2\frac{1}{2}$ inch ftop cock in the lead pipe, by way of regulation and convenience, as follows:

The little refervoir is fuppofed to be of fufficient fize to hold feeding water for one or more hours fervice at an extreme; it is then to be found, by a few days trial, what quantity of water continually running from the great refervoir, to be let out by a cock placed in a tree at the bottom, and then paffing through a common open caft trench into the little refervoir, will keep it fupplied during the twenty-four hours fervice, and in like manner, the ftop cock in the lead pipe muft be regulated fo as to fupply a quantity into the pan equivalent to the boiler's confumption, which being once found, will need little alteration, fo that when the engine ftops working, there will be nothing to do but to fhut the cock at the pan, and when it begins working to open it, the regulation of quantity being done by the ftop cock in the lead pipe; there will, however, be juft the fame attention neceffary in the keeper, to fee that the water in the boilers keeps right by the gage cocks, as in the common way of feeding, and to give the ftop-cock in the pipe a touch accordingly.

It may probably happen that one feeding fpout may run lefs water from the pan than the other, or that one boiler may raife more fteam than the other, in which cafe, if a

couple

couple of thin wedges are driven under the iron frame that supports the pan, on the opposite side of that which runs too little, the quantity may thus be adjusted betwixt them as nicely as you please.

If the intervals of working do not exceed half an hour, it perhaps may be found convenient to let the boilers feed at such a rate as to take in the water continually.; for though in the interval the water will go into the boilers almost cold, yet the fire that must necessarily be kept on, will keep the whole to a boiling state, that now only produces a loss of steam, and consequently of feeding water.

Lastly, the snift pipe, instead of being carried out of doors, may be returned, and come down into the pan, and its end being bored with several small holes, be immersed about four inches into the pan's water, but not so deep as to prevent its snifting; by this means the heat of the steam of the snifting pipe, that otherwise would be lost, will be added to the feeding water in the pans.

COMPUTATION OF QUANTITY.

A cube yard of water will last an engine of sixty-inch cylinder, in boiler's feed, full eighteen minutes, that is, suppose the engine to go twelve hours in twenty-four, at the rate of forty cube yards per day, a reservoir, therefore, of forty yards square, and two yards mean depth, will last the engine eighty days; and a reservoir of seventy-six yards long, and twenty-six yards mean width, will nearly do the same thing at five feet deep.

J. SMEATON.

Austhorpe, 8th October, 1779.

P. S. The small reservoir will hold twenty-four working hours water, at eight yards wide by ten yards long, and to draw off a yard depth of water; but it will be well to dig the small reservoir a foot deeper, near the mouth of the pipe, that the water may go in clear of sediment.

RAVENSBURN

RAVENSBOURNE ENGINE.

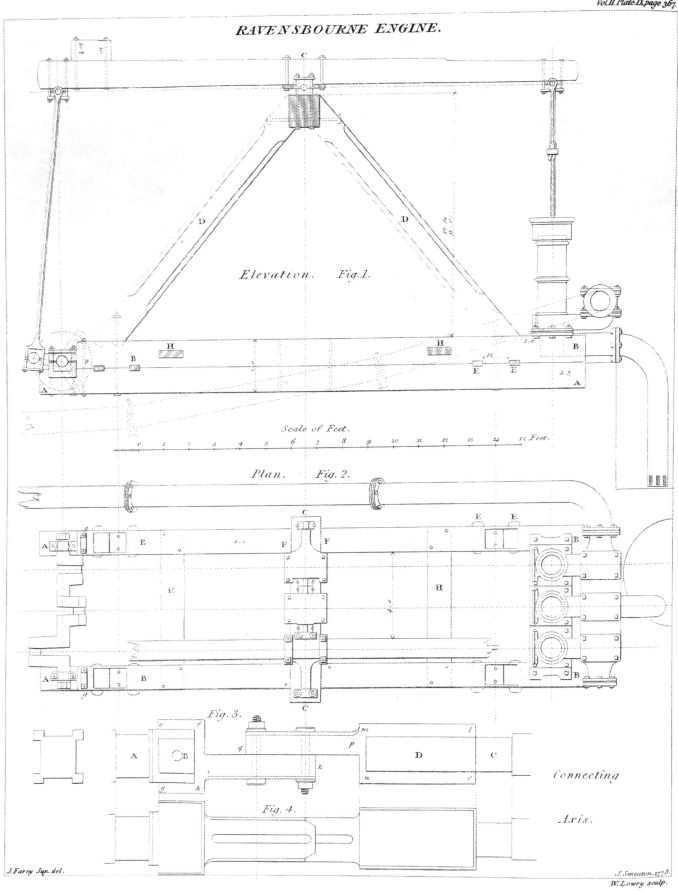

Elevation.   Fig.1.

Scale of Feet.

Plan.   Fig. 2.

Fig.3.

Connecting

Fig. 4.

Axis.

J.Farey Jun. del.

J. Smeaton. 1778.
W. Lowry sculp.

Published as the Act directs, 1812, by Longman, Hurst, Rees, Orme and Brown, Paternoster Row, London.

RAVENSBURN ENGINE.

EXPLANATION of the plan and elevation for the Ravensburn engine.

See the defign plates 9, 10.

THE two ground fills are intended to be laid upon brick walls, confifting of one or as many courfes more, as are neceffary to bring them up to a proper level from a folid foundation, or inftead thereof they may be founded upon pile heads, cut off to a proper level.

The weights upon the crank end of the regulating beams, are intended to overhaul the forcers and fill the barrels, fo that the crank rods will, together with the crank, be always bearing downwards; there will therefore be no ufe for under braffes for the crank rods, or upper braffes for the crank, otherwife than for fafety, they may therefore be made of yew or hard oak.

As there is intended no directing frame for the regulator's beams, their fteadinefs will depend upon their gudgeons, which, for this reafon, are intended to be as far diftant as poffible; the middle one therefore is made as long as can poffibly be admitted between the outward regulators; which fhortening the gudgeons of the outward regulators, on the fide next the middle one, are compenfated by a greater length on the outfides, and is extended beyond the length of the middle gudgeon, becaufe the bearings on each gudgeon of the outward regulators, not being equal, will have occafion for a greater length to keep them equally fteady. The gudgeons are caft upon broad plates, upon which the regulators are bolted down, without any perforation in the middle of the beams, the method of which will readily appear from the drawing, fig. 2, plate 9.

The regulators are fuppofed to be one foot high by nine inches thick in the middle, and to be made fo much bigger in height at the crank end, and fmaller at the forcer end, as will fuit the taper of the trees they are fawn from; they may alfo be made a little taper in the width if it be found fuitable to the times, viz. fuppofe ten inches wide at the crank ends, nine at the gudgeon, and eight inches at the forcer, I mean them however to be all three alike.

The centre or axis of each barrel, when fcrewed down in place, may be found upon the regulator when level, by laying a ftraight rule againft the infide of the barrels up to the regulators, and then finding the middle between, but the centre of the force rod braffes muft be fet four-tenths of an inch further from the regulators gudgeon than this middle point, becaufe the curve, defcribed by the centre point of the force rods,

will

will be eight-tenths of an inch out of a ftraight line, and by doing it this way, will never depart from the middle line, but four-tenths of an inch on each fide, which will avoid the neceffity of arch-heads and chains, as I had firft intended.

I propofe that every part of the engine may be put together, and in working condition, in the yard, before any part of it is removed to the engine houfe.

N. B. The nuts for fetting down the gudgeons upon their braffes, are intended to be hexagons, becaufe if fquare in turning round they will interfere.

Every gudgeon is intended to have its own feparate pair of braffes, but thofe between the regulators will be clofe together; the regulator in the plan, fig. 2. that is, furtheft from the upwright view, is fhewn in place as completed, but the ends broken off, the other two are fuppofed not in place, but the lower braffes in place ready to receive them.

The working barrels fhewn in plate 10, fig. 3, is a plan of the whole, and fig. 2, is the fection of the middle working barrel, as alfo of the horfe tree, with the fuction pipe, and of the faddle tree or cheft.

A B is the cylindrical part of the barrel, or bored part of the forcer, of ten inches diameter, below which it widens to

C D, the chamber for the fuction valve, and conveyances of twelve inches diameter.

E E fhew the flanch or fquare bafe upon which it ftands, being eighteen inches fquare.

F F is the upper face or feat of the horfe tree, which has three round holes of eight inches diameter, denoted by

G G, which are covered by the fuction valves.

H I is the communication pipe of the horfe tree, being eight inches fquare, and carries the water to all the three barrels.

K L is the fuction pipe and flanch, which is round, and of eight inches diameter, placed anfwerable to the middle barrel only.

M M N O O fhew the branch or conveyance pipe, one to each barrel, the entry M M and paffage to N are fquare, that is, eight inches wide by fix high, but gradually

gathered

DESIGN FOR THE WORKING BARRELS FOR RAVENSBOURNE ENGINE.

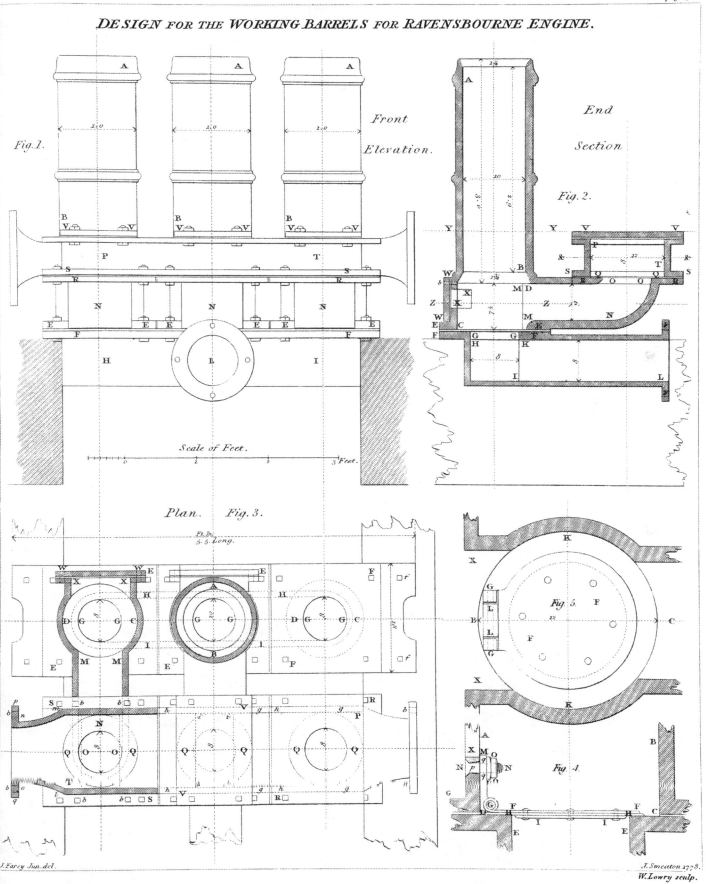

Fig.1.

Front

Elevation.

End

Section

Fig.2.

Scale of Feet.

Plan.   Fig. 3.

Fig. 5.

Fig. 4.

J.Farey Jun.del.

J.Smeaton 1778.

W.Lowry sculp.

Published as the Act directs, 1812, by Longman, Hurst, Rees, Orme and Brown Paternoster Row London.

gathered into the round hole O O of eight inches diameter, through which the water is forced into the cavity of the cheſt or ſaddle tree P T, twelve inches wide, and eight inches high, which, like the horſe tree, communicates with all the three barrels.

The hole O O is covered by the forcing valve, and this is in the middle of R R, a flanch of eighteen inches ſquare, upon the branch of each barrel, ſo that theſe three flanches of the three barrels form a ſeat to which the under face of the ſaddle tree applies itſelf, and fixes by the flanches S S, this under face being perforated in three places with a round hole of twelve inches diameter, marked Q Q, ſurrounding each valve, and concentrical with the hole O O.

Directly over each valve, in the upper ſurface of the cheſt, is a perforation of twelve inches length, equal to the width of the cheſt, and ſix inches wide, each covered with

V V, a flat lid ſerving to look at, and take in and out the valves ſeparately, without unſcrewing the cheſt, and with the ſame intent.

W W repreſent a door cloſing the opening X X, of nine and a half inches wide, and three inches high, by which the ſuction valves are got in and out.

So far the ſame letters explain both figures, but for a more full explanation of the plan fig. 3. obſerve that ff F F ſhew a part of the upper face of the horſe tree, without a barrel upon it, the part F F ſhewing the place for the baſe of the barrel, the dotted lines H I. H I. ſhewing the figure of the ſquare communication pipe under the ſame, the ſucking pipe lying directly under the branch pipe of the middle barrel, cannot here be repreſented, but is ſufficiently ſo in the ſection fig. 2.

In like manner R R ſhew one of the naked flanches of the barrel's communication pipe, cleared of the cheſt ; upon which the dotted lines g h. g h. ſhew the place of the ſides of the cheſt, with the manner of their being gathered into an eight inch round hole and flanch at each end, for joining the pipe of conduct at either end, the other being ſhut by a flat plate.

The middle barrel A D is ſuppoſed to be cut by a horizontal ſection upon the level of the line Y Y, fig. 2. conſequently the whole reſpecting it is ſhewn as put together with the lid V V in place, the dotted circle Q Q shewing the round hole in the bottom and i k i k the square hole in the top of the chest, that is shut by the lid at V V.

The near barrel C D X M is reprefented by a horizontal section through the middle of the chamber, according to the level of the dotted line Z Z in fig. 2. and the area S S is shewn as things would appear upon an horizontal section through the middle of the chest upon the level of the dotted line & & in fig. 2. but l m n o p q reprefent that end of the chest gathered into a circle and flanch as it will appear from above when finifhed.

N. B. The whole length of the horfe tree, and of the chest or faddle tree (each being in one piece) is five feet five inches.

The fmall fquares fhew the bolt holes, and the places of bolt holes fhewn by dotted lines are all marked at b.

In the fection fig. 4,

A B C D reprefent a fection of the chamber where the valves lie, being twelve inches in width.

E E is the infide width or diameter of the opening of the clack feat, being eight inches diameter.

F F is the upper plate of the valve, being about nine and a half inches diameter, but branching out on one fide, fo as to form a hinge or tankard lid joint at G.

H H is the leather of thick dintle, and ten inches diameter.

I I is the under plate to be rivetted by about fix rivets in the circumference, and one in the centre.

X is the perforation to introduce the valve covered by the clack hole door.

In the plan fig. 5,

K B K reprefents a part of the circumference of the chamber, on the fide where the hinge is applied, fo that the line B C in the plan will correfpond with the line B C in the fection.

G G

G G will be the centre, or axis of the hinge, whose middle part L L, is in breadth two and a half inches, and is formed from the upper plate of the valve.

G L G L unite in one above the joint, and rise up as in

The section.

G M is the rising part of the hinge by which it is fixed by

N N the bolt, and

O O, the nut; this bolt may be about five-eighths of an inch diameter, and pass through a round hole at p to be bored in the flat side of the chamber; but the chamfered head at N, being sunk in square with a chisel, will prevent the bolt from turning round while the nut O O is tightening, and the hole for this bolt must be bored sufficiently near the perforation X that its head may be covered, and secured from leakage by the clack-hole door.

The rising piece G M must be adapted on the outside to the curve of the chamber, but the inside flat; the hole in the rising piece, by which the bolt passes through it, must be at q q, a full eighth of an inch on a side wider than the bolt, that the valve may fairly seat itself, and find its own place, and will be firmly fixed by the counter plate R.

N. B. The center-pin of the joint must be full one-sixteenth of an inch less in diameter that the holes in the lid and ear-pieces, nor must the joint between the lid and ears be close, so that the valve may be at free liberty to seat itself according to the pressure of the water; and that it may not be confined from settling as the leather wears thinner; the rising piece must be pushed down as far as it will easily go, and then there will be all the liberty given that may be for the valve to settle without confinement; indeed it would be better if the holes in the joint were made a little oval, both in the lid and ears, the long way of the oval to be in a vertical position, and as the centre pin is expected to wear the most of any thing, it would not be amiss to make it of steel, but not tempered, so that a little brass ring or collet being put on at each end, they may be rivetted to keep it in place, in which it should have no confinement either endways or sideways.

The only thing to be attended to, in respect to the tightness of the joint is, that its

vacillation

vacillation fideways fhall not be fufficient to bring the under plate of the valve upon the feat, fo as to prevent the leather's coming to a fair bearing; but to prevent all poffibility of this, the under plates may be made a little oval, fo that fideways they may not be above feven inches in diameter.

The above particularly defcribes the valves for the barrels, the forcing valves will not differ, except from the circumftances of the chambers being fquare, and the doors being at the top, which will require the rifing piece to be flat, and the head of the bolt N N, inftead of being chamfered and funk in, to be a broad flat head, fo as to admit a piece of leather under it on the outfide to make it water tight.

J. Smeaton.

Austhorpe, 6th June, 1779.

SEACROFT

SEACROFT COKE FURNACE.

The REPORT of JOHN SMEATON, engineer, concerning the powers necessary for working a coke furnace at Seacroft.

THE whole rife, as per level taken by Mr. Eaftburn, from the tail of the colliery drain to the furface of the mill pond at a full head, is thirty-three feet four inches, and there is reafon to fuppofe, from obfervations fince made, that by purfuing Mr. Porter's drain, there will be a lofs of level of two feet; our neat difference to work upon will therefore be thirty-one feet four inches, upon which fall, to allow for a fufficient head and clearances, let us for the prefent fuppofe the water-wheel to be overfhot of twenty-eight feet high, and upon this height, by calculation drawn from my experience, to work a coke furnace roundly, that is, at a middling fpeed, will require 7,294 tons of water per day to be expended upon it.

I alfo calculate, that this quantity of water expended upon the prefent corn mill will grind corn at the rate of three bufhels, or a load per hour; it will, therefore, follow, that for fo much of the year as the natural fupply of water will keep the prefent mill going, fo as to grind at the rate of a load per hour, fo much of the year this natural fupply may be expected to work the furnace as above mentioned without any foreign aid, and this information the proprietors will the moft naturally get upon the fpot. From what has occurred to me upon the fubject, I fhould ftate it as follows:

That for five months in the year there will be a full supply of water to go on
continually at the rate above mentioned, which will give - - 5 months' water.
That for three months more there will be as much as will give two-thirds of the
quantity requisite, which will amount to - - - 2 months' water.
And that for the other four, there will at an average not be more than what
will cause the mill to go six hours in twenty-four, which will amount to 1 month's water.
_____
The furnace may therefore expect to have yearly - - - 8 months' water.
_____

Now, though the average of the four dry months is ftated at fix hours water, yet it frequently happens, that for three months together the natural fupply does not amount to above three hours water per day; fo that without fome fubfidiary power, this furnace muft undoubtedly blow out every fummer, and this fubfidiary power I advife to be a fire engine, and that of fufficient power to work the furnace at the above rate, inde-

pendently

pendently of the natural fupply, for when the natural fupply is too fcanty to work the furnace, the engine being fet a-going will work till the natural fupply has filled the ponds, and then the engine may ceafe working, and fave fuel till the ponds become empty again; or the natural fupply may be employed at the boring mill, even at the fcarceft time.

The whole quantity then, 7294 tons per day, I can engage to raife back by a fire-engine of no more than a thirty-inch cylinder, and this engine I can warrant will work with three cwt. of coals per hour, that is, feventy-two cwt. per twenty-four hours, of the quality of the late Halton Bright, that is, if we allow $2\frac{1}{4}$ cwt. the horfe pack weight to each corf, this will amount to two dozen and eight corves per day. The quality of the coals wherewith this engine will probably work is unknown to me; but in proportion as they are better or worfe engine coals, the confumption will be greater or lefs; but this engine would be worked with about eighty-fix cwt. of raw fleck per twenty-four hours, fuch as were ufed to be led from the fleck heap to the engine at Halton; and if we allow the fame weight and meafure to the dozen of fleck, as of coals, this will be no more than three dozen two corves of fleck per twenty-four hours, which, if laid down at the engine door, at 1 . 6d. per dozen, the engine will work at the price of 3s. 3d. per day in fuel, more or lefs, as the fleck (or fmall coal) will be procured and laid down at the engine door for more or than lefs 1s. 6d. per dozen.

Now, if the engine is worked four months in the year, at the rate of 3s. 3d. per day, the whole amount of the fuel will be only £19 16s. 6d.; but this likewife will be more, if the engine is worked longer than the whole four months, either on account of the defect of the natural fupply, or to keep the boring mill at work; or if done at many different intervals, an addition will be required to make the water boil each time of lighting the fire.

The engine will be attended by one man, when its work can be done in twelve hours, and by two men when working twenty four hours: when it wants leathering the wright of the works will be wanted to affift, and when not ufed, the engine-keepers will be employed in other labouring work.

J. SMEATON.

Austhorpe, 16th January, 1779.

DRAWING

DRAWING COALS BY HORSES AND BY ENGINE.

Comparative ESTIMATE of drawing coals by horfes, or by a coal engine worked by water fupplied by a fire engine.

ESTIMATE by horfes.

	£	s.	d.
The pit eighty-two fathom, to draw a corf in two minutes for twelve hours per day, will take nine horses at £25 per annum each horse, - - -	225	0	0
To two gin drivers, at eight-pence per day each, six days per week, is eight shillings per week, and fifty weeks per year, is - - - - -	20	0	0
To one horse keeper, at one shilling per day, the whole year, - - -	18	5	0
Charge of a pit drawn by horses, - - -	263	5	0

ESTIMATE by the water coal engine.

	£	s.	d.
Suppose the fire engine to cost £500, and the coal engine £300, in the whole £800, for which, allowing £10 per cent., will be per annum, - -	80	0	0
Coals, nine bolls per day, at four-pence, that is, three shillings per day, and for six days per week, eighteen shillings, and for fifty weeks per year, - - -	45	0	0
A boy to attend the coal engine, at nine-pence, six days per week, and fifty weeks per year, is 300 days, - - - - - - -	11	5	0
An engine keeper for the fire engine, at eight shillings per week upon the whole year,	20	16	0
The engine wright's attendance on the two engines, at sixpence per working day, 300 days per year, - - - - - -	7	10	0
Charge of a pit drawn by a water coal engine, - -	164	11	0

N. B. It is expected that the wear and tear of horfes and gins will be greater than that of the fire engine and water engines.

J. SMEATON.

Austhorpe, 14th August, 1776.

Mr.

Mr. WESTGARTH's ENGINE.

The REPRESENTATION of JOHN SMEATON, engineer, concerning the Hydraulic Engine, invented by Mr. WILLIAM WESTGARTH, of Coal Cleugh, in the county of Northumberland, for raising water by water.

Mr. SMEATON begs leave to advertife his friends and acquaintance, as well as thofe who are defirous of promoting the fuccefs of good mechanical inventions, that they will find the engine, the model of which will be exhibited by Mr. Weftgarth, to be highly worthy of their attention and regard, being, in his opinion, one of the greateft ftrokes of art in the hydraulic way that has appeared fince the invention of the fire-engine.

This machine is founded on one of the moft fimple principles of ftaticks, viz. that of a heavier column of water preffing up, or raifing a lighter; an idea which, though proportionably obvious as it is fimple, and therefore far from new, having been adopted by feveral able and eminent mechanics, yet it has been referved for Mr. Weftgarth to form the proper expedients, and obviate thofe difficulties which attended the practical execution of machines intended to work on this principle, and which contrivances are (as to my knowledge) not only new and peculiar to Mr. Weftgarth, but which alone have contributed to its fuccefs, and, therefore, though the whole compofition dependent thereon is highly deferving of a patent, or exclufive privilege, yet Mr. Weftgarth choofing rather to communicate his invention to the public, feems the more highly deferving of the public, and of the patrons and encouragers of arts for that very reafon, and he feems ftill the more worthy of proper notice, as he has not been hafty in communicating his invention till the value thereof has fully appeared by a number of engines built at large, and applied to real ufe.

Independently of the various drafts, experiments, and effays on models, which I am informed Mr. Weftgarth had in hand for years paft prior to his attempting a machine in large, I had the pleafure of feeing the firft complete machine of this kind at work for draining or unwatering a lead mine belonging to Sir Walter Blackett, at Coal Cleugh aforefaid, in the fummer of 1765, fince which time that machine has been fhewn to all thofe who had the curiofity to fee it; he has now erected four others in different mines in that neighbourhood, one of which I have feen, and all attended with

great

great fuccefs, as I have much reafon to believe, not only from the nature of the con-struction, but from the reports of thofe for whom they have been erected.

This machine is not only peculiariy adapted for the raifing of water by water for draining of mines, but the fame principle can readily be extended to the raifing of water for fupplying towns, gentlemen's houfes, &c. and univerfally for raifing water from any depth wherever a fall of water can be procured, and particularly in thofe cafes where the fall is great, that is, where it exceeds thirty or forty feet, it will (in my opinion) not only exceed all other known machines, in effect, but in fimplicity, and that whether the quantity of water that is to be applied be great or fmall; I, therefore, think that this machine is not only very curious in itfelf, but will be of great advantage to the public.

CORN MILL, TO BE WORKED BY A FIRE ENGINE.

To the Honorable the Commissioners of His Majesty's Victualling-Office.

GENTLEMEN,

IN compliance with your order of the 14th May laſt, deſiring me to give my opinion, which of the two methods I prefer of conſtructing a mill, to be worked by ſteam, (one by the intervention of water, the other by a crank), and the reaſon of that preference, and to give in a plan of ſuch a mill as I would recommend, of ſufficient power to grind 400 quarters of corn per week, I have fully and duly conſidered the buſineſs, and find that in point of quantity of coals to be conſumed for raiſing a power ſufficient to do this buſineſs, ſuppoſing a machine in either way perfectly underſtood, there will be no material difference; but in point of convenience and good effect in the practical part, it appears to me that the difference will be very conſiderable; for, in the firſt place, I apprehend that no motion communicated from the reciprocating beam of a fire engine can ever act perfectly equal and ſteady in producing a circular motion, like the regular efflux of water in turning a water-wheel, and much of the good effect of a water-mill is well known to depend upon the motion communicated to the mill-ſtones being perfectly equal and ſmooth, as the leaſt tremor or agitation takes off from the complete performance.

Secondly, all the fire-engines that I have ſeen are liable to ſtoppages, and that ſo ſuddenly, that in making a ſingle ſtroke the machine is capable of paſſing from almoſt the full power and motion to a total ceſſation; for, whenever the ſteam gets lowered in its heat and elaſticity below a certain degree, for want of a renewal of the fire in due time or otherwiſe, the engine is then incapable of performing the neceſſary functions for preſerving its motion. In the raiſing of water, (a buſineſs for which the fire-engine ſeems peculiarly adapted), the ſtoppage of the engine for a few ſtrokes is of no other ill conſequence than the loſs of ſo much time, but in the motion of mill-ſtones grinding corn, ſuch ſtoppages would have a particular ill effect.

When a water-mill ſtops for want of water, the motion is loſt by degrees, and the feeding of the mill-ſtones with corn depending on the rapidity of the motion, ceaſes alſo gradually, ſo that when the mill is ſet a-going again, the ſtones not being full-charged with corn, there is no impediment to their being ſet forwards, and in like

manner,

manner, when the mill is stopped by the shutting down the sluice, the miller, being apprised of his own intention, takes off the feed of corn, so that by such time as the mill stands still, the stones are in a great measure cleared ; but if a mill in full work were to stop in the time of a single stroke of the engine, it would stop with the stones full-charged with corn, for the miller could not be apprised thereof, so that before the motion could be renewed, the stones must be raised from their grinding bearings, in which case a quantity of half-ground meal would come through before they could be re-adjusted to their work, and which could never be done till they had acquired a regular speed, which would take up some time, when immediately deriving their power from the reciprocating motion of a fire-engine.

It is true that much care in the engine-keeper may prevent frequent stoppages, but no one can be expected to be so much upon his guard that this shall never happen, and if it were to happen but once in twelve hours, it would confuse the regular operations of the mill to such a degree, as to render it very disagreeable to those concerned in the working of it, as it would generally happen when they were not aware of it.

By the intervention of water, these uncertainties and difficulties are avoided, for the work, in fact, is a water-mill, and as in the construction here presented, there will be a sufficiency of reservoir or mill-pond, capable of keeping it going one minute withou sensible abatement, it seldom happens that if, by any inadvertence of the engine-keeper, the engine stops, but that in less than a minute, and generally in less than half a minute, he can set it a-going, so that the mill will regularly continue at work, and if any thing should go wrong with the engine for a greater length of time, as the mill will stop gradually, no particular derangement can happen, further than so much loss of time, as the miller will always be apprised thereof by the gradual loss of the mill's motion.

For these reasons, were I to establish a work of this kind at my own cost, I should certainly execute it by the intervention of water, and therefore must greatly prefer it.

In putting a design of this kind into actual execution, so as *bona fide* to dispatch 400 quarters of corn in a week, it is necessary to make a provision of power beyond the bare calculation, to allow for necessary and unavoidable stoppages, and to bring the water round into the reservoir again ; for this reason I have estimated the work to be dispatched in about twenty hours a day, and seven days to the week, in which case it

will

will be proper that the engine fhould be of fufficient power to raife 460 cube feet of water per minute, to the height of thirty-four feet.

The fize of the engine neceffary for this bufinefs, and the quantity of coals to work it, will be afcertained by Meffrs. Bolton and Watt, as well as the proper conftruction of the engine.

I am, gentlemen,

your moft obliged,

and moft humble fervant,

J. SMEATON.

Austhorpe, 23d Nov. 1781.

GOSPORT

GOSPORT WATER WORKS, &c.

To the Honourable the Commissioners of His Majesty's Victualling Office.

The REPORT of JOHN SMEATON, engineer, upon the state and distribution of the waters at the brewery at Weevill, near Gosport.

HAVING carefully viewed and examined the state of the waters at this establishment, in regard to their practicability of furnishing an ample, lasting, and certain supply, not only for the two breweries working there, and such further additions in that branch as may be thought necessary; but also for watering such large fleets of his majesty's ships as may have occasion to lie at Spithead; I have principally observed as follows:

That the new well lately dug there, furnishes the highest degree of probability, that from this alone the whole supply which can possibly be wanted, will be amply furnished, when a proper engine is fixed thereon, by which the water can be conveniently raised and distributed; for this reason I cannot hesitate to recommend to the board to construct such machinery, reservoir, and pipes, as may be capable of effecting this business, leaving whatever has been constructed before by way of furnishing a part of the water wanted, in its present state, so that if necessary, the same part may still be supplied by the same means as heretofore.

I understand, that so large a fleet as has lately been at Spithead, has been frequently in want of a supply to the amount of 200 tons of water per day; the brewery at present consumes to the amount of 85 tons, and if this quantity should be wanted to be doubled, the brewery would then take 170 tons, which, with the shipping, will create an occasional want of 370 tons per day.

At present there are no means of raising the water from the new well, but by a hand pump, which, when worked continually, has never been able to lower the surface above five or six feet, the total depth in water being twenty-seven feet. The quantity therefore that it may be capable of furnishing, by no means appears from this reduction, but may be judged of by a consideration of the following facts:

That when the auger by which the bottom was bored, pierced through the stratum of clay in which the well is sunk, into the stratum of sand which affords the water, at

which

which they arrived fooner than expected, the water came in fo faft upon them, that they were obliged to clear the well, and it was obferved, that within the fpace of fix hours, it rofe twenty feet perpendicular; the cubic content of which cavity being full forty-eight tons, it follows, that the average rate of fupply was full eight tons per hour, or at the rate of 192 tons per day; but as the utmoft height to which it will rife, but little exceeds twenty-feven feet, it muft be fuppofed that the rife of the fur-face was much flower, when the well had got twenty feet depth of water, than at firft, when there was fo much lefs preffure upon the rifing column; fo that when an engine is placed thereon, capable of clearing and keeping clear the well of water, that its influx would be at a much greater rate than the average quantity of eight tons per hour, above ftated.

The certainty of this increafe appears from the facts that attend the Cooperage well, from whence the fupply has heretofore been drawn, and which, in many refpects, appear fimilar to the facts that attend the new well.

When the water of the Cooperage well has been drawn out by the horfe machine (not long fince erected upon it) after ftanding fourteen hours it gets twenty feet depth of water, on fetting the machine to work, it draws out in three hours the twenty feet, together with the growth of water that is continually coming in while the other is drawing out, and which together amounts, by my computation, to forty-nine tons; after this, as I underftand the practice is to let all ftand for five hours, and then work the machine again as before, and in two hours they again drain the well, we may therefore fuppofe, that in two-thirds of the time they draw two-thirds of the quantity, and this will amount to $32\frac{2}{3}$ tons, all then again ftands for fourteen hours, including the night, and the well fills as before to twenty feet, in this way the quantity drawn per day, or twenty-four hours, will be $81\frac{1}{3}$ tons.

Now the cubic capacity of this well, filled in twenty-four hours, twenty feet high, being meafured, I make to be thirty-five tons, that is, it comes in at the average rate of two and a half tons per hour; but the quantity drawn out of the well, after ftanding five, and working two, in the whole feven hours, it affords, as before ftated, $32\frac{2}{3}$ tons, which will be at the rate of $4\frac{2}{3}$ tons per hour, that is, almoft double of the average rate of rifing twenty feet; fo that if inftead of working twice in twenty-four hours, as above fpecified, which, I fuppofe, in ordinary, may be both convenient and fufficient, yet, if upon an exigency the well were emptied four times in twenty-four hours, at regular and equal intervals, as the furface would then never rife above ten

feet

feet, I am satisfied it will afford an average of five tons per hour, that is, 120 tons in twenty-four hours, instead of 81 ¾ as at present.

Hence taking it for granted, that in like manner the New well will afford double the average quantity of water, if constantly kept low, to what it does in rising twenty feet, it will follow, that the New well thus worked, will afford sixteen tons per hour, a quantity more than equal to all the occasions that are computed to be wanted.

To render this supply still more ample, and therefore certain, as I apprehend that the influx of the water through the bore hole must be in part resisted by a column of sand that will necessarily introduce itself into the bore hole along with the water, I would advise, when a machine is established, capable of drawing out the water, to make four auger holes more, round the circumference of the well, which, giving the water more vent, will in effect lessen the obstruction, and probably increase the quantity of water, but in every event will render the supply more certain, as a single hole is liable to accidental obstructions, by extraneous matter getting into it, which may be lodged in the stratum of sand which supplies the water.

On making trial of the New well's water, I do not find the least difference between that and the Cooperage well, which I understand has been long used and approved, nor indeed do I find any material difference amongst all the waters of that part of the country; even the fine clear stream that falls into the head of Portsmouth harbour, at Fairham, is quite as far from the perfect softness of rain, or distilled water, as the New or Cooperage well; and the well called the Fortune, from whence the water is forced through pipes for the service of the town of Gosport, appears to me just the same. In reality, I apprehend the stratum of sand, which lies at different depths under a bed of clay, extends itself over a considerable part of that flat country, the water which it affords being nearly the same to all.

That the water of the New well and Cooperage well proceed from the same body or source, thus appears, for, notwithstanding that at the New well they pierced the water stratum to a less depth by twenty feet than at the Cooperage well, and though the rate of its influx is in very different quantities, yet the ultimate height to which they all rise is nearly the same, viz. to near upon the level of high water mark at common spring tides, and the ultimate height to which the New and Cooperage wells rise, is in fact, as near as I can measure them, precisely the same.

An

An engine therefore capable of raiſing 400 tons per twenty-four hours, out of the New well, will either perform the whole ſervice or drain the well, and if the New well ſhould happen to fall ſhort by reaſon of any unforeſeen intervening circumſtance, yet as the Cooperage well will independently furniſh 120 tons per day in addition, as has already been ſhewn, the machinery already upon it can be occaſionally uſed, or the two waters may be united under ground, according to the idea of Mr. Whitbread, by an adit or drift of communication.

That the drawing out of the water of one well, will not materially affect the quantity to be drawn from the other, appears from this circumſtance; that when I was there, the New well attained its utmoſt height when the Cooperage well was drawn empty, which, indeed, ſeemingly contradicts the idea of their having one common ſource; but it is to be conſidered, there is no open communication between them like that of a pipe, but only by the interſtices of a maſs of ſand common to both; each therefore gets its water from the pores of the ſurrounding maſs, and were there two wells at the diſtance of five yards, inſtead of 500 feet, as the caſe is, I doubt not but that the emptying one would lower the water conſiderably in the other, and though the water percolates very ſlowly through a bed of ſand, when the diſtance is conſiderable, yet as the water drawn from it can never riſe higher than the level of its ſource, it will ſtill follow, that equal ultimate heights point out the identity of the ſource.

At preſent the water is raiſed into a ſmall reſervoir upon the ſurface of the ground at the Cooperage well, from whence it runs through elm pipes, and ſupplies at pleaſure the quay reſervoir, from whence the water is taken into caſks for the ſhipping; when not wanted there, it runs into the new brewhouſe well, whoſe water, by an underground drain, communicates with the great reſervoir, and alſo with the old brewhouſe well; the ſurfaces therefore of theſe three are always upon a level, and the conveyance of water into the new brewhouſe well, is in effect conveying it into the reſervoir. From theſe wells and reſervoirs, the water is again raiſed by horſe machinery into the liquor backs of the old and new brewhouſes, to the height of twenty-five feet for the old, and thirty feet for the new brewhouſe.

Now I find, if the water be raiſed at the New well into a reſervoir or back, but ſixteen feet higher than it is already raiſed at the Cooperage well, that it will then run by pipes into, and ſupply the liquor backs of both the preſent brewhouſes, and a new one if required, without pumping, which will be a great eaſement to the horſes uſed there, as I underſtand a couple of horſes additional are put on whenever they raiſe the

liquor,

liquor, and I find that a proper engine conftructed at the new well, will perform the fervice mentioned, with two horfes at a time, the fame as ufed at the Cooperage well, and when the water is not wanted to be raifed into the new well refervoir for the fupply of the breweries, or otherwife, the water need not be raifed higher than the furface of the ground, where it will pafs off to the quay refervoirs as faft as it is raifed, or on fpe-cial occafion, by having ready filled the propofed refervoir at the new well, which I intend to hold about eighty tons; this may in fifteen minutes be transferred into the quay refervoir, fo that it is fcarcely poffible to conceive any emergency can happen, that the propofed eftablifhment will not be equal to; whereas at prefent, when the quay refervoir is emptied, and the Cooperage well alfo, there is no refource for water for the fhipping, but by raifing water from the great refervoir by a hand pump.

In this arrangement, the great refervoir may be made ufe of as a treafury of water for occafional fervice; and if therefore an intermixture of rain water with the fpring water, fhould be thought of fervice for the brewing, the furface water from the roofs of the whole buildings, as fuggefted by commiffioner Kirk, may be collected and carried into this refervoir, where it may be kept feparately, and ufed at difcretion.

J. SMEATON.

London, 5th Nov. 1779.

I expect that two horfes at the new well engine will perform the ordinary fervice I fee going on, but if the full quantity of 200 tons be required to be daily delivered, then to keep the fervice going during twenty-four hours, will require three fetts and one fpare horfe in cafe of accidents to the reft, that is, in the whole feven horfes.

If thought neceffary by way of eafing the horfes, a wind engine might be raifed upon the fame building, of fufficient power to perform the whole fervice, whenever the wind amounts to a frefh breeze.

GENERAL description of the horse engine, proposed to be erected upon the new well, at his majesty's brewery works at Weevil, near Gosport.

PROPORTIONS.

	Ft.	In.		
Mean diameter of the horse track, - - -	30	0		
Diameter of the great wheel in the pitch circle, - -	19	1	containing 144 ⎫	Pitch
wallower, - - - -	2	3½	17 ⎭	5 In.
fly wheel, - - - -	10	0		
Sweep for the crank, - - - -	1	6		
Diameter of the working barrel of the pump, - -	0	6		

THIS machine, according to the propofition contained in my report to the honourable the commiffioners of his majefty's victualling office of the 5th of November laft, to which I beg leave to refer, being intended to be made capable of raifing water, not only for the fupply of the two brewhoufes already eftablifhed there, and to fuch other brewhoufes as may be there erected, but alfo for the watering of fuch fleets of his majefty's fhips as fhall be occafionally affembled at Spithead, to the amount in the whole of 400 tons per twenty-four hours, I have endeavoured to affemble the following properties, that is, ftrength and durability of all the matter, materials, and fundamental parts, eafe of repair of fuch as are moft liable to decay by weather or wear, to have all poffible fimplicity, fo as to be as little fubject to be out of order as may be when fuddenly wanted upon an exigency, and the whole of fuch form and conftruction as to be capable of being adequately repaired, and kept in order by the proper artificers in the country, and laftly, to be capable of raifing all the water that can be raifed to the fame height by the number of horfes propofed to be employed; the conftruction I have defigned is fet forth in plate

With this view, to avoid an unneceffary number of moving parts, I have contrived the whole bufinefs to be done with a fingle pump of fix inches barrel, and to make a ftroke of three feet, and which, from the proportions given to the different parts of the machine, may be expected when the horfes are moving at a very moderate working pace, to make from fixteen to eighteen ftrokes per minute, which rate will deliver 100 barrels per hour, and if worked 24 hours, will amount to 400 tons per day, that is, on fuppofition that the fpring fhall fo faft fupply it, or if not, that an under-ground communication be made with the Cooperage well, or others that fhall be fufficient.

A fingle pump worked by horfes would of itfelf produce a very irregular motion, but this will be equalifed in great part by the application of a piece of caft iron, to the inner or crank end of the working beam, which being fuppofed of or about half the

weight

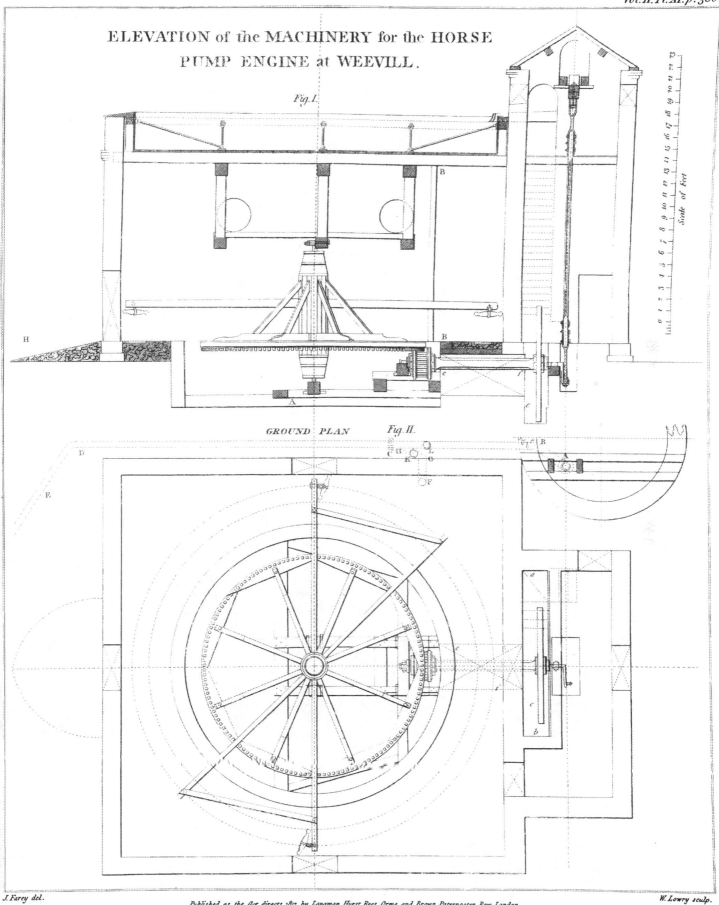

ELEVATION of the MACHINERY for the HORSE
PUMP ENGINE at WEEVILL.

Fig. 1.

Scale of Feet

GROUND PLAN   Fig. II.

J. Farey del.   Published as the Act directs, 1812, by Longman, Hurst, Rees, Orme and Brown, Paternoster Row London.   W. Lowry sculp.

weight or column of water in the pump, when the crank acts to draw water, it is helped by half its gravity by the weight, and when the fpear of the pump is defcending to fetch a new ftroke the crank is lifting the weight, fo that the inequality of the refift-ance being by this means reduced to half, the remainder will be taken off by a heavy fly of caft iron, applied to the fame axis as the crank, which renders the refift-ance to the horfes perfectly equal; this general idea eftablifhed, the machine is fo plain and fimple, that I apprehend the whole matter will be perfectly underftood by mere infpection of the defigns, with fuch fhort defcriptions as are fubjoined hereto.

It is neceffary however to obferve, that as it cannot be known for a certainty, what quantity of water the fpring may produce, its rate of producing water being inferred from the rate of its firft filling, and from analogy with the Cooperage well, it doubtlefs would be fatisfactory to know the rate at which it will nearly afford water, before the whole expenfe is gone into, and in this refpect, as a confiderable charge feems likely to attend its been emptied with common hand pumps, it cannot be done better than by the engine itfelf, and which may be afcertained as foon as the machine and pump are fixed, fo as to deliver its water at the furface, that is, before the refervoir is made, or pipes laid to convey it; and then there would be no expenfe incurred more than would be proper for a machine for the new well fingly, fuppofing it no more than equivalent to the Cooperage well in point of produce, fave that the building, ferving as a fhed and cover to the horfe track, might have been more flightly founded and built, and alfo raifed to a lefs height in cafe it were not intended to carry a refervoir of eighty tons upon the top of it.

I have fuppofed the refervoir to be lined with lead, this however may be left to choice, as, from calculation of the weight of the lead at 8lbs. to the foot, it does not appear that it will be much more expenfive than the rate of back making.

The new pipes laid to convey the water to be fix inches diameter, may be either of elm or of iron, for as they will pafs chiefly through the wafte, and not at all interfere with any prefent building, if laid with good elm they ought to laft full twenty years, and can eafily be come at to be renewed; and in cafe any new building fhould be wanted to be built acrofs them, iron pipes may then be laid down under the fite of fuch building before it is raifed.

It is further to be noted, that all the means of fupply that now fubfift will remain undifturbed, till it is feen whether the continuance of them be neceffary.

Austhorpe, Feb. 19th, 1780.   J. SMEATON.

Explanation

Explanation of plates XI. and XII.

Plate XI. fig. 1. reprefents the upright fection of the houfe and refervoir, and alfo of the out-jetty for the crank, working-beam, and ftairs, fhewing the upright elevation of the machinery.

A, walling for fupporting and keeping fteady the grand carriages and headftocks.— B B, pipe for conveying the refervoir water to the main.—C, reprefents a pit, whofe bottom muft be feven feet below the ground line B H, which is fuppofed to be level with the furface of the well kirb, and to preferve the neceffary width in the bottom, the outfide fets off for the length, a b muft be in part omitted on the outfide of the main well, as fhewn fig. 1.—e f in the plan and fection is an arched communication between the wheel pit and the pit C.

Fig. 2.—A B C D E, the mean pipe of conduct, five inches diameter from the pump to the joint C.—F is the perpendicular pipe from the refervoir to the ground under which it paffes through the foundation wall to G, and joins the main from the pump at H.—K L are two ftop cocks.—K is the refervoir ftop cock, which, being fhut, retains the water in the refervoir, while L being open can fend it as the engine draws it to the quay refervoir, or even to the liquor backs of the brewhoufe.—L is the pump ftop cock, which being fhut, will oblige the water to afcend to the refervoir in preference to all other places.

Plate XII. fig. 1. is a fection through the middle of the out-jetty and elevation of the building, working-beam, and pump, alfo a front view of the pump and corner of the building; in thefe figures the dotted fquares a b and c d, denote two pieces of oak to be walled flufh into the refpective faces of the building, of about 3½ or four feet long, each containing a mortife for a hook tennant to fix the ends of the crofs pieces, for fteadying the pump, frame, and ftage.—e f fhew the pit for the crank.—g h fhew the headftocks or carriage for the crank neck gudgeon.—W, a caft-iron weight to balance part of the weight of the column of water in the pump, to be eighteen inches long, ten inches wide, and nine inches high.

Fig. 2. is the defign for the pump, being a front fection of the brafs barrel and clack feat piece of caft iron.—A B, the brafs barrel.—C C, the leaden pump bore fcrewed down with a flanch thereupon, to be five-eighths thick at the length next the barrel, and

each

WEEVILL ENGINE.

End View of Weevill Engine.

Scale of Feet.

Fig. 1.

Fig. 5.

Fig. 3.

Fig. 2.

Fig. 4.

Fig. 6.

Fig. 7.

Fig. 8.

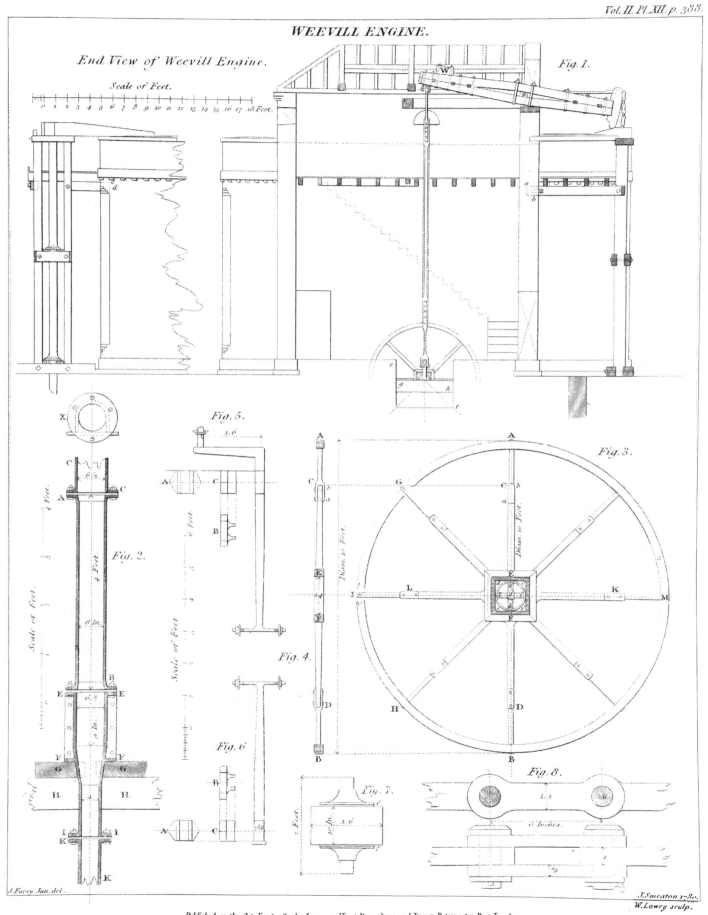

J.Farey Jun. del.

J.Smeaton 1780.

W.Lowry sculp.

Published as the Act directs, 1812, by Longman, Hurst, Rees, Orme and Brown, Paternoster Row, London.

each length gradually lefs, fo as to be three-eighths thick at the top.—E F is the clack door piece, whereof E E is the upper flanch, whereon the brafs barrel is fcrewed, and F F is the lower flanch, which refts upon the crofs planks G G, fupported on the beams H H, fuppofed double. The clack feat-piece terminates in the flanch I I, to which is fcrewed the leaden fuction pipe K K, the bottom of which muft be pierced with fixteen blaft holes, of one inch diameter.—X is the plan of the top of the clack feat-piece.

Fig. 3. fhews the conftruction of the fly wheel, whofe circumference is intended to be three inches fquare in caft iron, with eight round holes for fixing the arms, which arms are in part to be of caft iron, but joined to the ring by pieces of wrought iron, which will prevent all difficulty in the cafting, and at the fame time furnifh the means to bring it to its true centre ; fig. 4. is a fection of the fly wheel, which it is neceffary to confider along with the other, the fame letters denoting the fame parts in each. A B, the ring of the fly wheel, ten feet diameter. C D, two oppofite arms of caft iron, being eight in number, and all connected in one piece by the fquare E F, whofe internal opening is fourteen inches ; the end of each arm is to be perforated with a couple of holes of about three-eighths of an inch diameter, as at a b, by which the wrought pieces a b, C A, are rivetted and firmly fixed upon the caft iron ones, fo as to make them out to a proper diameter, and being brought to a centre, they are fixed in the circumference of the ring (the ends being fplit to receive a wedge as is fhewn at G & I) exactly like a trenail, which, being faftened, the furplus is chifeled off, and the circumference left even.

The only circumfpection that is neceffary in fixing the arms, will be (after they are all refpectively fitted and fet fair, the wrought to the caft iron ftumps) to get the two oppofite ones, as fuppofe, A C and B D into place, (the whole lying horizontally) and rivetted, but not wedged, which is not to be done till the laft, then one of the fide ones, as fuppofe I L is to be got in and rivetted, which done, that for the oppofite one at K M, the double part is to be made hot, fo that when offered to the arm, a few gentle blows of a hammer, fupported by an oppofite fledge, will bring it to fet quite true to the caft iron, when this may alfo be rivetted ; and in like manner the other four are to be got in and rivetted one by one, heating the double parts to make them fet true, if found neceffary ; laftly, the wedges are to be driven, and the whole, by the means above obferved, having no falfe bearings, the ftrength of the whole will be united, and the wheel may be moved at pleafure.

When

When the fly wheel is hung, the fquare c d, e f, of twelve inches, is fuppofed to be the end of the wooden axis (upon which alfo the wallower is to be fixed) the intermediate fpace being left for wood wedges to be hung, and brought into round and flat like a wooden wheel ; g, is fuppofed a fection of the iron fpindle, and h i k r are ftaple cramps driven in between the fpindle and the hoop to prevent its getting loofe, the fame being alfo fuppofed at the other end where the wallower is hung.

Fig. 5. fhews the crank gudgeon; fig. 6. fhews the wallower gudgeon, with their bearing braffes. The neck of the crank gudgeon is not propofed to be fteeled, but made of good tough iron well hammered, and fmoothly turned, and left of the full fize of $2\frac{5}{8}$ inches diameter, fo that being feven inches in the length of the bearing, the wear will be very inconfiderable ; the wallower gudgeon will be eafily underftood by the figure, the bofs or head of the end being intended to counteract the tendency of the wheel, to drive the wallower from it.

A A, are the plans of the refpective bearing braffes; B B are the uprights thereof feen endways of the gudgeon, and C C are the uprights as feen crofsway of the gudgeon.

N. B. The two gudgeons fig. 5 and 6, according to the defigns, are fuppofed to be of one intire piece or bar, tapering gradually from $2\frac{5}{8}$ at the crank end to $2\frac{3}{8}$ fquare at the wallower end, put through a wooden axis bored like a pipe, and opened at each end to a fquare, to admit of wedges, and afterwards to be fecured with ftaple cramps, as defcribed for the fly wheel but as the fmith may not have a convenience to turn it in one piece on account of the length, being fourteen feet hree inc hes inthe fpindle part, it may be made in two pieces, each properly centered and turned, and then fhut together in the middle, and brought to a ftraight line ; but if this fhould happen to be attended with difficulty, they may be made feparately and laid into the wooden fhott, with a fcrewed T at the tails, as fhewn in the defigns, and fecured with ftaple cramps as already fpecified.

Fig. 7, is the plan of the caft iron gudgeon for the working beam, of the thicknefs of three inches, except that f f reprefent two fillets on the upper and under fide, to keep the gudgeon immoveable, between the two beams, of which the working beam is propofed to be compofed, and which is firmly clipped in by the fcrew ftumps fhewn in Fig. 1, the gudgeons, if not turned, muft be fmoothed with a file. Fig. 8. fhews

DESIGN FOR A STONE BEER VAULT FOR WEEVILL BREWERY.

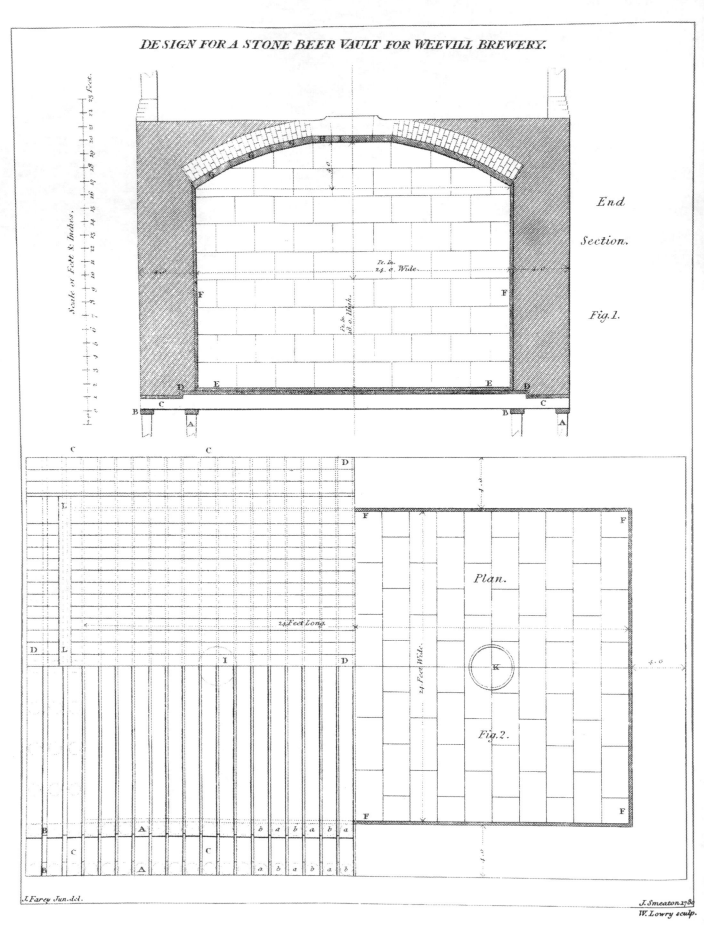

End

Section.

Fig. 1.

Plan.

Fig. 2.

J. Farey Jun. del.

J. Smeaton 1786

W. Lowry sculp.

Published as the Act directs, 1812, by Longman, Hurst, Rees, Orme and Brown, Paternoster Row, London.

ſhews the deſign for the links of the pump chain, on a larger ſcale, which will be ſufficiently explained by inſpection.

N. B. By way of giving a durability to the wallower, I would recommend the rounds to be of caſt iron, turned or well ſmoothed with a grindſtone, by this means the part moſt ſubject to wear will laſt many years, and always be in order for work.

WEEVIL

WEEVIL BREWHOUSE.

DESCRIPTION of the design for a stone beer vat for his majesty's brewhouse at Weevil.

Plate

THIS vat is propofed to be forty-two feet long and twenty-four feet wide, infide meafure, and eighteen feet depth from the crown of the vault to the floor; the arch rifing four feet, which vat will contain upwards of 2000 barrels; the vat to be four feet in thicknefs, and every where lined with Elland Edge flags.

The top of the vault being fuppofed to be even with the furface of the adjacent ground, the bottom will be required to be funk confiderably into the ftratum of blue clay that is every where met with in this diftrict under the upper ftratum of gravel, which is generally about nine feet thick, which clay itfelf being every where inpervous to water, a great part of the folidity of the work will depend upon an intire exclufion of water from the bottom, or upon preventing the ill effects of fuch as may unavoidably happen.

Section

A A, denote piles driven into the bottom of about eight or nine inches diameter at the heads, and five or fix feet long, more or lefs, fo as when driven down, to go pretty ftifly into the ground; the tops of trees of their natural taper, and pointed, will very well ferve this purpofe, and beech as good as any, there being two to each end of each beam.

B B, fhew the ends of ftring pieces of fir, four by twelve, going lengthways of the vat, and lying fair upon the pile heads after being levelled.

C C, are tranfverfe beams croffing the vat and its foundation walls, of one foot fquare, and at diftances of about four inches, difpofed as per plan.

D D, is a floor of two inch fir plank, running lengthways upon the tranfverfe beams, which is to be very carefully caulked, and that it may be feen to be free from leakage,

this

this floor, where not neceſſarily covered by the walls, is to be left uncovered till the laſt, and where defective, amended, after which it is to be covered with

E E, a paving of large four inch Elland Edge flags, well jointed and bedded in terras or pozzelana mortar.

The ends of the tranſverſe beams, at one foot from the inner face of the walls, are to be taken down three inches lower than the reſt, not only to give a firm footing to the wall, and prevent its being preſſed in before the Elland Edge floor is put in, but by breaking the joint of the walls upon the floor, will be likely to be the moſt effectual means of preventing the tranſpiration of water from without through that joint.

The walls are to be built of ſtrong well burnt brick, carefully walled on both ſides, and the inſide grouted full, courſe by courſe, with terras or pozzelana mortar, and alſo as the walls are carried up, they muſt be very well and carefully plaiſtered with a good coat of the ſame mortar, to prevent in every poſſible degree the reverting of water through the walls from the outſide inward, when the vat is empty, and thereby throwing off the inward lining of the ſide walls with Elland Edge flags of two inches thick F F, which are to be applied after the outwalls are built and ſettled.

As it is very difficult to make any lining of thoſe flags ſtick to the roof of the vault after it is built, I propoſe to compoſe it originally of a ſegment of a polygon, com-poſed of thick Elland Edge flags of ſix inches, G G G H, which will be ſufficient to ſtep four inches upon the wall at each ſpringer, and leave two inches of projection to cloſe in, and fix the top courſe of Elland Edge flags of two inches thick, F F, which are to be applied after built; the middle pieces are ſuppoſed to be broader than the reſt, in order to admit the man-holes to go through the entire ſtone as repreſented at I.

Plan.

One half of the length is ſhewn as when completed, the arch being ſuppoſed unput on, or removed; the other half is ſhewn, on one ſide the naked tranſverſe beams reſted upon the piles and ſtring pieces, the other with the timber flooring ready to re-ceive the walls.

The ſame letters of reference denote the ſame members in the plan, as before de-ſcribed of the ſection, beſides which the dotted circle I, repreſenting one of the man-holes, K being in place of the other, fhews a kind of pan of caft iron of one inch thick, and three inches deep, its upper border to be even with the ftone floor, and compofing a part thereof; its ufe will be that in emptying the vat by a pump, it will ferve as a pump to drain the whole floor; underneath the other man-hole may be a fimilar one, if thought neceffary, and I would recommend two openings in preference to one, becaufe the vault can be more eafily ventilated, and either or both can be fhut, if occafion require, by laying a border of clay round them, and a flat ftone upon them.

K L reprefents a riband let one inch into the planking, and projecting upwards three inches, to ferve the fame purpofe at the ends as the letting down of the beams for the fides.

Precautions in founding the floors.

As a principal concern will be to found them clear of water, the clay muft be uncovered to a fufficient breadth to fink the reft of the foundations therein, leaving a border by which the water iffuing from the gravel may be collected in one place, and there pumped out continually, without letting it run down into the cavity in the clay, which it is expected will afford no water, and it may be confidered whether it may not be lefs expenfe to form a drain for it into the new well, the water being continually kept down there for the neceffary purpofes of brewing, &c. than keep a pump going continually for the purpofes abovementioned.

When the piles are driven, and the ftring pieces eftablifhed, the clay muft be carefully beaten under them, and up to them, and to prevent them from rifing, they fhould be trenailed at every other pile, as a, a, a, a, a, &c. and as the tranfverfe beams are brought on, they muft likewife have the clay beaten under them, and to prevent their rifing, fhould be trenailed to every other pile at b b b. b b b, &c. fo that every beam will have a trenail at each end, and before the floor is laid, the interftices between the beams fhould be beaten up with clay, endeavouring to make all completely full if poffible, without ufing fo much force as to draw the trenails. Laftly, in laying down the planking, they fhould each be progreffively under beaten with clay, fo that all may be full, and as much as poffible exclude the admiffion of water from fhowers or otherwife; and round the whole border of the timber-work, the clay being beaten up to a level with the floor, the brick-work may then be eftablifhed thereon.

Precautions

Precautions in raising the brick-work.

Care must be taken after raising a competent number of courses, and performing the terras plaiftering outside as before mentioned, to pick up and cleanse all the scraps of brick, lime, &c. from the clay bottom, and carefully fill in behind the wall with clay in its natural or middling state of stiffness, so as to make all so close together, as to exclude the water, and that rather by treading than ramming, for by violent ramming in of more stubborn matter, the wall might be bilged if twice as thick as prescribed; this is to be done as high as the natural stratum of clay reaches; after which a coat of clay of three or four inches thick behind the terras plaifter will be sufficient, with common filling, ramming the materials pretty tight behind the springing of the arch, after the arch is turned.

When the arch is completed, the Elland Edge lining of the side walls may be proceeded with, to be flushed upon the back, and in part jointed with terras mortar, to be finished in the face with cement of Mr. Whitbread's composition, and lastly the floor, taking care to stop all transpirations in the caulking or planks, if any appear.

The beft lime for the purpose will be had from Watchet in Somerfetfhire, upon the Briftol channel.

The beft compofition two parts lime in flour, one part of pozzelana or terras, and one part clean fharp fand, well beaten to a tough pafte, and the grout to be made (not from the raw materials, but) from the beaten mortar diluted with water to a proper confiftence when well ftirred.

J. SMEATON.

Austhorpe, 26th Auguft, 1780.

CHIMNEY

CHIMNEY WINDMILL.

EXPLANATION of the designs for Chimney Windmill.

PLATE XIV. exhibits a vertical fection of the whole mill. Plate XV. figs. 1 and 2 are defigns for the whips, fig. 4, is a plan of the upper kirb, and moveable part of the kirb, and fig. 3, reprefents part of the cap feen from behind.

A B, plate XV. fig. 1, is the mafter whip, reaching eight inches beyond the centre. a b c d fhew the mafter tenant, fhouldered two inches on the under fide, and two inches on the upper, which will leave it five inches in thicknefs. C D and E F, are two whips that are mortifed on to the mafter tenant, and butt upon it, in the line a b, and alfo mitre with each other according to the line B e. G H and I K are two whips that are brought on after the three former are in place; they mitre into the angles of the former upon the lines f a g, h b k; their tenants are fhaped according to the line f l m g, as relative to the whip G H, and the other in a fimilar manner. Thefe tenants are fhouldered two inches upon the upper face, and four inches upon the lower face, and are three inches thick.

N. B. That part of the mafter tenant that is mortifed away for the tenants of the whips G H, I K muft be continued to the end of the mafter tenant; that is, the fpace a m n muft be continued out to d o, by which means the mafter whip, as well as any of the reft, can be drawn out fingly, to be repaired in cafe of accident.

Fig. 2. fhews the manner in which the whips are to be ftirruped down upon the horns, being a fection feen fideways, wherein A B is the mafter whip, C D the fame whip, as that in the plan fig. 1. marked with the fame letters; f h c d, is the mafter tenant, and i k the place of the tenant of the whip I K, but the mortife to be cut out of the mafter tenant to the end at d.

The proportions of the wheel work are as follows:

The fails are thirty-four feet nine inches long from the centre, and fix feet wide acrofs the face of each fail; the brake wheel M, plate XIV. has feventy-two cogs at five inches pitch; the wallower N, which it turns, has thirty-one, at the fame pitch; the main fpur wheel O, feventy-two cogs, at four inches pitch; the nuts or lanterns

P P,

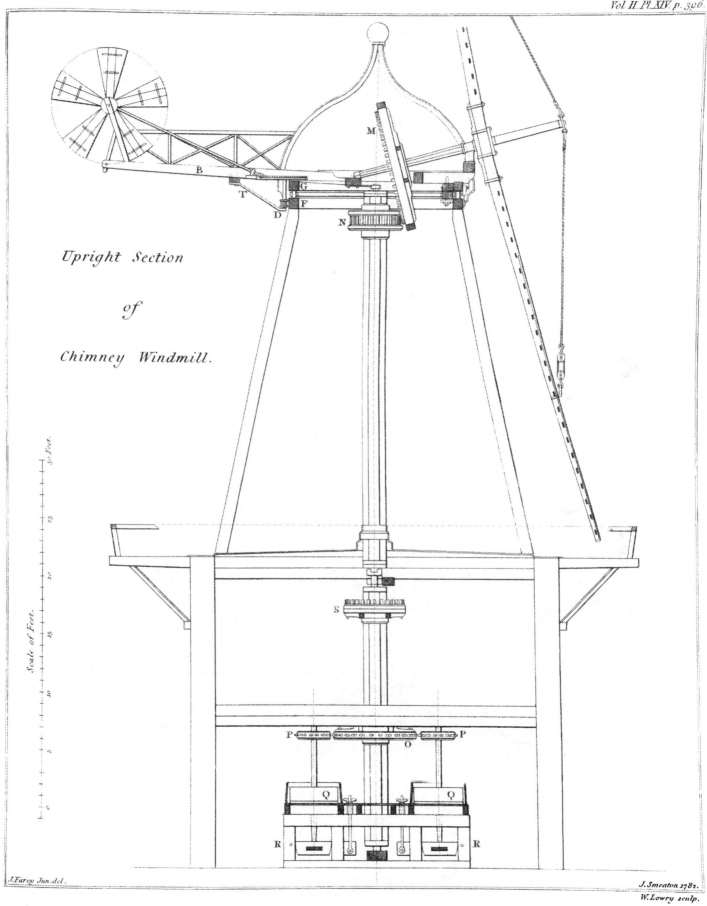

B

T
D
G
F
N
M

S

Upright Section

of

Chimney Windmill.

P P
O

Q Q

R R

J.Farey Jun. del.

J.Smeaton 1782.

W.Lowry sculp.

Published as the Act directs, 1812, by Longman, Hurst, Rees, Orme and Brown, Paternoster Row London.

Scale of Feet.

30 Feet.
25
20
15
10
5
0

CHIMNEY WINDMILL.

Design for joining the Whips *in the Center, and manner of fixing the same to the Horns.*

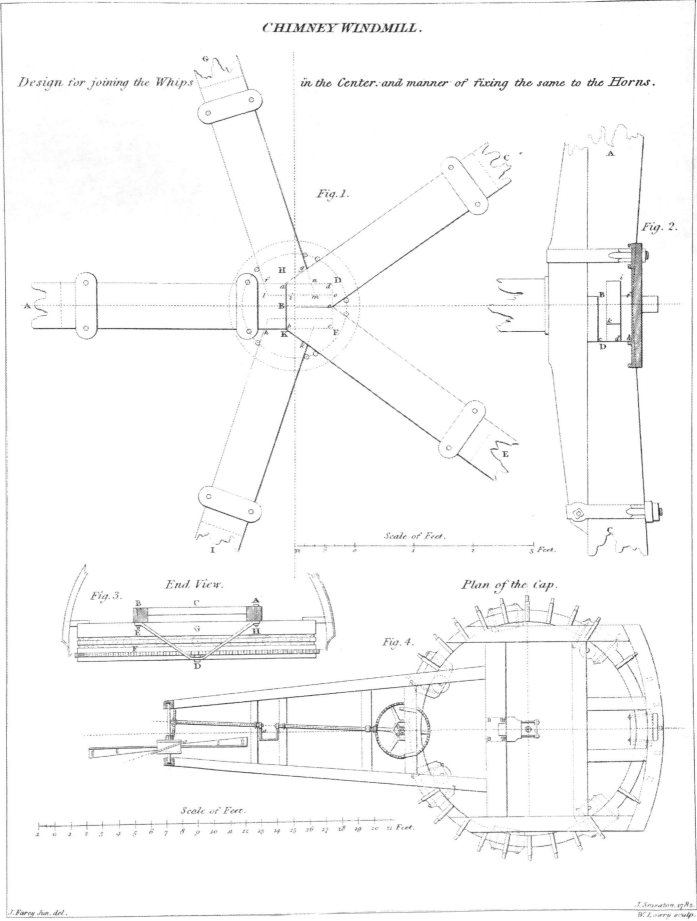

Fig. 1.

Fig. 2.

Scale of Feet.

Fig. 3. End View.

Plan of the Cap.

Fig. 4.

Scale of Feet.

J. Farey Jun. del.

J. Smeaton. 1782.
W. Lowry sculp.

Published as the Act directs, 1812, by Longman, Hurst, Rees, Orme and Brown, Paternoster Row, London.

P P, for the two pair of four feet blue ſtones Q Q, contain twenty-ſix cogs, at four inches; the nuts or lanterns, for the two pair of five feet ſtones, twenty-eight cogs, at four inches; theſe laſt are not ſhewn in the deſign, being behind the others, with which they form a ſquare, upon the Hirſt R R, one pair of ſtones being ſituated at each angle. The ſwimming wheel S, which is for driving the ſmall geer, has fifty-ſix cogs, at 3¼ pitch.

In fig. 3, of plate XV. which is an upright of the hind part of the cap A B, are ſections of two tail trees, which ſupport the tail vanes; C D, the perpendicular line, being the centre line of the upright ſpindle; G, part of the upper kirb; F, part of the lower ditto; H D E, the two iron arms or braces, which, at D, ſupport the braſs box for the lower gudgeon of the pinion, which turns the cap round, the box is received in a broader and ſtronger part of the iron brace; the box and gudgeon, and conſequently the lantern are prevented from ſhunning their work, by a back brace ſhewn particularly at T, plate XIV, which, by a wedge at its extremity, can be adjuſted ſo that the lantern may be brought to a proper depth in the dead wheel fixed round the lower kirb F of the cap.

The numbers of the iron wheels for turning the cap are as follows:

The dead wheel upon the lower kirb F, is made in ſixteen ſegments of fourteen teeth each, making 224; the lantern pinion working therein ſix leaves; the crown wheel upon the top of the main lantern ſpindle three feet ſix inches diameter, number, ninety-ſix teeth; the nutt or pinion turning it ſix leaves; the contrate wheel at the top of the oblique ſpindle, that is driven by the ſail wheel pinion, two feet diameter, number ſeventy-two; the pinion upon the tail vane ſpindle, to drive the above number thirteen.

The axis of the tail vanes is intended to be a little oblique to the middle line of the cap, as ſhewn in this plan, that the main ſails may be canted a little from the wind's eye, the right way, and the long oblique ſpindle being divided near the middle by a handle, (the arm A releaſing at pleaſure by a joint), the mill may be turned in a calm by hand.

OIL.

OIL MILL, AT HULL.

EXPLANATION of the design for an oil mill, for Mr. Mayson Wright, to be worked by water raised by a fire-engine, at Hull, by J. SMEATON.

Fig. 1. ſhews the water-wheel in front, with its penſtock.

Fig. 2. repreſents the general deſign of the work, the water-wheel being ſeen end-ways.

Fig. 3. repreſents the main ſpur-wheel in front, with the manner of fixing it upon the ſquare flanch upon the iron axis.

> N. B. The whole of the machinery being drawn true to a ſcale, the explanation will be confined to ſuch things only as do not obviouſly appear from the ſeveral drawings or meaſures thereon.

Further explanation of fig. 2.

The axis of the water-wheel is ſuppoſed to be parallel to the length of the houſe, and the plane of the wheel to be parallel to the croſs wall that ſeparates the engine and refining-room from the great building; this view being ſeen from the ſide next Hull. So much of the floors and interior walls of the main building as will interfere with the work, is to be removed, and it is to be noted, that by lengthening or ſhortening the tumbling axis, the ſtones may be ſet at any diſtance, at pleaſure, from the water-wheel.

The head-ſtock at a may be reſted upon the ſaid croſs wall, if it proves ſufficiently founded, and of ſtrength to break a hole therein, otherwiſe a new head-ſtock wall had better be erected parallel thereto.

b, ſhews the ſection of a beam to reach acroſs the building, in order to ſupport the upper gudgeon of the axis of the ſtones, and c is a rider bolted to it, to carry the gudgeon of the tumbling axis.

The

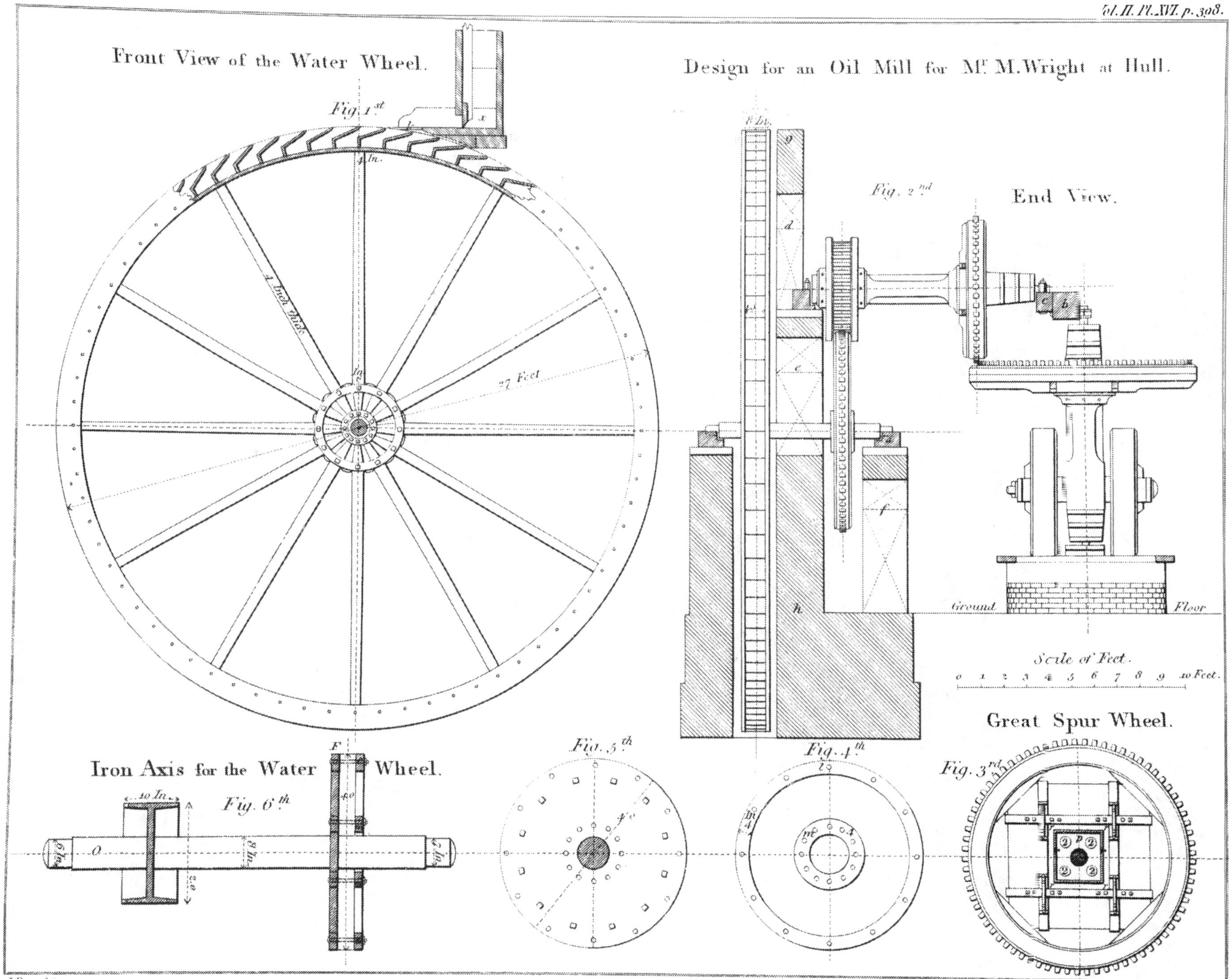

Front View of the Water Wheel.

Fig. 1.st

A Back Slide

4 Inc.

7 Feet

Iron Axis for the Water Wheel.

Fig. 6.th

Fig. 5.th

Fig. 4.th

Fig. 3.rd

Great Spur Wheel.

Design for an Oil Mill for Mr. M. Wright at Hull.

Fig. 2.nd

End View.

Ground

Floor

Scale of Feet.
0 1 2 3 4 5 6 7 8 9 10 Feet.

L. Berry del.

W. Lowry Sculp.

Published as the Act directs, 1812, by Longman, Hurst, Rees, Orme and Brown, Paternoster Row London.

Vol. II. Pl. IV. p. 308.

The unshaded spaces d e f, are supposed arched openings in the walls, of three or four feet in width, the lines whereon d e f are respectively marked are supposed in the spring of the arches; and it is to be noted, that in some convenient place of the water-wall g h, an opening is to be left near about the height of the centre, on one side or other, in order that the shield boards of the wheel, next the wall, may be fixed and afterwards repaired. The raising of the water-wall to the height represented will be necessary, in order to prevent the splash of the wheel from annoying the inner works, and also to support the penstock on cross beams rested thereon. The height of the bed stone may be at pleasure higher or lower, by varying the length of the axis of the stones to suit it, as also the height of the ground floor, if it happens to be a little higher or lower, respecting the surface of the water in the well, than here drawn. The press works are here omitted, as nothing is intended out of the usual course, Mr. Wright himself being the best judge of the conveniency of their situation.

Further explanation of fig. 1.

This view of the water-wheel is supposed to be taken on the contrary side of the wheel to that where the works lie, and shews the manner of fixing the arms to the iron axis, being bolted on between the great flanch on one side, and two strong cast-iron rings on the other; the intermediate spaces being filled up by drift wedges or chocks, the angles of which being somewhat obtuse, they are hindered from starting by wooden pins or keys, that stop against the inside of the great ring, and go through the chocks and the flanch. The side of the wheel descending will, therefore, be to the west, in case the stones are chosen to go round with the sun, and the penstock is supposed to be supplied by a square wooden pipe or trunk, proceeding from the reservoir, and in the direction of and parallel to the axis of the water-wheel, and consequently will enter the penstock sideways; the clear opening of this trunk in the inside is shewn by the space x, here represented on the far side, but in reality entering by the near side, which in this section is supposed to be removed. At k, in the bottom of the sheet of the penstock, is shewn the edge of an iron plate, intended to make the water deliver clean, and convey it nearer the centre of the wheel, without losing height; the width between the two cheeks of the shoot, whereof k marks the far one, is eight inches, and this is the breadth of the iron plate, being somewhat rounding at the extreme and most projecting in the middle. When the water has passed the middle of the wheel at bottom, it is to be conducted by a quarter-round turn, through the cross wall into the well, and from the centre of the wheel should gradually deepen from one inch, that the bottom of the con-

duit

duit is fuppofed to be covered with water to fix or feven inches, or even twelve on entering the well.

Further explanation of fig. 3.

The centre circle is a fection of the iron axis of eight inches diameter, the four circles round it are holes through the flanch of the main cogg wheel; the fquare furrounding that is a hollow fquare of caft iron, two feet outfide and ten inches in length, fupported in the middle of its length by the faid flanch; the fquare fpace between the faid iron fquare and the arms of the wheel, fhews what is occupied by the blocking and wedging, being three inches on each fide, in order that the arms may be feparated to the diftance of two feet fix inches. The plates fhewn upon the croffing of the arms, are upon the fide that is cut, in order to bind each firm together, the plates therefore on the far fide lie the contrary way; the reft of the conftruction of the wheel is by backs and facings, in the ufual way.

The other figures exhibit particular defigns of the caft-iron axis, and the parts relative thereto, wherein

Fig. 6, fhews the axis itfelf, with the round flanch F, and its two rings, as alfo the fquare flanch G, and its box in fection.

Fig. 4, fhews the two rings, l the bigger, and m the lefs, relative to their fections, marked with the fame letters in fig. D.

Fig. 5 is the great round flanch feen upon the flat, wherein the holes are thofe for the keys or pins to pafs through to keep the chocks and wedges in place.

The fquare flanch and box, as feen end-ways, of the axis, are reprefented in fig. 3, wherein p is the fquare flanch, q four holes through it, principally to facilitate the cafting, r r r r is the edge of the fquare box, whereon the main cogg-wheel is wedged, in the fame manner as upon a fquare wooden axis.

GENERAL OBSERVATIONS.

The quantity of water furnifhed by the fire-engine being at prefent no more than 274.3 wine gallons per minute, the work done thereby can be expected but little to

exceed

exceed the power of two horfes yoked at once, that is, to turn the ftones fomewhat better than four turns per minute, with a full charge as in horfe works; but as it is faid that ftones of this kind do their work in the beft manner when they are made to revolve about ten turns per minute, and as they cannot be expected to revolve at that rate with the prefent power, it is moft likely Mr. Wright will think proper to increafe his power; I have, therefore, thought it good to adapt the numbers in fuch a manner as to be fuitable when a power fufficient is procured, to make the ftones revolve their full pace of ten turns per minute; and in the mean time I would propofe to diminifh the charge or quantity of feed put on at a time, till the ftones will make fix or feven revolutions per minute; this, perhaps, may not conduce to increafe the produce of the work beyond the common charge at the horfe pace, though, if any difference, I expect the way propofed will be for the better. But I apprehend it will be neceffary, on account of giving a proper velocity to the water-wheel, for, in order to give the ftones ten turns per minute, it is neceffary that the water-wheel fhould revolve flower than the ftones; and, therefore, according to the proportion affigned, when the ftones revolve at four turns per minute, the water-wheel will be under $2\frac{1}{2}$ turns, a pace fo flow that I can hardly expect it to move regularly or equally forwards; it will, therefore, be proper to diminifh the charge till the wheel moves regularly and takes its water agreeably, which I expect will be when the water-wheel makes about four turns, and, confequently, the ftones about fix or feven. But, in cafe Mr. Wright fhould fully refolve not to increafe his power, fo as to make the ftones revolve nine or ten turns with a full charge, there is nothing to be done but to make the wallower and fpur-wheel upon the horizontal or tumbling axis, of equal numbers.

To increafe the power fo as to make the ftones revolve ten turns per minute, with a full charge, and to do work in proportion, will require an engine that will raife $2\frac{1}{2}$ times the quantity of water that the prefent engine can do, an improvement that is totally out of nature to make with the prefent engine; but, with an engine properly conftructed, of twenty-five inches cylinder, I can undertake to raife to the height of thirty feet 1000 gallons, wine meafure, per minute, with the expenfe of 1 cwt. 3 qrs. coals per hour; and I do apprehend, that if a new engine were fet about in conjunction with the mill, or proper conveniencies left for placing it, that fuch an engine would be made for £350, or thereabouts.

J. SMEATON.

London, 15th March, 1776.

LEAD HILLS WORKS.

The REPORT of JOHN SMEATON, engineer, upon the state of the powers employed, and capable of being employed at the Susannah Vien, at Lead Hills, for unwatering the low workings thereof.

AT the requeſt of the court of directors of the Scots Mines Company, on the 27th and 28th of Auguſt laſt, I carefully viewed every thing both above ground and below, that I conceived could conduce to give me a thorough knowledge of the circumſtances attending the Suſannah Vien at Lead Hills, ſo far as related to the drainage of the water from the low ſoles or low workings of that vien, both with reſpect to its preſent and future proſpect.

The principal circumſtances attending this mine in the reſpects mentioned, I found as follows :

But I muſt firſt premiſe, that the ſeaſon preceding my being there was remarkably dry and warm, and which, as I was informed, had been very particularly the caſe in the quarter of Lead Hills, and, indeed, the remarkable dryneſs of the mooriſh ſoil that there abounds, in a great meaſure ſhewed the ſame thing, the drought being continued to, and during the time I ſtaid there, which was till Monday the 30th of Auguſt.

The loweſt level that now drains this mine, which iſſues out near the company's ſmelt mill, is called the Poutchill level, and by it iſſued in that dry ſeaſon near upon the whole water that is collected in the diſtrict, its quantity was then conſiderable, amounting by my gage thereof to no leſs than 205 cube feet, that is, to twenty four hogsheads, or ſix tons per minute.

There is now bringing up a low level at the random of twenty-two fathoms below the Poutchill level, which, coming from a lower part of the burn or rivulet, by which the waters of Lead Hill make their departure, will flank the vien, and enter it near to a right angle ; this level is carried directly towards that part of the vien where the deepeſt workings are, ſo that when this level is finiſhed, or brought up to the vien, the whole of the water, or nearly the whole, that now makes its departure by the Poutchill level's mouth, can be let fall into the low level.

Of

Of the water that now iſſues from the Poutchill level mouth, a part of it is gathered upon the ſurface, being the collection of a great number of ſprings and waters iſſuing from old upper workings, which are collected by a water-courſe that extends upon the ſurface ſeveral miles in length, and is ultimately conducted to the top of a ſhaft near the mouth of Porto Bello level: it there falls down by different ſtages to the amount of twenty-nine fathoms, before it reaches the top of the wheel of the preſent bob engine, which is employed in raiſing ſuch water from the depth of thirty-one fathoms, as has been unlooſed in the mine below the random of Poutchill level.

This ſurface water I gaged, as I then found it near the ſhaft, where it goes underground, and found it amount to ſeventy-ſeven cube feet per minute.

The whole water iſſuing by Poutchill level, nor this part of it in particular, was never known to be leſs than it was at the time of my view, and which, on that account, was very fortunately timed; yet this ſurface water gave power and velocity to the preſent engine more than ſufficient to draw the water, it being upon the ſnore, or drawing air, ſo that it did not fill its barrels by about one quarter part, as I judged it.

The water ariſing underground is pretty conſtant; yet, like moſt other ſprings, is increaſed by the winter's rains, and the ſurface water much more; for the diſcharge of the Poutchill level in the month of April, being meaſured in a ciſtern, amounted to more than double the quantity now aſſigned by me, the operations confirming each other; but in general, when the water underground is increaſed, the ſurface water is ſtill more ſo, ſo that the engine moving with greater velocity, is generally competent to draw its water, except during the froſts in winter, which often continue long there, when the ſurface water becomes frozen; for though it in general proceeds from ſprings, which coming out at forty-five degrees of the thermometer, may be called warm, when compared with froſt; yet by running for ſo many miles in an open leet, expoſed to the froſty air, it at laſt ſubmits to the cold, ſo as to prevent any water from getting home to the engine, the conſequence of which is, that during thoſe froſts, which I am informed continue frequently for four months in the winter, the workmen are totally driven from the low workings.

This engine has a twenty-four feet water wheel overſhot, and draws a column at two lifts of thirty-one fathoms, the bores of the pumps being $7\frac{1}{2}$ inches working barrels; it makes a five feet ſtroke, and when I ſaw it, made $5\frac{1}{2}$ ſtrokes in each barrel

per

per minute; which, allowing for air then drawn inſtead of water, amounts to 380 cube feet per hour, that is, to about 11¼ tons, which, together with the ſurface water from the tail of the wheel, is delivered into the Poutchill level, and makes a part of the water thence iſſuing.

Below the level of the engine foot, the works are carried down in two ſeparate places, from one of which they are extended to the north, and from the other to the ſouth, there being betwixt the two a check in the vien, where the two checks are ſo cloſe together as to make a water tight joint, ſo that it acts as a dam, and prevents the water of one ſet of workings from affecting the other, while both can be cleared by the engine; this check thus keeping the workings ſeparate, is a real convenience, as either may be worked without drawing the water out of the other.

One of theſe workings goes down eleven fathoms, and the water thence ariſing, which is by far the moſt conſiderable, is drawn by hand pumps. This work was full of water at the time I was there, the pumps not being at work; but as the water was running out at the top of the pump, which I was informed was as near as poſſible the ſame in quantity as if drawn by the pump, it would therefore be the ſame to my meaſure thereof as well as the engine, and which I found to be 206 cube feet per hour.

The other was clear of water, which produced much leſs, and was drawn by tubs from the depth of fifteen fathoms, the amount of which was then at the rate of ſeventeen cube feet per hour, ſo that the whole growth of water ariſing below the engine pump foot, amounted to 223 cube feet per hour, which, deducted from what the engine then drew, viz. 380 feet, will leave 157 cube feet per hour, ariſing between the Poutchill level, and the engine pump foot.

When I was there they were preparing to put in an additional engine, whoſe water wheel was of thirty feet diameter, and eighteen inches in the ſole; with this engine they propoſe to ſink thirty fathoms below the preſent engine pump foot, that is, fifteen fathoms below the deepeſt workings at preſent in this vien. This engine is to receive the power by placing its wheel above the preſent one, ſo as to occupy a part of the twenty-nine fathoms that the ſurface water has been deſcribed to deſcend from the ſurface to the top of this preſent wheel; and after delivering its water from the tail of the new wheel, it is then to be re-conducted upon the preſent wheel, and do its office as at preſent; but the pumps which are intended to be of ſeven inches bore, are to be

placed

placed in a pump funk by the fide of the other, and to raife, not only the water that is raifed from the low foles by the hand pumps and the tubs, but fuch as may arife in going ftill deeper, to the depth of thirty fathoms below the prefent engine foot.

Having now defcribed the prefent ftate of things, I proceed to give my opinion thereon.

I underftand from all hands, that the putting in of the prefent engine is to be confidered as a temporary expedient to draw the water now drawn from the low foles by the hand pumps and tubs, till the low level is got up to the vien, which may poffibly take three or four years; after that, we are to confider what kind of power this will afford us, when the water iffuing from the Poutchill level can be let fall as a power from that level twenty-two fathoms down to the low level, the draft from the low foles being then relieved by twenty-two fathoms in the perpendicular rife; for then the deepeft of the prefent workings will be only twenty-four fathoms under level, and the pump foot of the intended new engine will be no more than thirty-nine fathoms, which to the prefent level will be fixty-one fathoms; that is, whether there is fuch a kind of probability of then wanting a fufficient power from this fall of water as to make it neceffary now to prepare conveniences for the erection of a fire engine.

		Cube feet per minute.
The quantity ascertained by me to proceed from Poutchill level, the 27th September, 1779, - - - - - - -		205
The surface water ascertained the same day for turning the engine, -	77	
Also, what the engine drew from the low soles, 380 cube feet per hour, that is, $6\frac{1}{3}$ cube feet per minute, which, to avoid fractions, call -	6	
		83
Will remain water raised in the mine and its communications, that will not be subject to frost, - - - - - - -		122

Which quantity, of 122 cube feet per minute, being let fall twenty-two fathoms, and acting upon an engine properly conftructed to be acted upon by the weight of the whole column, will be equivalent to the power of a fire engine of thirty-fix inches cylinder, and this in its weakeft ftate, and no advantage fuppofed to be received from the furface water, which, if we take into the account, will, for the greateft part of the year, be equivalent to a fire engine of forty-five inches cylinder; but the conftant power equivalent to a thirty-fix inch cylinder would raife double the produce afcertained of the

the prefent engine from a depth of 128 fathoms below level, which is 104 fathoms deeper than the prefent loweft workings.

It is probable, that in increafing the extent of the prefent workings of the low foles along the vien, more water will be cut, and as they go deeper, ftill more; but if all the water that can be collected at thirty fathoms below the level now bringing up (that is, fix fathoms below the prefent loweft workings), is there intercepted, and raifed to that level, without being fuffered to go down, it is probable, that the water raifed under that level, will be confiderably lefs in quantity than what is now raifed by the prefent engine; for, after a certain depth, depending on local circumftances, it is a general obfervation among all miners, that the deeper you go, the lefs water you raife, fo that it is highly probable that a lefs power than is equivalent to a fire engine of thirty-fix inches cylinder, will clear this vien of water to as great a depth as it will carry lead.

It is remarked by Mr. Milne, who was fo obliging as to furnifh me with copies of Mr. Sterling's letter to him of the 9th of January, 1779, and his anfwer thereto of the 10th of February following, that the principal appearance of water was at the iffuing of the Poutchill level, from whence he concludes, that the principal part of the water that has been raifed in thefe mines is now dropped down to that level, as being the loweft point to which it can defcend, which obfervations I muft confirm; he, therefore, concludes, that nature following the fame courfe, when the low level is brought up, and the vien opened, the greateft part of all the water will alfo find its way down to that level, and of confequence there will be an end of that power, arifing from the defcent of the Poutchill level water into the low level, which was infifted upon as a fufficient power by Mr. Sterling, and the effects of which have been more particularly calculated and afcertained in what precedes; and this being fuppofed to be the cafe, as the ground to be worked upon is very hard, and will take feveral years to fink a perpendicular fhaft from the furface down to the deepeft and moft valuable part of the prefent workings proper for a fire engine to be placed upon, it is now time to begin fuch a fhaft, that it may be ready when it fhall be wanted.

This, undoubtedly, is a queftion of a very ferious import, and demands a critical inveftigation how far the fubfidence of the prefent water is likely to take place; becaufe, if it does in the manner and degree fuggefted by Mr. Milne, there is an end of all reafonings and calculations about its power and effect. 1ft. That in laying open

the

the vien, the water raifed therein has in a great meafure got down to the Poutchill level, being now the loweft level to which it can defcend, is certain : but I apprehend that this has been owing to there being no motive at the various times at which the fprings have been cut for keeping the water up, but it has been fuffered to defcend to the low level as the fhorteft way of getting rid of it, and the water rifing in a great variety of places, it is now impracticable amongft old workings to collect it at a higher random.

2dly. I underftand that a great part of the water that now iffues at the Poutchill level mouth is not the produce of the Sufannah vien, but proceeds from the old works of feveral other viens that have been unwatered by a crofs level taken out of the Poutchill, and which now, and for a courfe of years paft, have iffued their water invariably here, and confequently can be fubject to no change by the future workings of the Sufannah.

3dly. The Sufannah having been worked as far into the mountain as the boundary, the whole of the fprings rifing in or above the level of the Poutchill are collected therein, and could be kept upon that level by an artificial bottom, though the prefent natural bottom were cut away; it is therefore only fuch fprings as arife in the very bottom of the level that have paffages under the floor of that level, and inverted fyphon-wife, proceed from higher fources; I fay it is only fprings of this kind that can be affected by under-workings, and which kind of communications we muft apprehend can fcarcely exift, except near the boundary, where the ground is whole in the direction of the vien; for, as the ftrata ftand edgeways here, or parallel to the checks of the vien, very little water can make its way crofsways of the checks : it follows, therefore, that very little of the prefent water would be let down, though the vien were worked to the extent of the boundary under the random of Poutchill level, provided the bottom were left whole, or made good artificially, where any chafm or opennefs might occur.

4th. The prefent low level being carried flankways to the vien, there is very little chance of its cutting any water which has communication with the water that now makes its way into the Poutchill level.

From all thefe confiderations taken together, I am induced to think that proper care and attention being paid to the fubject, it will only be a very fmall proportion of the water now iffuing from Poutchill level, that will get down to the low level, when brought up to the vien, except by paffages cut for the purpofe; but from

what

what has been eftablifhed, even if half of it were to get down to the level, yet the other half would be a power fo refpectable, as to be capable of drawing as much water as now arifes below the Poutchill level from a depth of 128 fathoms, and double that quantity from half the depth, which would ftill be forty-four fathoms below the prefent low workings, and will doubtlefs laft feveral years; and even this power will be of that confequence where coals are at a confiderable diftance in a mountainous country, that it will be very well worth while to employ a power of water as far as it can be employed, and build a fmaller fire engine in aid of the water engine, whenever that fhall become overpowered.

But here it may be faid, that fince the placing of a fire engine underground may be impracticable, or at leaft inconvenient *, for the reafons given by Mr. Sterling, and as it will doubtlefs take feveral years to fink a perpendicular fhaft from the top of the mountain, in cafe a fire engine fhould become neceffary, contrary to our prefent fpecu-lations, what is then to be done? Is a valuable mine to ceafe work till a fhaft can be funk? I anfwer no.

The argument here is to time, not expenfe; becaufe, if a fhaft could be funk in the fame time that the engine was building, though the expenfe of the fhaft would be the fame, it would, doubtlefs, be proper to defer fo as to begin both together, therefore faving the expenfe of the fhaft. I would fooner undertake to put up a fire engine of any fize that might be required, without either placing it underground, or finking a fhaft from the furface, than to place it upon a fhaft fubject to the expenfe, though it could be done in a day.

In fhort, I would place my engine at the tail of the low level, and work my pumps in a pump put down to the low workings by a fliding fpear through the low level, and fuch a pump muft, in effect, be funk for placing the pumps of fuch an engine or engines as muft be ufed, previous to the execution of any engine upon the twenty-two fathom fall.

It may be now time to fay what kind of a water engine I would recommend for this new fall of twenty-two fathoms.

I believe I may lay it down as a certainty, that no engine has yet been invented that furpaffes the bob engine for the miner's ufe, when well proportioned and conftructed,

and

* A fire engine deep underground has been in use some years past in the coal mines of Whitehaven.

and in such situations as it is adapted to, but as, particularly underground, it would grow very cumbersome, if erected with a water wheel exceeding thirty feet, it follows, that in the height of twenty-two fathoms, it would take four such water wheels and engines, one above another, to exhaust the power, or, in other words, to receive the full benefit of this fall, and which altogether would become a very complex and expensive assemblage of machinery.

Now, as it frequently happens in mines, that very great falls can be obtained, and but small quantities of water, then, as has been said, the bob engine taking in but five or $5\frac{1}{2}$ fathom with convenience at one stage, will be subject to a considerable expense in the erection, and do but little business; whereas, if the power of four, five, or six such wheels could, (where there is fall enough) be united in one machine, this would avoid not only great expense in the erection, but the complexity, weight, and incumbrance of several machines, and therefore it has long been a *desideratum*, which several have formerly attempted in different ways, but without particular success; this proposition, however, was about fourteen years ago happily accomplished by the late very ingenious Mr. William Westgarth, at Coal Cleugh, in Northumberland, to the lead mines of which place he was principal agent for the late Sir Walter Blackett.

This engine acts upon a simple statical principle, not unlike the fire engine, substituting the weight and pressure of a column of water, for the pressure and weight of a column of the atmosphere, so that as the column of water that may press upon one end of the balance beam may be of any perpendicular height, a fall of twenty, or even sixty fathoms, may be taken to one machine, and as it is not requisite that this fall should be in a perpendicular line, but acting through a pipe, may be bent in any angles that suit the place. This machine very particularly adapts itself to the frequently occurring circumstances of lead mines.

The machine mentioned has been constantly at work ever since its erection, and I believe still is, without ever being subject to any material disorder, other than common repairs. Some others were also erected in the same neighbourhood by the same ingenious person, some of which are still working, and the others not, owing to the mines ceasing working, or the water being relieved by bringing up a deeper level *.

* One upon this principle I have myself erected that has now been at work ten years, and is likely to continue.

It is, therefore, Weſtgarth's ſtatical engine, mentioned in the papers before me, that I would adviſe to be erected upon the twenty-two fathom fall, which working with the quantities of water ſpecified, is capable of producing the effects I have calculated upon.

This engine has alſo this further advantage; like the water wheel, it has, according to its ſize, a certain rate of working, which it cannot exceed, but will perform in proportion to any leſs power applied; ſo that if it is calculated to go ſixteen ſtrokes per minute with a given quantity of water, and this be reduced to one-fourth of the quantity, it will not ſtop, but make four ſtrokes per minute.

I come next to make mention of the engine now at work, and that erecting: the preſent working engine at the time, and for the purpoſe it was erected is not amiſs, the ſize of the water wheel being properly adapted to the quantity of water, and the pumps to both, and I may ſay the ſame thing of that now erecting: it were, indeed, to have been wiſhed, that in the order of things, each engine had been in the other's place; for then, as the new one will be the more powerful, each would have had a quantity of work to do, proportioned to its powers, and the preſent one, which is the weaker engine, on account of the leſs diameter of its water wheel, would have had only its water to have drawn from the low ſoles, which, as has been ſhewn, is leſs by 157 cube feet per hour than what is drawn by the preſent engine from its own pump foot; but the fact is, it is not eaſy, but indeed very troubleſome, to change the ſituation of ſuch an engine underground, without ſtopping the works, and as long as the old one can manage its water, it is no fault that the new one ſhould be ſtronger; for, however powerful, it can draw no more water than there is to draw, and as it appears from what has already been ſtated, that the preſent engine was capable of drawing about one-fourth more water than there was to draw, it is to be hoped, that in the interval of the low levels coming up, there may not be more water cut in addition to the preſent, by an extenſion of the work of the low ſoles, than what may ſtill be maſtered by the preſent engine; yet, if this ſhould happen, there will be wanted a ſecond new engine in aid of the preſent one.

Some little aid may be indeed drawn from the new engine, or by ſhortening the preſent engine pumps, and lengthening the other; but as this will take up time, during which no water can be drawn, I would rather in that caſe recommend the putting in of one of Weſtgarth's engines, that would at once do the buſineſs of them

all,

all, and if neceffary, exhauft the whole power of the furface water that now falls in the whole thirty-three fathoms, of which the power of twenty-eight fathoms is now loft, and after the new wheel is put in, above twenty will ftill remain unemployed; indeed, had I been confulted before any material progrefs had been made with the new water wheel engine, I fhould have clearly advifed one of Weftgarth's, which, when erected, would have put the prefent engine out of employment. The advantage of fuch a powerful engine would have been great, even fuppofing the prefent machines to continue capable of maftering the water, becaufe it would much fooner have cleared the mine after a ftoppage by froft.

J. SMEATON.

Austhorpe, 16th October, 1779.

MILL FOR GRINDSTONES.

REMARKS concerning the mill for turning grindstones, and grinding blacking.

AS it was defired to adapt grindftones to the new boring mill, and thinking them better to be feparate, I adapted the defign immediately to the fame head and fall of water, and I apprehend, that at all common times, there will be water enough for the whole, efpecially when the fluice is open to receive water out of Stenhoufe dam-head, but as water, when there is plenty, may as well be taken from the forge dam-head, as any other, the defign of this mill will equally apply itfelf to either, only converting the breaft wheel into an overfhot. However, for the fingle purpofe of turning grindftones, the prefent plan will have the preference, in going more fteady. Iron rings will not be needed for this wheel, wooden ones, of the common fcantlings, viz. about fix inches fquare, will be fufficient.

As the fame patterns for the wheels and nuts as in the gun mill are propofed here to be made ufe of, I apprehend they will be done cheaper than of wood, but if otherwife, wooden wheels of equivalent numbers will do; alfo, if the practice of making iron axes for the water wheels, be found to recommend itfelf fo far as to be equally cheap with wood, they may alfo be ufed here.

J. SMEATON.

Austhorpe, October 15th, 1770.

N. B. I fuppofe one pair of four-feet common mill-ftones, cut with furrows, as for corn, will be fufficient for grinding the blacking, but as it will be very little more ex-penfe by lengthening the tumbling axis, and the hirft, to make provifion for two pair, it may be advifeable to do it, as there is no faying what purpofes may occur to employ them.

Upon the whole it feems advifeable to let all ftand as they do, till the new mill be built, and from its performance the conveniences of the reft will fhew themfelves.

BOCKING

BOCKING FULLING MILL.

DIRECTIONS for converting the larger water wheel of Mr. Nottidge's fulling mill at Bocking, so as to drive two stocks by geer.

1st. THE water wheel muft be turned about fo as to go the contrary way round, which may be executed three ways, the whole wheel and axis may be lifted out together, and the axis changed ends; fecondly, if the wheel is made with clafp arms it may be un-wedged, the axis drawn out, the wheel lifted out and turned, and the axis put in again the fame way as before; or thirdly, the fhrouds being taken off and changed fide for fide, the buckets may then be reverfed, all the reft remaining as before.

2dly. The water wheel muft be raifed three inches and a half higher than at prefent, fo as to have no more than two inches and a half tail water in dry feafons, and muft be fet as much down ftream as poffible.

3dly. It muft be fitted up with a breafting fweep of wood, brick, or ftone, from the centre of the wheel to the bottom, then carried level three or four feet, and then re-conciled by a flope with the prefent floor.

4thly. The water muft be made to go on to the wheel in a reverfed direction to what it now does, which can be done without any material lofs of fall. The trough at the proper point is made to divide, and its bottom to proceed on the fame level as at prefent, on each fide of the wheel, and from the trough of union the bottom rifes five inches into the fhuttle trough, the two being reconciled by a flope. The water wheel will then ftand as follows:

	Feet	Inches
Tail water in dry feafons - - - -	0	$2\frac{1}{2}$
Height of the water wheel - - -	12	0
The furface of the water at a full head above the top of the water wheel	1	$1\frac{1}{2}$
The furface of ditto above the fhuttle trough bottom - -	1	$3\frac{1}{2}$

5thly. A fpur wheel, of forty-eight cogs, at a five inch pitch, muft be put on to the eaft end of the water wheel axis, to drive a lanthorn or wallower of twenty-three rounds,

put

put upon the end of a tumbling axis lying upſtream, which axis you have in your power to make of what length you pleaſe, to ſuit your convenience. In this you place your feet anſwerable to your ſtocks, according to your own method. Upon this axis, to give it a pleaſant and ſteady motion, muſt be placed ſomewhere, a loaded wheel or fly, which the width between your upſtream and down ſtream topſetts will not admit to exceed five feet five inches diameter, and which will moſt conveniently be made of caſt iron, of the ſquare of five inches; it will weigh about half a ton, and you may have it caſt all in a piece. The ſpur wheel and lanthorn I ſhould alſo have had of larger dimenſions if room would have allowed; but to make them wear longer and more kindly, it would be of advantage to put in a double row of cogs, which is eaſily done by putting on a facing on each ſide: this however regards only the length of time that the geer will laſt.

The beſt place for the loaded fly is between the two ſtocks; it ſeems to me that you may diſpoſe of the ſpur wheel and lanthorn in the vacancy next eaſt of the wheel, and as near as may be to the wheel; then to take off from the weſt ſide of the place for the two ſtocks by a new tranſverſe timber, ſo much as only to leave room for one ſtock; this will leave you a ſufficient vacancy between the cog-wheel and the ſtock, to run the loaded fly on the eaſt ſide of the moſt eaſternmoſt tranſverſe timber, to put in a new tranſverſe on the eaſt ſide of the loaded fly, and another as far eaſt as will take in a place for the ſecond ſtock, and this will ſtill leave a conſiderable vacancy at the eaſt end.

Reſpecting matters of convenience, in regard to the placing of the ſtocks, you muſt judge for yourſelf, and in regard to the thickneſs and ſtrength of the parts, your mill-wright will, I doubt not, judge properly, but you muſt ſtrictly adhere to the proportions I have given, becauſe the performance depends upon it, and which done, you will find that the mill will not only go with much leſs water than in now does in dry ſeaſons, but will alſo bear much more tail water in time of floods.

You muſt however remark, that where the gate draws ſo ſhallow, it is required to draw higher (to let through the ſame quantity of water) than where it draws lower.

J. SMEATON.

Austhorpe, 27th June, 1772.

FLINT

FLINT MILL AT LEEDS.

DIRECTIONS for weathering the Sails of the Flint Mill at Leeds.

THE lattice of the fails being compofed of twenty-five bars in the fcheme or fcale of weathering, No. 1 reprefents the angle of weather of the point bar; No. 2 that of the 2d, and fo on increafing to the 20th bar, the 21ft being the fame, and then gradually diminifhing to the 25th, which is fuppofed to be at four feet diftance from the centre of its weather, the fame as the 16th.

N. B. If the angle for each bar is taken off from the fcale by a bevil, between the line called the parallel of the axis, and the refpective number of the bar, and this bevil applied to the leading fide of the whip, the mortife being fet off on each fide the whip, will give its proper angle.

The weather of the leading board is to be fomewhat different to that of the lattice for the fail cloth, viz. the leading board from the 25th down to the 12th bar, the board, &c. is to be nailed upon the leading ends of the bars produced as ufual; but upon the leading end of the point bar, or No. 1, there is to be applied an angular piece or wedge, whofe thicknefs at the head is one-eighth part of its length, and this being applied with its point next the whip, and fixed upon the face of the bar, will give the board a degree of weather at the point, more than the cloth by the angle of this wedge, this being fuppofed fixed, the reft of the bars Nos. 2, 3, 4, &c. are to be furnifhed with wedges growing gradually thinner, fo as to end in nothing at the 12th bar, reconciling the 1ft and 12th, upon which 12th, as before mentioned, the board fixes as common without any rifing.

It will be neceffary to attend minutely to thefe directions for the weathering of the fails and boards, the effect depends upon it, and however uncommon they may appear, the fuccefs will follow.

N. B. It will be proper to weather the face of the whip correfpondent to the bars, from the point to the 17th bar, and then gradually to lofe it in a fquare for the fake of ftrength. All other fides of the whip to be kept fquare as ufual.

J. SMEATON.

Austhorpe, 29th April, 1774.

PUDDING

PUDDING MILL.

The REPORT of JOHN SMEATON, engineer, upon the powers of the Pudding Mill, in the parish of St. George, in Surry, near Blackfriars Bridge, ·for raising water for the service of the neighbouring parts of the town, and other purposes.

THE power of the pudding mill depends in a great meafure upon the tide water that is taken from the river Thames into certain ditches or drains, that in a manner encircle this parifh as well as interfect it in certain directions; this tide water is taken in by means of a draw gate fluice, which, when in fpring tides it has flowed to a certain gage mark upon a poft placed in the mill pond for that purpofe, the fluice is put down to prevent the waters rifing to fuch a height as would annoy the houfes and premifes of the inhabitants; the water being thus fhut in the ditches to a certain height, it there remains till the water of the Thames has reflowed to three hours ebb, or thereabouts, when the floor or apron of the mill conduit begins to be fufficiently clear for the water wheel of the mill to begin to work, and then the mill gate or fluice being drawn, the water fo pent up in the aforefaid ditches, is gradually let down upon the water wheel of the mill, and turns it for a greater or lefs time in proportion as the water has occafion to be expended at a greater or lefs rate, but in any rate in which it can be fuppofed to be confumed in doing bufinefs, it will or may be fpent before the action of the wheel comes to be obftructed by the rifing water of the next tide, there being an interval of about fix hours in every tide in which the water contained in the ditches may be expended in working the mill: this is properly what is called a tide mill, and its power or quantity of bufinefs to be done thereby on each tide, will depend upon the quantity of water that can be pent up and expended, together with the perpendicular fall that can be given to it; but as both thefe will be variable in different kinds of tides, that is, of fpring and neap, it is proper to obferve, that from the gage mark upon the poft to the bottom of the mill conduit, there is a fall of feven feet, but that upon the 8th of June laft, being two days after the moon's quarter day, and confequently dead of neap, and at a time of the year when the tides ran fhorteft, the utmoft flow of the tide was eighteen inches below the gage mark, and confequently from the furface of high water to the bottom of the mill conduit, there were no more than five feet and a half: And I further obferved, that when the water was reduced one foot below the high water mark of this tide, the ditches were in a great meafure emptied, very little water remaining except in the little pond contiguous to the

mill

mill, from whence it may be concluded, that very little bufinefs of any kind can be expected to be done when the tides are in this ftate. I fhall therefore make my computation upon the bufinefs that may be expected to be done at fuch tides as will fill the ditches up to the gage mark, and from the obfervations I made on view of the ditches, and of the plan of the parifh of St. George, furnifhed for my infpection by Mr. Burrows, and alfo from obfervations which I have fince got made by Mr. Holmes at my defire and direction, I have good grounds for a computation, the refult of which is, that fuppofing the ditches filled to the gage mark, and the water required to be raifed forty-five feet, 500 hogfheads, wine meafure, of water, may be raifed per tide, and confequently double that quantity in two tides, being a little more than twenty-four hours; but fuppofe, that at an average, 500 hogfheads be raifed per twenty-four hours for the fervice of that part of the town, this quantity, at the average price we are paid at our works at Deptford and Greenwich, fhould bring in water rents to the amount of £190 per annum; the principal difficulty that will attend this tide mill as a water work, will be this, there will be about five days every fortnight at the dead of neaps, during which the fupply of water will be fo trifling that all the principal families will be under the neceffity of having a ciftern or refervoir, capable of holding five days ordinary confumption of water, for, in regard to extraordinary confumptions, and among the lower clafs of tenants who cannot be at that expenfe, as the times will be always known before hand when the fupply will be defective, viz. one day before the quarter day of the moon, the quarter day itfelf, and three days after, they will be led to contrive accordingly.

It is at prefent impoffible to make a proper eftimate of the expenfe, till a proper plan to proceed upon is made out, and the extent of the fervice known, but to give fome idea thereof, in order that its eligibility may be judged of, I fuppofe that a capital of £1500 will be required to be laid out upon thefe works, viz. £500 in building a proper engine and repairing the fluice and mill gates; £500 more in making a proper refervoir, for taking in the Thames water as clean as poffible, for the engine to raife, and in the erection of a ftand pipe, and fuppofing one mile of five inch pipes to come to, all expenfes included, five fhillings per yard, this will be £440, but as it is probable, that neither a mile of pipe of fo large a bore as five inches will be wanted, nor yet that a mile in length will be fufficient, we may reckon the article of pipes at £500 at leaft; the attendance and annual repairs may be ftated at £80 per annum, and if £20 per annum be fet off for the rent of the mill in its prefent condition, there will then remain £90 per annum, in return for the capital.

VOL. II.   3 H   In

[418]

In order to judge what value the mill may be of, for the common purpose, I have also computed what its produce might be expected to be in grinding wheat, and making flour, the result is, that when the ditches are filled to the gage mark, it may be expected to grind twelve bushels per tide; but to allow for deficiencies of neap tides, we will rate its performance at sixteen bushels, or two quarters, per twenty-four hours; this, at 2s. 6d. per quarter, will be £91 5s. per annum, and to repair and alter the premises to this effect, will take a capital of at least £200, supposing all the machinery and materials upon the premises, to go along therewith; the miller's wages and repairs of the mill cannot be rated at less than £60 per annum, and if £20 be set off for the rent of the premises as they now are, there will remain only £11 5s. per annum, in return for the capital.

It is impossible to judge with exactness, but the above statements are the most probable that I can form; from whence it appears, that unless it be judged adviseable to go into the scheme of a water work, it is not likely to fetch more than £20 per annum rent, or even so much, unless you can let it to some neighbouring manufacturer in the dying, scouring, fulling, or iron branches, to whom, as a particular and local convenience, it may be worth much more than to be employed in a business, where the mills in the country are upon the same footing in point of situation.

Lastly, I must point out, that from appearance the ditches not only seem a good deal filled up and obstructed, but I apprehend are curtailed in dimensions by the new roads, and new buildings, so that in fact, the reservoir or mill pond of the Pudding mill contains less water than heretofore, and though it is probable that some advantage may be got to the mill by clearing the ditches, yet as this advantage will be purchased by the expense that will probably attend the clearing or getting them cleared, I have contented myself with reckoning upon what is now subsisting.

J. SMEATON.

Newcastle, 16th August, 1775.

P. S. The Falcoln iron foundery in that neighbourhood employ horses in boring and turning cast iron work; it seems therefore to me, that this is a thing that would suit them.

DALRY

DALRY MILLS.

REPORT of JOHN SMEATON, engineer, upon the memorandum of Mr. John Russell, of the 16th July, 1771, concerning his mills, called the Dalry Mills, near to and west of Edinburgh.

ACCORDING to a level taken by Andrew Meikle, there is a fall of twelve feet three inches from the furface of the water at a full head, to the bottom of the new wheel race, and it is probable that nine inches additional head may be gained at a moderate expenfe, fo as to make in the whole thirteen feet of fall. Upon this certainty of twelve feet three inches, and probability of thirteen feet, I would advife an overfhot wheel of eleven feet diameter, which, with the conftruction I propofe, will work very well with a head of one foot three inches, and even with one foot, or, occafionally, with nine inches; but if the probability of raifing the prefent full head nine inches is thought great, to give more compafs, I would advife the wheel to be ten feet nine inches only.

A water-wheel of this kind fhould, upon this head and fall, with the water I faw, dif-patch at the rate of 2½ bolls of wheat per hour, and five Winchefter bufhels, fuppofing half the water taken to the oat mill and fnuff mill.

I would propofe the overfhot wheel to go the fame way round as the prefent intended breaft wheel, and from the bottom of the wheel up the breaft, to the height of the axle, to work a true curved fall of ftone work, and above the axle to rife perpendicular. This part of the work will, therefore, be obliged to be done up anew, by a frefh lining of aifler. The great cog wheel not to exceed eight feet diameter, out and out, fo that in cafe the water wheel is eleven feet, the under-fide of the cog wheel will be eighteen inches above the bottom of the race, and if ten feet nine inches, then it will be 16½ inches, which I apprehend will be fufficient to keep it clear of the water in ordinary times, and in fpeats, if made of and geered with oak, it will go wet, as many mills are obliged to do conftantly.

If this general idea is approved of, it will be neceffary for me to give a defign for the water wheel and trough, and fhall alfo fend the proper numbers to accompany it; but, if it is refolved to go with the breaft wheel of fixteen feet, as now defigned, I fhall, if required, give the proper numbers to accompany that.

An

An alteration in the oat mill and fnuff mill will, undoubtedly, be neceffary, in order to allow water to the wheat mill in dry times, as thefe mills, as they now ftand, occupy the whole water; and though no advance of rent can be expected upon thefe mills, how ever good and valid they could be made, yet the lefs water they are made to ufe, in doing their prefent bufinefs, the more can be fpared to the wheat mill, which will proportionably advance its value.

To do this in the moft effectual way would, doubtlefs, be to rebuild the oat mill in the manner I recommend for the wheat mill, with a double motion, and the fnuff mill can be driven from the fame wheel, as propofed by Mr. Meikle; and were the premifes mine, I would certainly do it in this way, as the oat mill and fnuff mill would go together with two-thirds of the water they can otherwife be made to do with, but if Mr Ruffell thinks this will be launching into too great an expenfe, it will certainly be a great improvement to place the two wheels to take the water from each other, as they will then both work with the fame water; but in this cafe, when one of them wants to work alone, it will confume the fame water as if they both worked, and therefore will greatly diminifh the fupply that at thofe times might be taken to the flour mill; and as the race of the lower mill muft be funk, together with all the machinery, which, when moved, will be found to want many repairs, and perhaps feveral parts new, this will be fomething towards building a new mill.

The wafte of water, when one of the two prefent mills is at reft, will be prevented by joining the two falls into one wheel, as propofed by Mr. Meikle, but, unlefs a cor refpondent alteration is made in the infide, there will be more water confumed in turning the two mills together, than when they take the water from each other, and to fink the wheel and make a correfpondent alteration, will again fomewhat approach towards a new mill, and, after all, you will be but poffeffed of an old mill, continually requiring half as much more water as if conftructed upon the beft principles; that is, in one cafe you may fpare about one half the prefent water to the wheat mill, in the other cafe two-thirds, and the difference will be an addition to the neat profit of the undertaking, and you will have your mills all new.

If Mr. Ruffell will be pleafed to fix upon his plan, Mr. Smeaton will endeavour to advife him how to execute it in the beft manner, but to draw a fcheme for every poffible mode of execution, would require more time than Mr. Smeaton can poffibly allot to this bufinefs.

J. SMEATON.

Edinburgh, 28th Auguft, 1771.

MEMORANDUM

MEMORANDUM for Mr. Smeaton, concerning John Ruffell's mills, called Dalry Mills.

BY level taken by Andrew Meikle, millwright, this day, he reports, that from the furface of the water, ten inches lower than the back fluice to the bottom of the wheel race of the new intended flour mill, is eleven feet one inch of fall, and when the water is gathered by ftopping the fluices of the mill, it rifes fourteen inches higher in the dam.

Andrew Meikle has drawn a plan of the intended flour mill. The houfe is building, but no part of the mill work is cut out. Mr. Smeaton is defired to confider Mr. Meikle's plan, and to make fuch alterations as he fhall think proper.

The fnuff mill is let to Nov. 11th, 1773, at £40 per annum, and the oatmeal mill to the fame term at £30 per annum; it is not probable that they will let much higher when the leafe is out.

The conftruction of the new flour mill renders it neceffary to make fome alteration on thefe two mills. Andrew Meikle propofes to make one water wheel ferve both of them. Mr. Smeaton will pleafe to confider, whether this may be right, and if he thinks the propofal right, he will give his directions, whether it is to be done by an overfhot wheel or in the common way, and what dimenfion and conftruction the wheel fhould be of.

The difficulties that occur to Mr. Ruffell in making thefe alterations are, that the wheel muft neceffarily be placed fo low, that it may create a confiderable additional expenfe by rendering it neceffary to take down and rebuild the fide walls of the mill next the water.

2dly. Though the prefent oatmeal mill may poffibly admit of finking the wheel, yet the foundation of the fnuff mill, being confiderably higher than the oatmeal mill, it may not admit of finking the machinery within the fnuff mill, fo as to anfwer to the machinery of the outer wheel of the oatmeal mill when funk; if, therefore, any thing can be contrived, fo as to prevent the taking down and rebuilding the fnuff mill, it will be a confiderable advantage.

3dly. Another inconvenience arifing from having one outer wheel for both fnuff and oatmeal mill is, that the mills cannot be let to different tenants; it is, therefore, to be
confidered,

considered, whether it be practicable to have two water wheels for these two mills, both in one line after the other? but if, by dividing the fall, this should be inconvenient, the circumstance of letting the mills to different tenants is not to be minded.

Mr. Smeaton will also consider whether the troughs which is the conduit which leads the water from the sluices to the head of the fall, should be built of stone or made of timber, as Mr. Russell will follow his direction as to that matter.

Mr. Smeaton will please to consider the whole, and give such directions, in writing, as he shall think proper, so that Mr. Meikle may execute whatever plan he proposes.

The REPORT of JOHN SMEATON, engineer, upon the construction of Dalry Mills, belonging to Mr. John Russell.

IN case you can make an addition of six or eight inches or thereupon to your present head and fall of twelve feet three inches, your water wheel may be eleven feet diameter, but in case you cannot, and must stand as you now are, it will be prudent to make it no more than ten feet nine inches diameter, but should you happen to get more than six or eight inches head addition, for every inch addition above six, you may add one inch to the diameter of the water wheel above eleven feet.

FOR THE WHEAT MILL.

Water wheel, in diameter,	11 feet.
The main cog wheel,	54 cogs, at 5 inch pitch.
The great lanthorn,	19 do. at do.
The spur wheel,	62 at 3½ inch pitch.
The lanthorn on the spindles,	13 at ditto.
The diameter of the French stones,	4 feet.

N. B. The water wheel is intended to go eight turns per minute, so that Mr. Meikle can suit his barley mill motions accordingly. I apprehend the following will suit.

The main cog wheel,	54 cogs, at 5 inch pitch.
The great lanthorn,	17 do. do.
The spur wheel,	68 do. at 3¼ inch pitch.
The lanthorn on the spindles,	11 do.

The

The same motions and a water wheel of a similar conftruction will anfwer for the oat-meal mill as for the wheat mill; but I fhould advife to put up a pair of ftones for grinding oatmeal of about four feet fix inches diameter, with lanthorn on the fpindle of fifteen, keeping the five feet ftones with the thirteen lanthorn for fhelling only; an axis laid oblique to the horizon from the main cog wheel will turn the fnuff mill.

As the main geer will be often wet, it will make it wear much longer to have caft-iron rounds in the great lanthorns, particularly of the wheat mill, which will come to the greateft preffure; if they are made fmooth by grinding, particularly where the cogs take, they will not deftroy the cogs at firft. This is frequently done in our beft mills in this part of the kingdom.

I apprehend that three dreffing mil's will be fully fufficient, but the manner of proceeding in making wheat into flour in Scotland is fo very different from what is ufed here, that Mr. Meikle will be better able to advife you upon this head than I can; it is good not to have more machinery at firft than you want, but to make provifion for what by experience you will find to wanted.

J. SMEATON.

Austhorpe, Nov. 27th, 1771.

KILNHURST

KILNHURST FORGE HAMMER MILL.

See plate XVII.

Fig. 1. is an elevation of the machinery in front.

Fig. 2. an elevation fideways; fhewing the water wheel and conduit.

The following are thofe parts which are not rendered evident from an infpection of the defign: a, is the breaft board of the fhuttle; b, is a piece of an iron plate nailed on to the nofe or edge of the fhuttle, to make it deliver the water in a clean fheet.

A, is a pump, worked by a tappet in the water wheel axis, to keep the cog pit free from water, when the tail water is above the bottom thereof.

The water wheel is fifteen feet diameter out and out, and fix feet wide with thirty-fix floats; the water line, when at a full head, is two feet above the crown of the fall D, and five feet ten inches above the bottom of the wheel race; the outfide ftones both upftream and downftream of the fore bay D, or breaft or fall of the wheel, to be jointed with cement of terras, and the infide with good mortar, mixed with forge fcales or minion; the flope B is formed by an apron of earth fpread on the mafonry.

The open wheel E, or the water-wheel fhaft within the houfe, to be ten feet fix inches in diameter, with feventy-two cogs, at $5\frac{1}{2}$ inches pitch, and the wallower or trundle F, which it turns, to have twenty-five rounds at the fame pitch, and will be 3 feet $7\frac{3}{4}$ inches diameter; the fly wheel G, is to be eleven feet fix inches outfide diameter, being furrounded by a caft iron ring eight inches by fix in fection; the iron ring H, with four cogs (defcribing a circle of four feet eight inches diameter), is to be fixed on the end of the fhaft of the fly wheel, for lifting the hammer I; the gudgeon for the end of this fhaft is to be caft upon a crop d, fig. 1, the back face of which is to be made one quarter of an inch broader than the fore face, that when the wood is wedged round it will be grafped, and held in like a dovetail, exclufive of the four fcrews; but if this is not preferred, a common gudgeon may be applied to the fame ring.

The

Design for a HAMMER MILL, at Kilnhurst Forge, Yorkshire.

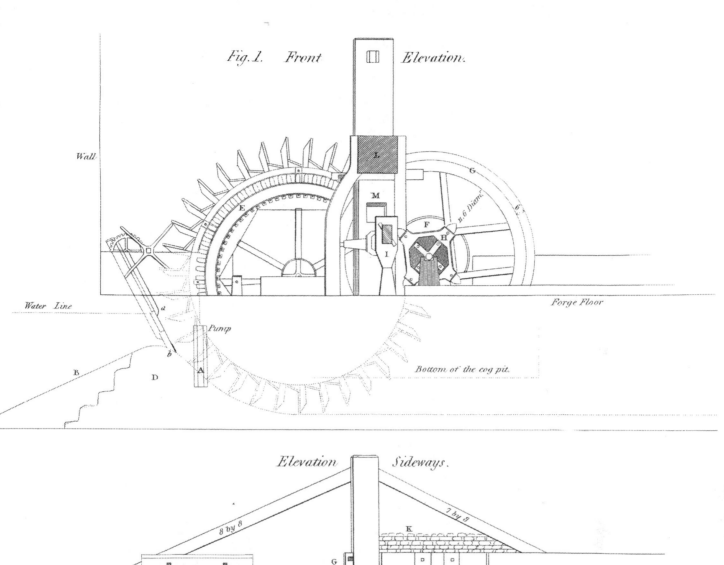

Fig. 1.   Front   Elevation.

Wall

E

L

M

F

H

I

G

n.6 Diam.

6

Water Line

a

Pump

b

B

A

D

Forge Floor

Bottom of the cog pit.

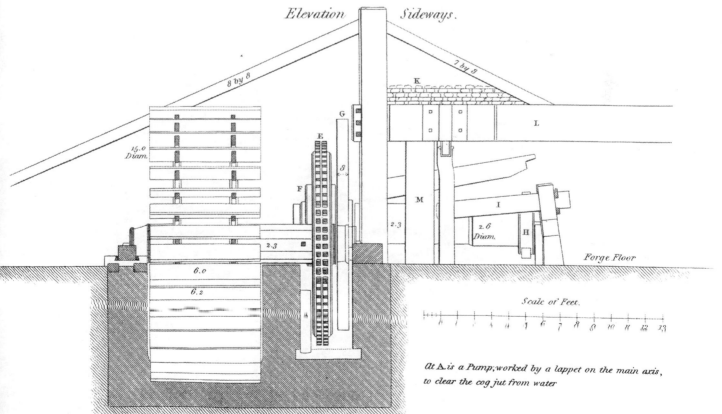

Elevation   Sideways.

8 by 8

7 by 8

K

L

G

E

F

M

I

H

15.0 Diam.

8

2.3

2.3

2.6 Diam.

6.0

6.2

Forge Floor

Scale of Feet.

At A is a Pump, worked by a lappet on the main axis,
to clear the cog jut from water

M. Farey delin.   London, Published by Longman, Hurst, Rees & Orme, 1810.   Lowry sculp.

The reft of the machinery, being the framing or harnefs for the hammer, is fuppofed to be fubject to the corrections of the forge carpenter, and are only fhewn in the defign to mark their relative places.

K, are pigs of iron, to ferve as a weight upon the drome beam L, and puppet poft M.

THORNTON WATER WHEEL.

See plate XVIII.

EXPLANATION of the design for the water wheel for the upper paper mill at Thornton.

THE water wheel is intended to fit its fweep and fides of the conduit, the fame as if it were a breaft wheel.

The outfide rings and outfide of the fhrouds are intended to be flufh.

If the arms cannot be got curved, as reprefented, they may be made ftraight, according to the dotted lines.

a a—Reprefent ftraight pieces of hoop iron, about $1\frac{1}{4}$ inch, or $1\frac{1}{2}$ inch broad, nailed on, in order to ftrengthen and preferve the edge of the bucket boards.

A—Reprefents one of the two fliding fhuttles, where note, that in order to make the fame water tight

b b—Reprefent a piece of leather about two or three inches broad, to fecure the joins between the fhuttle and the bottom of the trough.

E—Reprefents the end of a piece of leather about $2\frac{1}{2}$ or three inches broad, and of the fame thicknefs with the former, which, being faftened to the tail of the fhuttle, flides along with it, in order to keep it upon a parallel.

f—Reprefents the end of a piece of leather fixed to the fhuttling forehead of the fhuttles, in order to fecure the joint between the forehead of the fhuttles and the floping board D ; and in order to make it deliver the water fair, the lower part of the leather is intended to fally below the wood about half an inch for the length of the open fpace between the parts c c, where it will be clear of the leather firft defcribed.

g g—Reprefent an iron plate, which will not only fecure the fharp edge of the floping board D, but by turning $1\frac{1}{4}$ inch of the breadth of the lower part of the plate into an horizontal direction, will make the ftream nearer horizontal, and caufe the point of the bucket more freely to enter the ftream.

h—

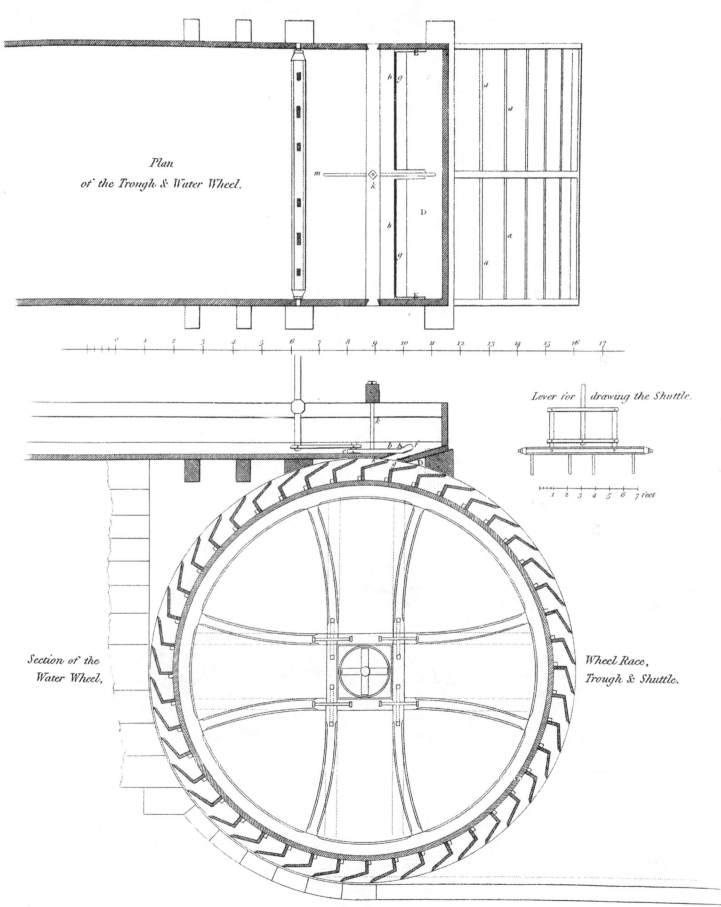

Plan of the Trough & Water Wheel.

Lever for drawing the Shuttle.

Section of the Water Wheel.

Wheel Race, Trough & Shuttle.

J.Farey delin.t   *Designed by J.Smeaton.*   *Engraved by Wilson Lowry.*

London,Published by Longman,Hurst,Rees & Orme,1810.

h—Reprefents the end of a piece of leather, to fally or hang down a little below the wood, in order to prevent the water, when going out in a full bore, from following the bottom of the trough.

N. B. Befides the above defcribed, bits of leather fhould be nailed to the lower fide of the parts c c, to hang down a very little, to make the ftream deliver clean at its ends.

k—Reprefents an iron bolt going through the bottom of the trough between the two fhuttles, in order to prevent the bottom from fagging by the weight of the water thereupon, and to prevent the fhuttles from rubbing againft the bolt; a fmall fillet m, muft be nailed to the bottom to keep them feparate.

Each fhuttle is drawn by two parts, to keep their motion parallel, and that one fhuttle may not draw before the other, or one fide of the fame fhuttle before the other, by the twifting of the roll, fpringing of the leavers, &c. it is propofed to be framed according to the upright thereof, which is drawn to a leffer fcale.

WINLATON

WINLATON BLADE MILL.

See plate XIX.

THE general defign for the blade mill at Winlaton is as follows:

The water wheel is to be overfhot, eight feet wide, and thirteen feet high, and its bottom even with the furface of the adjacent river. It is intended to receive its water from the head, in the fame direction as at prefent, by a kind of mechanifm, which is particularly defcribed hereafter, the water is made to fhoot backward, and to turn the wheel round the fame way as at prefent.

The houfe is fuppofed to ftand as at prefent, or, if pulled down, to be erected of fuch dimenfions as are beft adapted for its purpofe; each tumbling axis may be made longer or fhorter, or to carry three ftrap wheels, if required.

As the water wheel axis, according to the above dimenfions, will ftand about four feet eight inches lower than it now does, if worked by a fingle geer, as at prefent, it would be neceffary to lower the floor of the houfe and the grindftones in the fame meafure, which might be inconvenient on account of the floods. By introducing the upright axis, the ftrap wheels will be about one foot higher than they now are; and, if productive of any advantage, may be made ftill higher, by placing the horizontal wheel higher on the upright axis.

Though the machinery will be fomewhat more expenfive on account of a double motion, yet it will be more lafting, as it will go more fmoothly, and with lefs friction.

This mill will go with at leaft half, if not one-third, of the water now ufed to do the fame bufinefs, and will perform with full effect, till the water is funk about fixteen inches under head; and then, if wanted by the ftilt mills or forges, the grinding mill muft ftop.

The wheel, though laid fo much lower, will in times of flood perform nearly as long as the prefent wheel; that is, it will perform with about four feet tail water, what would make about one foot in the prefent wheel; and, as the floods continue but a fhort time, it would be wrong to lofe an effential advantage for the fake of a few hours in the year.

As

Design for the MACHINERY for a BLADE MILL at Wadeton.

Fig. 1.
Elevation.

Fig. 2.
Front Elevation.

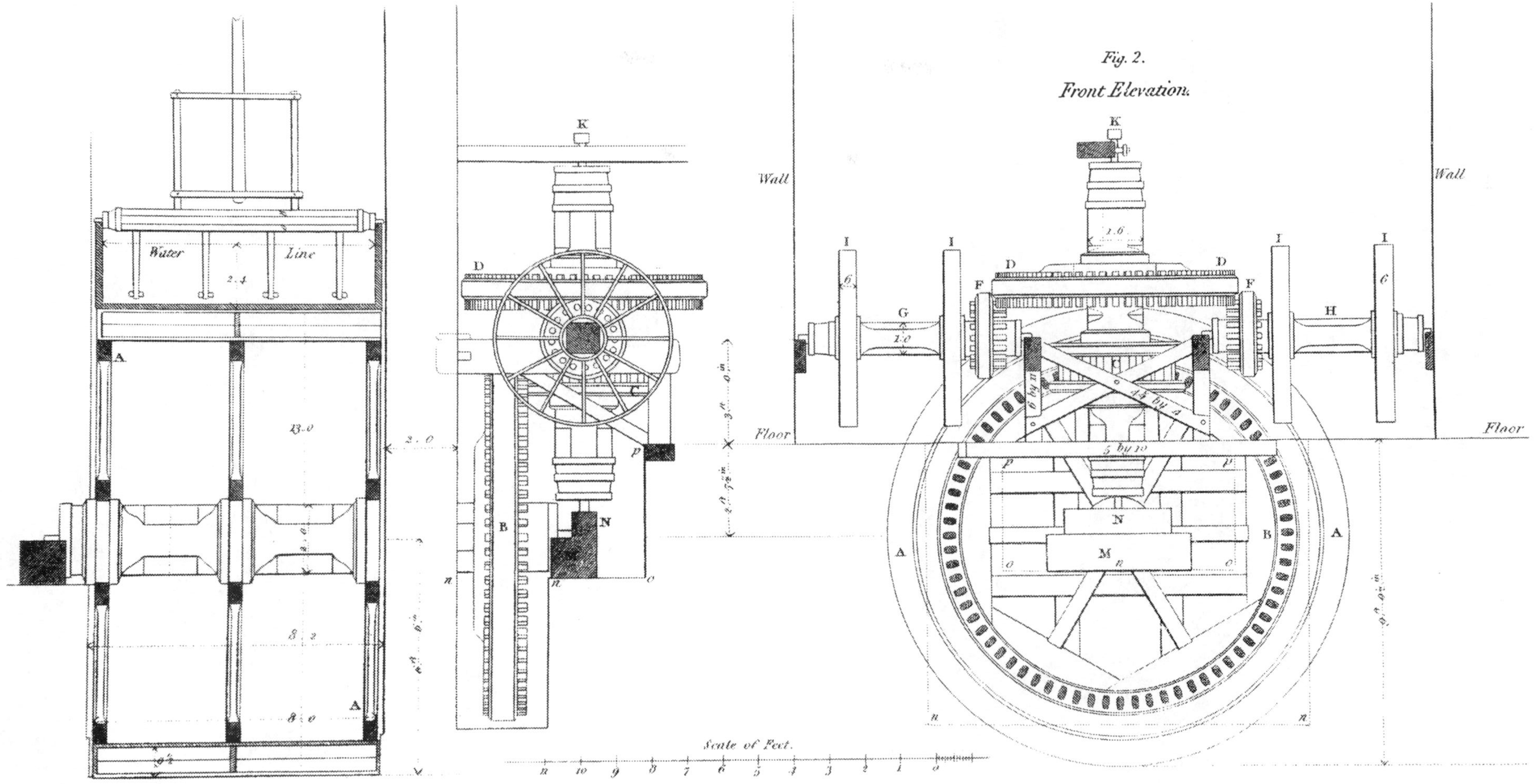

Scale of Feet.

Reduced from Mr Smeaton's drawing by I. Farey.

Designed by J. Smeaton.

London, Published by Longman, Hurst, Rees & Orme, 1810.

Engraved by Wilson Lowry.

Vol. II. Pl. XIX. p. 428.

As all the designs are drawn true to their respective scales, they will in a great measure explain themselves.

Fig. 1, is an elevation of the machinery for Winlaton Blade mill, with a section of the wheel and trough, and upright of the lever for drawing the shuttle.

Fig. 2, is a front elevation of the machinery seen from within the house : the design sufficiently explains itself, and the following are the proportions :

The water wheel A A is thirteen feet diameter out and out, and has forty buckets; the water line, when at a full head, is supposed to be two feet four inches above the top of the wheel.

The pit wheel B,	72 cogs,	at 5 inches pitch.
The great lanthorn C,	25	at 5 inches.
The horizontal wheel D,	60	at 4 inches.
The small lanthorn F F,	17	at ditto.

The wheels or runs I, I, I, I, which carry the straps for turning the grindstones, to be five feet diameter, and the length of the axis G, H, upon which they are fixed, to be adapted in length to the width of the house, and the wheels are to be set at such distances upon them as are most convenient for the disposition of the grindstones.

K is a square head upon the upper gudgeon of the vertical shaft, by means of which works may be turned in the upper room; n n, shew the lower cog pit for the main cog wheel, and o p, the upper cog pit or platform for the headstock M, and rider N, which are to be bolted together, to support the gudgeons of the water wheel and the upright axis.

The brasses for the gudgeons of the two lanthorns F F, must be rather deeper than the half circle, especially towards the side that the drift of the mill tends to.

SIR

SIR L. PILKINGTON's FLOUR MILL.

Plate XX, fig. 1, is a plan of the Hirft for Wakefield flour mill.

Fig. 2, is an elevation, and fig. 3, a feparate plan of the bridges and brayer for fup-porting the fpindle foot: the defign fufficiently explains itfelf, to be framed in a fyftem of triangles, which are well known to be the ftrongeft figures poffible, and do not depend upon any mortifing, becaufe the angles, if united by bolts, fo that they cannot feparate, will be far ftiffer than any mortifing or cramping can make a fquare frame. A B C are three beams, forming the upper frame of the hirft, and D, fig. 2, is the edge of a fimilar triangle, which lies upon the mill floor, and fupports the former by the upright E E, and crofs braces, F F, fig. 1, are beams framed acrofs the angles of the great triangle A B C, forming fmaller triangles, upon which the ftones G G are laid; the gudgeon of the great fpur wheel I, is fupported in a bearing fcrewed to the beam H; the bridges and brayers are fupported on three points, by mortifes in the upright timbers E E, and are adjuftable at one joint by a fcrew which bears the weight of the ftone; fig. 3, fhews this plainly, and alfo the wedges which adjuft the pofition of the fpindle foot. The proportions of the mill are as follows:

The water wheel twenty feet diameter; the difference between the head and fall eight feet at a full head; and then the water ftands two feet deep, over the crown or fall; the wheel has thirty-fix floats, and is feven feet wide.

The great pit wheel eleven feet diameter, eighty-four cogs.

The great lanthorn, three feet fix inches diameter, twenty-four cogs.

The wheels are in the ground floor: the upright fhaft goes through the firft floor, and turns the dreffing machines in the fecond floor, where the hirft is, it carries

The fpur wheel I, eight feet diameter, eighty-four cogs.

The pinions L, one foot eleven inches diameter, and twenty cogs; the ftones are four feet diameter.

EXPLANATION

Design for a FLOUR MILL, erected at Wakefield Yorkshire, for Sir Lionel Pilkington Bart. 1754.

Plan of the Hirst.

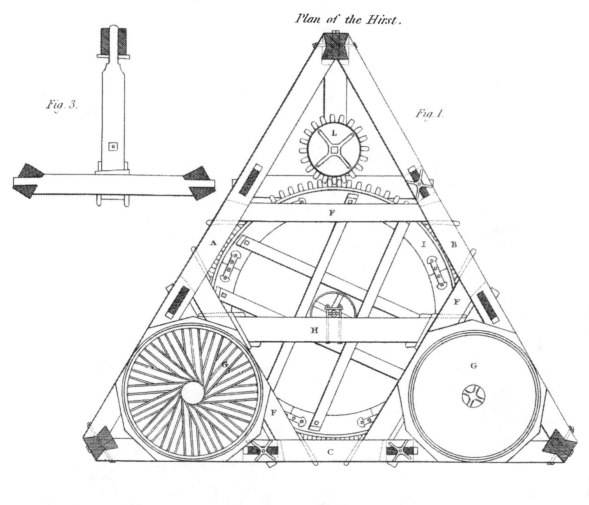

Fig. 3. Fig. 1.

L
F
A
I
B
F
H
G
G
F
C

Fig. 2.
Elevation.

G
G
C
L
E
L
E
E
I
E
D

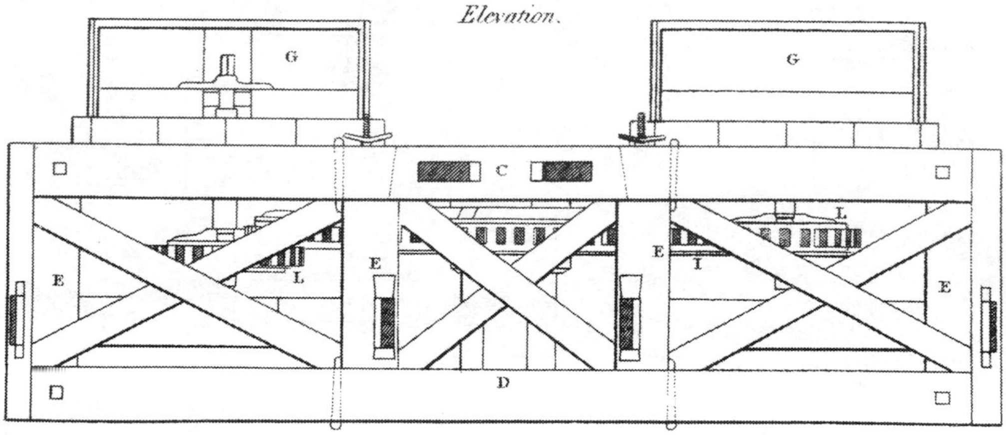

Scale of Feet.

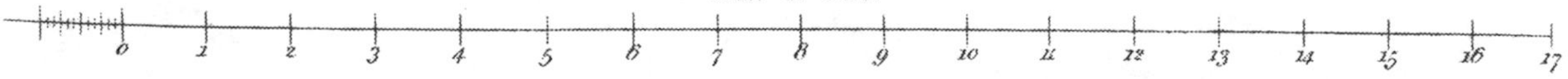

0
1
2
3
4
5
6
7
8
9
10
11
12
13
14
15
16
17

Designed by J. Smeaton.

...rey delin.

London. Published by Longman, Hurst, Rees & Orme, 1810.

Lowry sculp.

WATER ENGINE, Erected for Lord Irwin at Temple Newsham, Yorkshire

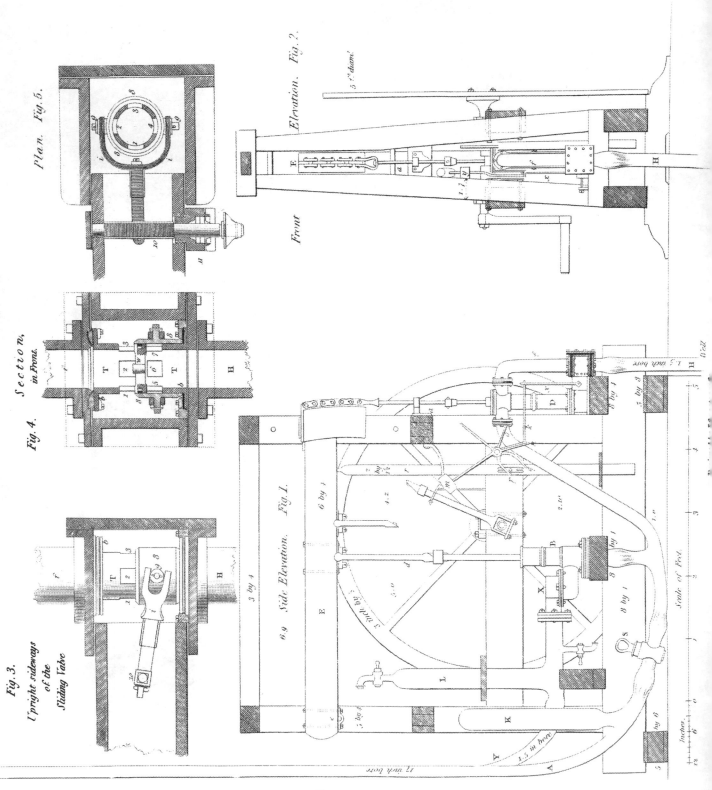

Plan. Fig.5.

Section, in Front.

Fig.4.

Fig.3.

Upright sideways of the Sliding Valve

Elevation. Fig.2.

Front

Side Elevation. Fig.1.

Wall

EXPLANATION of the designs for a water engine erected for Lord Irwin, at Temple Newsham, in 1770.

See plate XXI.

Figs. 1 and 2, are elevations of the whole engine, and figs. 3, 4, and 5, enlarged sections of the sliding valve, which regulates the admission of the water into the working cylinder, at proper intervals, to cause the motion.

The pipe A, which supplies the engine with water, is one-seventh of an inch bore, the perpendicular descent from the mouth, or entry at the spring of supply, is fifty-four feet, and the length 400 feet; the water is conveyed from the engine by a pipe H, which has a fall or descent of fifteen feet from the engine to the surface of the water in the pit or well into which it delivers, and has a stop cock, which regulates the discharge of the water. This descending column of sixty-feet of water is employed in working a pump, which throws back part of the water to a reservoir by a pipe Y of ascent, about 900 feet in length, and one-fifth of an inch bore; the delivery of this pipe at the top of the reservoir is eighty feet above the engine, consequently, twenty-six feet above the level of the head or spring of supply. The pipe A conducts the water to the pump B by one branch, and to the top of the cylinder D, which works the engine, by the other. The cylinder D is of brass, truly bored, and furnished with a solid piston, whose rod passes through a close stuffing box in the cylinder lid where leather is packed round it so closely, that no water can leak by it; the upper end of the piston rod is keyed into a small box, which connects it with an iron rod a, sliding through a guide, to make it more steadily; E is the working beam, moving round a centre at e; it has an arch head at the outer end, which is a segment of a circle struck from the centre e; the arch receives a chain by which the piston rod is suspended, and therefore has a vertical motion; d is the pump rod, jointed to the beam, and moving up and down with it; the forcer of the pump B is fixed to it at the lower end; f is a pipe, which forms a communication between the top and bottom of the cylinder, it leads down into a chest, upon which the cylinder is placed, and to which it is open at bottom; the chest is composed of plates screwed together, as shewn by the sections figs. 3, 4, and 5, which also explain the construction of the sliding valve contained within the chest; in these figures T is a short cylindrical pipe fixed across within this chest, and communicating with the pipe f at top, and the pipe H

(which

(which conveys the water away from the engine) at bottom; this short pine has a water-tight division, W, fig. 4, in the middle of it, so that there is no direct passage through it from f to H, but there are four square holes 1, 2, 3, 4, made in the pipe, at equal distances round it, above and also below the division, as shewn at 5, 6, 7. fig 4; a cylinder or tube of brass 8, 8, is fitted upon the cylindric pipe, and slides up and down upon it, being packed with leather round the edges of the middle partition near 1 and 3, fig. 4, that no water may escape between the two cylinders; the sliding cylinder 8, is just half the length of the other, and when it is pushed down, as in the figures, it covers and stops the four holes 5, 6, 7, in the pipe T, which are below the division W, and opens the four holes 1, 2, 3, above the division, allowing a passage from the pipe f, to the bottom of the cylinder D, fig. 1; on the contrary, when the slider is pushed up, the upper holes 1, 2, 3, are closed, and the lower ones 5, 6, 7, opened, making a passage from the cylinder bottom into H; the sliding valve 8, besides the leather packing, which surrounds the partition W, has leather placed at the top and bottom of the fixed cylinder T, at b b, and the ends of the slider are pressed upon these leathers to make a tight joint when up or down, but the cylinder fits as tight as it can be made; independent of these leathers, the sliding valve has a pin 9, projecting from each side of it, which are included between clefts made at the end of a forked lever i i, moveable on an axis 10, which passes through the side of the chest in a collar of leather 11, fig. 5, and has a long lever x, figs. 1 and 2, fastened upon it. By moving the upper end of this lever towards the engine, the sliding valve will be raised up, and by moving it in a contrary direction the valve will be pushed down; k is a small iron rod, jointed to the upper end of the long lever by one of its ends, and the other is suspended by hooks from a spindle, turning upon pivots supported by the framing; this spindle has several levers upon it similar to the working geer of a steam engine, as follows: m n o is a three armed lever, the arm m has a weight at the end, and is called the tumbling bob; n and o are two other arms, made in the same piece with m, these two latter arms strike against a pin fixed across in the end of the rod k, which is forked to admit the pin, and the arms n o act within the fork; p and q are two crooked levers, by which the spindle is moved, as handles; these levers are struck by pins fixed in a wooden rod r, which is jointed to the working beam, and moves up and down with it; y, fig. 2, is a piece of wood fixed to the upright beam of the frame, having pins projecting from it, which catch the tumbling bob m, and prevents its moving too far; S, is a stop cock in the main pipe, which regulates the quantity of water coming to the engine, and, consequently, the velocity with which the engine will work.

To

To defcribe the operation of the engine, fuppofe every thing to be in the pofition of the figures, and the pipes, cylinder, and pump full of water, the fliding valve is down in the pofition of fig. 3, and therefore forms a communication from the top to the bottom of the cylinder; in this ftate, the preffure of the defcending column of fifty-four feet of water is equal upon both fides of the pifton of the cylinder, and therefore has no operation either way, the communication with the pipe H being ftopped; fo that the water cannot efcape through it; but the preffure operates upon the lower furface of the forcer d of the pump rifing freely through the lower valve; this preffure being unbalanced, raifes up the working beam, and with it, the pifton of the cylinder, and the rod r; this, as before mentioned, has pins in it, one of which is feen in fig. 1, to be juft meeting the arm or handle p, which it raifes, lifting the tumbling bob m, and turning the axis with all its levers, when the engine arrives at the top of the ftroke, the tumbling bob paffes the vertical point, and falls down on the other fide of the centre; the arm o now ftrikes the crofs pin in the end of the rod k, and by drawing it, moves the lever x, and lifts up the fliding valve; this clofes the communication between the top and bottom of the cylinder, and opens a paffage to the pipe H, permitting the water to pafs through the pipe into the well, and thus get away from the engine, removing the preffure of fifty-four feet from beneath the pifton, though leaving it ftill acting at top of the pifton, and adding to it the twelve feet from the engine to the bottom of the well, in confequence of this column being fufpended in the pipe H; this unbalanced preffure caufes the pifton to defcend, bringing down the end of the beam and pump rod d with it; the valve in the bottom of the pump now fhuts, and the water in the pump being preffed by the pifton, opens the other valve at X, and goes up the pipe Y, to the refervoir, lifting a column of water of eighty feet.

When the engine gets to the middle of its ftroke, a pin in the other fide of the wooden rod r, and therefore not feen, takes the lever q, and forces it down, raifing the tumbling bob m, at the fame time; by the time the pifton arrives at the bottom of the cylinder, the tumbling bob is brought paft the vertical pofition, and fuddenly overfets by its own weight into the pofition of fig. 1; the lever n, now runs againft the pin acrofs the end of the rod k, and fhoves it from the engine, moving the long lever x, of the fliding valve, and the fhort lever i, fig. 3, down juft in the pofition of the drawings; this clofes the four lower holes in the fixed cylinder, and prevents the water going down the pipe H, and at the fame inftant opens the four upper holes, forming a communication between the top and bottom of the cylin-der. The preffure of fixty-fix feet, which caufed the pifton to defcend, is now removed, or rather balanced by caufing it to act equal beneath the pifton, and the column of water of fifty-four feet coming down the pipe A, forces open the lower valve of the pump (the valve at X clofing and taking the bearing of the column of eighty feet), preffes the under fide of the pump bucket, and raifes it up, as before defcribed, moving the beam and pifton with it, there now being an equal preffure, both above and below the pifton, it will be moved up eafily. When the pifton arrives at the middle of its ftroke the preceding operations are repeated, the pin in the rod r takes the lever p, and rifes with it till it arrives at the top of its ftroke, when it again paffes the vertical pofition, and inftantly falls over into the pofition firft defcribed; the lever o, taking the end of the rod k, and pufhing it towards the engine, raifes the fliding valve. opens the paffage to the pipe H, and the whole column of fixty-fix feet now preffes upon the pifton, and forces it down as before defcribed, overcoming a column of eighty feet upon the pump, though the diameter of the pump is larger than that of the cylinder; this happens from the chain of the pifton acting upon a much longer lever than the pump; K and L are two air veffels upon the pipes A and Y.

The proportions of the cylinder, pump, &c. are as below.

	Ft.	In.
The diameter of the working cylinder,	2	2
The length of its ftroke,	10	0
Distance from the centre of motion e, or radius of the lever it actuates	6	0
Perpendicular height of the column of water which preffes upon the piston of the cylinder when defcending 54 + 12,	66	0
Diameter of the pump barrel,	0	8
Length of its ftroke,	5	1
Distance of the pump from the centre of motion, or length of lever which works it,	30	6
Perpendicular height of the column of water preffing upon the forcer of the pump to raife it,	54	0
Ditto, raifed by the pump,	80	0
Difference, which is the height to which the pump actually raifes the water above its source,	26	0

LONG

ELEVATION & PLAN of a FIRE ENGINE, for raising Water for the WATER COAL GIN, at the Prosperous Pit, LONG BENTON COLLIERY.

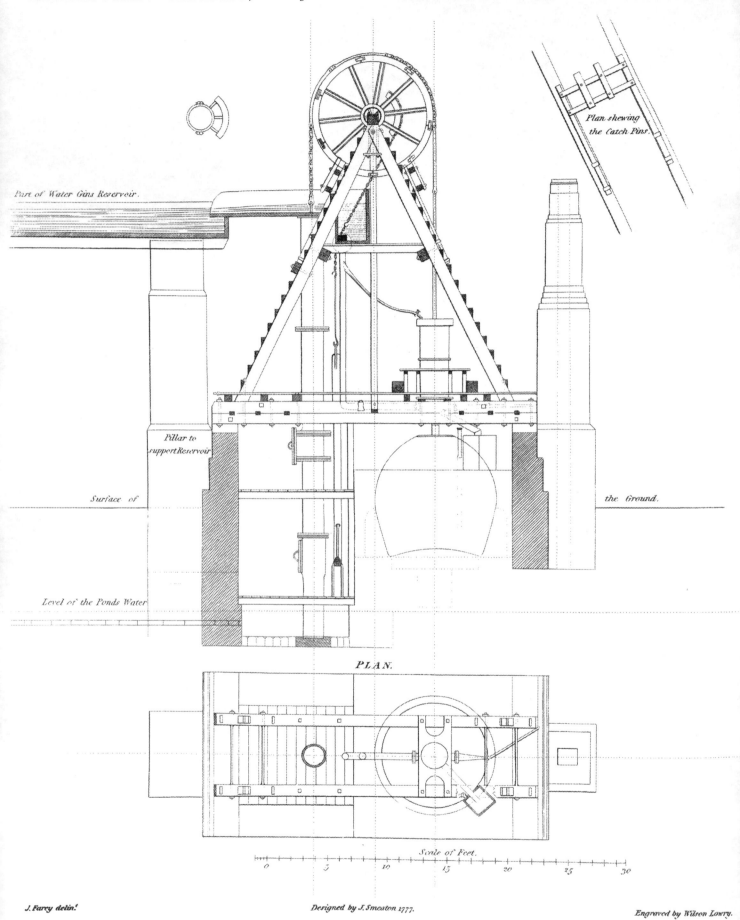

Plan shewing
the Catch Pins.

Part of Water Gins Reservoir.

Pillar to
support Reservoir

Surface of   the. Ground.

Level of the Ponds Water

PLAN.

Scale of Feet.

0   5   10   15   20   25   30

J. Farey delin.   Designed by J. Smeaton 1777.   Engraved by Wilson Lowry.

London, Published by Longman, Hurst, Rees & Orme, 1810.

LONG BENTON ENGINE.

To Mr. Walton,

Dear Sir,   Buxton, 2d July, 1777.

I NOW proceed to point out such things in the designs for the fire engine and water gin as may moft ftand in need of explanation, and firft with refpect to the elevation of the fire engine, you will perceive that I propofe that the hand geer for working the engine, fhall be down nearly upon the fame level with and abreaft of the fire door, fo that the perfon that feeds the fire can hand the engine without going up ladders or fteps; the particular method of communicating the hand geer with the regulator and injection cock I fhall fully explain by a particular draft, and fhall fend you alfo a draft for the regulator, which, for this fmall engine, will be beft wholly of brafs. The iron injection pipe, caft with the cylinder's bottom, I propofe to make really a part of the injection pipe, and to make the communication between the iron pipe and injection pipe by a fhort pipe of wood, with a brafs cap driven into it.

The upper ftage is intended to fupport the injection ciftern, which takes its water from, and fills to the fame height as the furface of the water in the lander trough, all that is not taken for injection going upon the wheel. I propofe a common lead pump occafionally to fill the injection ciftern till the water is delivered at the main pump head; but this ciftern fhould be lined with lead, to avoid all leakages, and to give liberty to the motion that the lander trough will have upon the great pump without affecting the injection ciftern; I propofe the communication to be made by means of a leather pipe to go through both, and to be turned and nailed flanch ways upon the infide of both; alfo the joint to be made good between the lander trough and the refervoir of the gin by a ftrip of leather nailed to both; the man that works the common pump to ftand upon the fame ftage as the injection ciftern; the contrivance for fixing the fprings upon the main triangle legs, and of getting them clofe enough to take the catch pins, without increafing the length of thofe pins, will afford an addition of fpring; but not being of ftrength fufficient to receive a dead ftroke of the engine without danger of breaking, they are propofed ultimately to bank upon two crofs pieces of half a balk, fix by twelve, mortifed and bed-fcrewed into the triangle legs, which will alfo firmly join the legs near the top; the fpring frames fhould be fhort of banking by an inch, upon thefe crofs pieces, which may be regulated by a fmall block of wood in the middle, to fill any part of the fpace between the crofs bars of the fpring frame and the faid crofs pieces. Upon the tops of the triangle legs, where

they

they are jointed in one, reft the head-ftocks, of caft iron, which are each to be bolted down by a couple of perpendicular bolts, as are fufficiently explained in the upright. Thefe iron head-ftocks are to contain the brafs cods for the gudgeons, and it is to be noted, that the cods are to take quite half the circle of the gudgeons, to prevent the danger of accident by the wheel's jumping out of its place upon a dead ftroke upon the bankings. The length of the cylinder in the clear will be fix feet fix inches, and the length of the ftroke, with the fprings, down to their ordinary bearings, to be fix feet; but I do not expeft the working ftroke of the engine to be more than five feet eight inches, and to go $14\frac{1}{4}$ ftrokes per minute. I propofe the whole of the great frame, as fhewn in the feparate fketch thereof, to be of fir; the breadth of the rim of the great wheel outfide, where the chains work, to be twelve inches, and the diftances of the two chains, middle and middle, to be about $7\frac{1}{4}$ or $7\frac{1}{2}$ inches; the heads of the pins to be outfide.

Refpecting the water gin, the principal thing that will want explaining is the conftruction of the refervoir and method of drawing the water upon the wheel. I fuppofe the column raifed by the engine to be thirty-four feet, thirty feet whereof is given to the wheel, three feet head, and a foot allowed for clearance of the wheel at bottom, and the return of the water from the pump's head into the refervoir. To conftruct the refervoir to hold three feet deep of water would be to burthen it with an unneceffary load, becaufe it never can be drawn off within one foot of the bottom, on account of the waters not having in that cafe a fufficient velocity upon the wheel; I propofe, therefore, that the refervoir be conftructed to hold but two feet, or rather two feet two inches water at a full head, and to make good the reft by fcrewing a trough of ten inches wide, and about five feet long, up againft the bottom at the end next the wheel, and in order to let the water down into it, four of the bottom planks to be pierced in the middle of each, with openings of fix inches wide, and ten long, equal to the width of the trough below; the laft to contain the penftock, fliding in grooves cut in each fide of the under troughs, for drawing on, and fhutting off the water from the wheel, which is propofed to be regulated by the machine itfelf at proper times for the meeting and landings of each corf; this will be done by this penftock, as to the reducing or fhutting off the water; but as the quantity to be drawn upon the wheel muft be exactly regulated by the quantity drawn (in the whole time of bringing up a corf) by the engine, this, I propofe to be done by a fmall regulating fhuttle outfide the head of the refervoir, which, when once adjufted, is to be fcrewed faft; it is not to flide in a groove in the trough flides, but to be leathered fo as to fill it fideways, and

make

make a water joint; this explanation, together with what will appear by infpection of the draft, will, I believe, be fufficient; the upright and fide view of the ftage w ill, I imagine, be fufficient to explain the manner of ftriking the great wallower out of the geer, and of the application of a convoy to the circumference of one of the great wheels.

In ordinary working, I do not expect that any thing further will be neceffary than that, as foon as the corf appears, the bank's man lays hold of it in order to ftrike it, and as foon as the boy in the cabin upon the ftage fees the bank's man have hold of it, or on notice from the bank's man, by means of the tackle, he hauls the great wallower out of the geer; he is prevented from ftriking it immediately into the oppofite wheel by an iron pin that fits two holes through the piece that contains the long mortife, in which the upper part of the upright lever flides, that fupports the wallower's gudgeon; when the pin is in one hole, it fuffers the upright lever to come into a perpendicular pofition, but no further; and when in the other hole, it does the fame thing when hauled out of the geer the contrary way; before the boy hauls into the geer he takes out the pin, and if after it is in, he puts the pin into the fame hole, the lever being then got to the contrary fide thereof, it will keep the wallower in its geer; but I expect in common the boy will, when grown expert, rather hold the wallower in its geer, by keeping the tackle tight, than handing the pin for any other purpofe than that of preventing the levers paffing the perpendicular, which will always be neceffary. In this manner I expect the operation will go on, when the people are expert in the ufe of the different tackle; but, to prevent accidents before they are grown expert, as well as inattention afterwards, I propofe two convoys, one upon the great wheel in the cabin, fo that if the water wheel happens to have got too much way after the water is taken off, to ftrike in again upon the other fide; he can curb the motion, or ftop it if neceffary, and if the corf happens to be got too high before the boy ftrikes out, as indeed is liable often to happen then the convoy upon the long axis near the barrel being, by means of a cord in the management of the bank's man, he can, on arrival of the corf, clap on his convoy, and prevent the weight of the corf from overhauling the barrel with too much violence; I alfo have a contrivance of a double ratchet wheel, which, for the fake of farther prevention of accidents, can be applied to the barrel, if thought neceffary, or found fo, by which the barrel will always be prevented from running back till the bank's man releafes the catch; but this, I think, will fcarcely be neceffary, as I hear they manage at Griff, without any other convoy, except one, applied to the great wheel. The whole is calculated to draw a corf complete in two minutes at Long Benton.

The

The wallower gudgeon is to reach through the upright lever and end, in a square of about two inches, to which is to be applied the work for governing the shuttle; this is to be a wrought gudgeon, steeled and turned.

As I found, on adjusting my work together, that it would require a very high pair of shears, to draw the bucket out of the pump head, (without unjointing the rod), which would give the whole a top-heavy and cumbrous appearance, I concluded that the easiest way would be to introduce a bucket-door piece, so that, by occasionally hooking a tackle block upon the engine's wheel, to one side or the other, either the piston might be drawn, or the bucket raised to the bucket door, without any additional parts for that purpose. I would advise that the joint of the bucket shank, with the spear, be made with a couple of bolts and nuts, rather than an off-take joint that fixes with a collar, because the nuts can be more easily undone than a hammer used in so confined a place.

I am, dear sir,
Your most obliged servant,
J. SMEATON.

I forgot to say in proper place that the reason why the real head of water would be three feet two inches, when only three feet are allowed, is, that the bottom of the trough would lie two inches below the bottom of the wheel.

LIST of mills executed by Mr. SMEATON, from a paper in his own hand-
writing.

Vine Elms china mill, wind.
Bafhing wood mill, wind.
Halton flour mill, water.
Wakefield ditto, water.
————— oil and wood mill, wind.
Heath engine, water.
Ridge wood mill, water.
Honeycomb groat mill, water.
Brittain furnace, water.
Carron, ditto, No. 1, water.
————— ditto, No. 2, water.
————— boring mills, water.
————— clay mill, water.
Waltham Abbey powder mill, water.
Hounflow Heath ditto, water.
Buffey mill, flour mill, water.
Dalry mills, flour mill, water.
Worcefter Park powder mill, water.
Thornton paper mill, water.
Bocking fulling mill, water.
Kilnhurft forge, water.
————— flitting mill, water.
Horfley wire mill, water.
————— tilt mill, water.
Knouchbridge flour mill, water.
Stratford ditto, water.
Colchefter fulling mill, water.
Holling's mill, ditto, water.
Kefwick grift mill, water.
Hounflow Heath copper mill, water.
Stratford engine, water.
London Bridge engine, water.
Alfton grift mill, water.

Woodhall

Woodhall grift mill, water.
Whittle ditto, water.
Throchley ditto, water.
Scremerston ditto, water.
Wanasworth flour mill, water.
Carshalton oil mill, water.
Hull ditto, water.
Cardington flour mill, water.
Leeds pottery mill, wind.
Deptford flock mill, water.
————-- engine, water.
Welbeck engine, water.
Thoresly ditto, water.
Wanloch Head slide engine, water.

END OF VOL. II.

S. Brooke, Printer, 35, Paternoster-Row, London.

Printed in the United States
By Bookmasters